Dimensions of Darwinism

This book is published as part of the joint publishing agreement established in 1977 between the Fondation de la Maison des Sciences de l'Homme and the Press Syndicate of the University of Cambridge. Titles published under this arrangement may appear in any European language or, in the case of volumes of collected essays, in several languages.

New books will appear either as individual titles or in one of the series which the Maison des Sciences de l'Homme and the Cambridge University Press have jointly agreed to publish. All books published jointly by the Maison des Sciences de l'Homme and the Cambridge University Press will be distributed by the Press throughout the world.

Cet ouvrage est publié dans le cadre de l'accord de co-édition passé en 1977 entre la Fondation de la Maison des Sciences de l'Homme et le Press Syndicate de l'Université de Cambridge. Toutes les langues européennes sont admises pour les titres couverts par cet accord, et les ouvrages collectifs peuvent paraître en plusieurs langues.

Les ouvrages paraissent soit isolément, soit dans l'une des séries que la Maison des Sciences de l'Homme et Cambridge University Press ont convenu de publier ensemble. La distribution dans le monde entier des titres ainsi publiés conjointement par les deux établissements est assurée par Cambridge University Press.

Dimensions of Darwinism

*Themes and Counterthemes in
Twentieth-Century Evolutionary Theory*

EDITED BY

MARJORIE GRENE

CAMBRIDGE UNIVERSITY PRESS

Cambridge

London   New York   New Rochelle
Melbourne   Sydney

& EDITIONS DE LA MAISON DES SCIENCES DE
L'HOMME

Paris

Published by the Press Syndicate of the University of Cambridge
The Pitt Building, Trumpington Street, Cambridge CB2 1RP
32 East 57th Street, New York, NY 10022, USA
296 Beaconsfield Parade, Middle Park, Melbourne 3206, Australia
and Editions de la Maison des Sciences de l'Homme
54 Boulevard Raspail, 75270 Paris Cedex 06, France

First published in 1983

Printed in the United States of America

Library of Congress Cataloging in Publication Data
Main entry under title:
Dimensions of Darwinism.
Includes index.
1. Evolution – Addresses, essays, lectures.
I. Grene, Marjorie Glicksman, 1910– .
II. Title.
QH366.2.D55 1983   575.01'62   83-1795
ISBN 0 521 25408 6
ISBN 2 7351 0060 X (France only)

CONTENTS

v

PART IV. SOME CONTEMPORARY ISSUES: THE SYNTHESIS RECONSIDERED

CONTRIBUTORS

RICHARD M. BURIAN Department of Humanities and Communication, Drexel University, Philadelphia, Pennsylvania 19104

STEPHEN JAY GOULD Museum of Comparative Zoology, Harvard University, Cambridge, Massachusetts 02138

MARJORIE GRENE Department of Philosophy, University of California, Davis, California 95616

ANTONI HOFFMAN Wiejska 14 m. 8, PL-00-490 Warsaw, Poland. (Present address: Department of Geology and Geophysics, University of Wisconsin, Madison, Wisconsin 53706)

WILLIAM C. KIMLER Department of History and Section of Ecology and Systematics, Cornell University, Ithaca, New York 14853

BERNARD NORTON Department of History of Science, University of Leicester, Leicester LE1 7RH, England

D. S. PETERS Forschungsinstitut Senckenberg, 6000 Frankfurt/M1, West Germany

WILLIAM B. PROVINE Department of History and Section of Ecology and Systematics, Cornell University, Ithaca, New York 14853

WOLF-ERNST REIF Geologisches Institut der Universität, D-7400 Tübingen 1, West Germany

BERNHARD RENSCH Zoologisches Institut der Universität, 44 Münster, West Germany

RUPERT RIEDL Institut für Zoologie der Universität, A-1090 Vienna, Austria

JOHN MAYNARD SMITH Department of Biology, University of Sussex, Falmer, Brighton, Sussex BN1 9QG, England

JOHN R. G. TURNER Department of Genetics, University of Leeds, Leeds LS2 9JT, England

ACKNOWLEDGMENTS

Our thanks are due to the Werner-Reimers Stiftung for their support and to the Maison des Sciences de l'Homme for including the present volume in their series. We regretted Professor Rensch's absence from our discussion, and regret also the pressure of work that prevented Dr. von Wahlert from preparing a paper about the ecological approach to evolution that he finds in the work of Klaus Günther and in the early work of Willi Hennig. Something of his intent may be gleaned from the lectures of Hennig, Günther, and von Wahlert himself in R. Siewing, ed., *Methoden der Phylogenetik* (*Erlanger Forsch. B*, 1971).

For my own part, I should also like to express my gratitude to a number of people for advice about the organization of the meeting: in particular, Stephen Jay Gould, Jonathan Hodge, William Provine, Adolf Seilacher, Roswitha and Wolfgang Wilkschko, and the late William Keeton. The editorial staff of Cambridge University Press have been consistently helpful in seeing the manuscript through the press.

M.G.

INTRODUCTION

MARJORIE GRENE

At the suggestion of Professor Wolf Lepenies of the University of Berlin, and under the sponsorship of the Werner Reimers Stiftung, a conference was held on August 27 and 28, 1981, at the Foundation's headquarters in Bad Homburg in West Germany, on the topic of twentieth-century evolutionary theory. Our leading concern might be formulated in terms of the question: How complete and how stable is, and has been, the evolutionary synthesis, or "neo-Darwinism"? A small meeting, with fourteen invited participants (of whom thirteen were present), the conference made no claim to be in any way exhaustive either of the historical or the philosophical issues involved in our subject matter. As the meeting was conceived, the primary focus was to be historical; but both the interests of the participants and the nature of the history we were concerned with made it inevitable that we carry our discussion forward to contemporary issues.

The essays that have resulted from our discussions speak to a number of very different aspects of our general question, let alone its answer(s). And to a thoroughly historicist philosopher of science, such as the organizer of the meeting and editor of this volume, this is as it should be. Let me fill in this general boast a little before taking up a few aspects of the particular questions we dealt with. I shall not, even then, affront the reader by summarizing each essay in turn; I trust that they are readable in themselves.

First, on the question of philosophy of science in general, let me make a confession of faith. By now, I believe, what used to be called the "received view" in philosophy of science may be decently buried and, indeed, forgotten. It was a view that divorced its analysis of science entirely from science as history, science as human activity, science as search. By now, thanks to historians of science, and to some philosophers and reflective scientists, we may reflect, more constructively, I believe, on science as a nexus of activities, of "practices,"

to use A. C. MacIntyre's term. Practices are characteristically located
in certain segments of certain societies, making certain kinds of epistemic claims in accordance with the patterns of thought of their
various disciplines or subdisciplines at a given time and stage of development, and authorized by the standards of the society that permits and sustains them (MacIntyre, 1981; cf. Brown, 1977).[1] This is
not to say that philosophy of science is thereby reduced to the sociology of science. Philosophers are interested not just in putting one
darned thing before or after another, nor in making "pure" causal
connections whether in psychoanalytical, Marxist, or other historio-
graphic fashions. We are interested in the nature of epistemic claims;
but we recognize that these are always embedded in a history from
which they take their shape and, to some extent, their content (Grene,
1976, 1978b; cf., e.g., Kitcher, 1982).

Moreover, history is always partial and fragmentary, characterized,
like everything alive, by what Merleau-Ponty called "patterned
mixed-upness" (Merleau-Ponty, 1968, p. 176). There is no end to its
complexity, and no way, on principle, to be sure what is and is not
illuminating. That statement may be partly an excuse for the very
assorted content of this volume – and of our discussions at Bad Homburg – but it is also, in my view, a reflection of scientific-historical-
philosophical reality.

If, further, an unresolved, and possibly irresoluble, plurality of perspectives is appropriate to metascientific questions in general,
whether they are (primarily) historical or (primarily) philosophical,
so much the more emphatically does this hold for our present subject
matter, the recent history of evolutionary theory. Darwinism has been
correctly described as a hypertheory, or supertheory that serves to
tie together a variety of scientific disciplines within biology (Tuomi,
1981; Wassermann, 1981). Ethology, paleontology, taxonomy (of the
evolutionary variety), and even, remotely, biochemistry depend at
various but crucial points on its organizing power. But perhaps because of that very comprehensive and regulative status, Darwinian
theory has been, since the beginning, the subject of unending debates, reinterpretations, and revisions. In the discussion that is here
partly recorded, we concentrated on two major themes or clusters of
themes: questions about the history of the modern evolutionary synthesis and questions of contemporary history centering on the problem of the role and limits of adaptation as an explanatory concept.

---

[1] For a statement of the standard, and, as I believe, mistaken, view that it is
science that should contain and legislate for the self-understanding of society,
see, e.g., Simon, 1982.

In fact, these questions too are closely interconnected; they are both touched on, directly or indirectly, in all four parts of this collection.

Historians of science often distinguish between internal and external causes of scientific change. The question may be raised, however, whether these are different in kind or only in degree. If science is viewed as a network of practices, dependent on the standards, concepts, and procedures of each subdiscipline, but authorized by and, to this extent at least, dependent on standards and beliefs of the wider society of which each discipline forms a part, there is no sharp separation, in principle, between broader social concerns and those of a given scientific practice. Granted, it is essential to the existence of any science that it observe its *own* ethical principles: truthfulness, accuracy, the demand for repeatable experiments, special criteria of experimental design, and so on. It is essential to the practice of any science, for example, that respect for truth take priority over ambition, vanity, or greed. But motives other than truth-seeking do, of course, play their part and so do the larger interests of the society as well as of the individual. Bernard Norton's essay about Fisher's interest in eugenics seems to illustrate a purely external relation of personal interest to scientific inquiry. What Fisher accomplished in evolutionary theory appears to have no connection with his early interest in eugenics, whatever we may think of it from our present perspective. And Wolf Reif refers to Beurlen's Nazi affiliation as having no effect on his scientific work. Perhaps. But Rupert Riedl mentions in passing a widely acknowledged case of the interaction of social and scientific beliefs: the relation of the Victorian middle class to Darwinism. Admittedly, social Darwinism cannot be fathered on Darwin himself nor yet on the theory of natural selection as a theory of organic evolution. Nevertheless, the recurrence of organic and social Darwinism as a team in the nineteenth century, in the early twentieth (with Sumner and company), and more recently in some of the offshoots of sociobiology, if not in the fountainhead itself, attest to the probable existence of some essential connection between the two. And more generally, it seems to me, a good case can be made for the existence of something about the Victorian state of mind that makes "utility," whether in Bentham's social thought or in the adaptationist bent of selection theory, seem self-explanatory.[2] Adaptations, like utility generally, are means and explain what they explain not in themselves but in relation to ends. But is the end of survival really adequate for human as well as organic history? Whatever the answer, the point

[2] See, e.g., the suggestions made by Philip Sloan (1981), about the broader influences on Darwin's thought.

here is just: There is something teasing about the way in which both utilitarianism and selection theory (in the strictest, or perhaps most reductive form) appeal to thinkers of a certain cast of mind. In any case, the question here is, rather, whether there is any such intrinsic connection or hint of an intrinsic connection in Fisher's case between his social theory and his genetical theory of natural selection. I don't know.

Another seemingly external factor in the history of science, however, surely does count as "internal" as well, and serves to show how difficult such a distinction may be to make, that is, the matter of national styles. Science is allegedly, and up to a point even is, international. Yet Darwinism has had notoriously different destinies in different European countries (Glick, 1972) and, as this collection partly shows, has continued to do so in this century. Despite Haeckel's popularization of nineteenth-century Darwinian theory, the synthesis took root only very partially in the twentieth-century German-speaking world. Rensch is, I believe it is fair to say, the only major German contributor who remained in Germany, and indeed his contribution to this volume illustrates scientific development in its genuinely international form. The story Wolf Reif tells is different: Notwithstanding Heberer's anthology of essays by subscribers to the synthesis, in German-speaking countries evolutionary thought, and the style of biological argument, retained, on the whole, its heavily morphological commitment. The largely English-language "modern synthesis," on the other hand, entailed at least in some of its bearers a positively antimorphological component: Witness, for instance, the remark of Mayr quoted by Riedl, and see also my comparison of the arguments of Simpson and Schindewolf (Grene, 1974).

Granted, the situation here is complicated and has grown in complexity in recent years with the growth of cladism, which derives from the intensely Germanic phylogenetic taxonomy of W. Hennig (1950). Cladism itself boasts of its anti-Darwinian bent and, from the point of view of orthodox evolutionists, appears to be something of an alien excrescence. Still, it is an increasingly vocal and increasingly prestigious self-proclaimed school; it may well contribute to a new "new synthesis" that will eventually overcome the less than global perspective of the present – or recent – orthodoxy. Indeed, even within the orthodoxy of the synthesis, Mayr has suggested a way to use fruitfully the approaches of all three contemporary taxonomic schools (Mayr, 1981, 1982a). For the relatively simple point I want to make here, however, that is by the way. What needs to be noticed, as against the overabstract conception of science as above and beyond any culture that carries it, is the presence in any given case of scientific thought of what Fleck calls a "thought-style" expressive of a "thought-complex" – and such styles or complexes, in turn, are carried and shaped by local, even linguistic, traditions (Fleck, 1981). As both Reif's narrative and Riedl's presentation of his own position make plain, there is a motif in German-language evolutionary thought that derives not only from a stress on morphology as such, but, acknowledged or not, from the heritage of Goethe and the *Urpflanze*. Similarly, as I have found in talking with followers of sociobiology, there is a lingering commitment to Newton's "hard, solid impenetrable particles" as well as to Newtonian linear causality in some versions of present Darwinian thinking. Some of Gould's adjurations to evolutionists to recall the importance of the constraints that limit the plasticity of natural selection at any given juncture in evolutionary history, or his references to the "science of form," represent, in part, efforts to overcome this parochial character (Gould, 1980*a*,*b*,*c*; Gould and Vrba, 1982). But the diversity, and even contradiction, certainly has existed and still exists, as Riedl's essay in this volume amply testifies.

Moreover, the difference is not only one of style in any superficial sense. It represents, in my view, a puzzling difference of substance as well, and the two, indeed, are intrinsically connected. Ernst Mayr has established as canonical the distinction between typological and population thinking in biology (Mayr, 1982*a*). The Continental, specifically the German-language, strain à la Beurlen–Schindewolf and Remane–Riedl clearly remains on the typological side. On the other hand, the development of the view that "species are individuals" has lately characterized population thinking to the point of denying the need ever to notice *any* similarities whatever among living things. Taxa are to be taken purely as lineage taxa, and that, said Ghiselin and Hull, is that (Ghiselin, 1974; Hull, 1976, 1980*b*). Yet, carried to its logical conclusion, this view seems to undercut the very starting point of any biological science, including the theory of evolution. How does one tell which "individuals" (in the everyday sense) are parts of which larger "species – individuals" except by noticing some kind of likeness among some and not others? David Hull has faced this issue by admitting that here, as in other areas, theory just does have to go against common sense – and so much the worse for common sense.[3] Yet what Riedl calls our "hereditary common sense" about natural kinds does seem, at some level, to underlie the practice of even the most theoretical biology. Doesn't one need some judgment of what something is *like* in order to notice either homologies or analogies? (cf., e.g., Pantin, 1954; Grene, 1976; Wiggins, 1980). Perhaps

---

[3] Personal communication; cf. Hull, 1980*b*.

judgments of "this such" are only everyday starting points, to be left behind, like physicists' everyday judgments, in the more sophisticated statements of a developed evolutionary science. But, then, how could Mayr have been so persuasive in the debate over numerical taxonomy by insisting on the importance of the weighting of taxonomic traits by the experienced taxonomist? (quoted by Riedl, this volume; cf. Mayr, 1981, 1982a; Pantin, 1954, 1968). Again, there seems to be some minimal, almost foundational, contribution here that some arguments drawn from a rigorous selection theory want us not only to overlook but to abandon. As against this extreme conclusion, Stephen Jay Gould's adjuration to "restore the organism to biology" seems reasonable (Gould, 1980c). Yet the morphological tradition generates its own puzzles. When I read Riedl's essay, or Hennig's original (1950) treatise on *Phylogenetic Systematics*, I find myself entering a different world, where what might have been modest methodological suggestions are somehow assimilated to an alien metaphysic. That I know this "other" scientific style fairly well and have even tried (vainly) to introduce it to the English-speaking world, by no means lessens that sense of strangeness (Grene, 1974, chs. 16–19).

Of course this is an oversimplification. As I have already pointed out, the cladistic school, who boast direct descent from Hennig, are English speakers (Hennig, 1966, introduction), and there are certainly non-English speakers attached to the synthesis, as not only the case of Rensch, but the essays of Peters and Hoffman attest. Reif's contribution is especially interesting in this connection, because it records the power of the morphological tradition without itself exemplifying its peculiar tenets. And further back in history there are also all sorts of interactions (see, e.g., Smith, 1982). Nevertheless, it seems undeniable that some kind of linguistic-cultural-scientific difference of style exists and needs to be reckoned with in any adequate account of the conceptual structure of recent evolutionary theory. The French case, it should be added, is different again; until recently, at least, it has represented a more finalistic than morphological resistance to Darwinism (Gavaudan, 1967; Boesiger, 1980; Limoges, 1980). But since we had no French participant in our discussion, only the German–English problem could be raised here.

In more obviously "internal" respects also, the recent history of evolutionary theory again presents pertinent conceptual complexities that show up with a new clarity in the light of controversies now current. Some of these are discussed explicitly in Part Four of this collection. As I have already remarked, we had intended to focus on the question of the role and limits of adaptation as an explanatory concept. Although this has been a recurrent question in evolutionary theory, it is the recent discussion of "adaptationism" that has brought it once more to center stage. Clearly – and Darwin himself was well aware of this – the theory of natural selection is a theory of organic change with respect to adaptation and only adaptation. It is a theory of the mechanisms by which organisms have become fit, that is, likely to leave offspring, in the environments in which they find themselves, environments which, indirectly, by the opportunities for new life-styles they offer, help to bring them about. But can "adaptation" do all the work? What other concepts belong to the core of modern evolutionary theory? The history of biology, Ernst Mayr writes, is not so much the history of theories (let alone of "facts") as it is of concepts (Mayr, 1982*a,b*). But concepts – and perhaps, especially, evolutionary concepts – have a way of expanding by cannibalizing other concepts that ought also to have a role in the whole explanatory framework. "Adaptation" seems to be particularly susceptible of such abuse. For example: the theory of natural selection is a two-step theory; there is random variation – plenty of it – and inheritance of those available characters that happen to prove slightly better adapted in given circumstances than the available alternatives. So undirected variation, basically a kind of randomness, not *only* selection, is an essential ingredient in the process. Moreover, biologists must recognize, if they think about it, the constraints imposed by past development – constraints of form, behavior, and physiology – within which variation and selection must take place. Differences in "tempo and mode" of evolution, too, may complicate the story. As the evolutionary synthesis developed, however, the conceptual plurality of the first – and founding – versions receded; and as Gould puts it, the synthesis hardened. Earlier versions, as the arguments of Provine and Gould demonstrate, had been less thoroughly "adaptationist," more tolerant of factors other than "means-to-survival" in the evolutionary story. In response to the hardening of the synthesis, the question of just how much work adaptation does or ought to do becomes a pressing one, especially in the light of proliferating challenges, both esoteric and exoteric, to the Darwinian tradition. Gould and Provine speak to this question from an historical perspective, and Burian and Peters address themselves more directly to the conceptual issue. The essays of Hoffman and Maynard Smith also bear indirectly on this problem, and, in fact, all three biologists directly dealing with current issues (Hoffman, Peters, Maynard Smith) defend the central adaptationist program. Kimler's essay, however, alongside Turner's, helps to complicate the picture. Turner, a student of P. M. Sheppard (Sheppard, 1958) would like to find mimetic theory leading directly into the synthesis in its strictest form (Turner, 1981) but acknowledges a recurring schism in evolutionary thought. Kimler presents convincingly a more complicated picture, arguing that an ecological perspective was important in this field and at the same time that ecological thinking, though part and parcel of Darwin's biological thought-style, played no major role in the development of the synthesis. Thus, rather than a single line of thought leading to the synthesis, we find a complex network of interacting interests and influences in which at some points, though not always or necessarily, the selectionist emphasis, or overemphasis, on adaptation assumes a dominant part. And it is this complexity that is spilling over in the many-sided debates about evolutionary explanation that are conspicuous today.

We touch on only a few aspects of this complex debate in our collection. It may be worthwhile, however, in conclusion, to separate schematically some of the issues that have arisen in the recent literature. Questions about the adequacy of the synthesis are often taken, on principle, as challenges to Darwinism. Of the many lines of debate now current, however, it seems to me, only two, or perhaps better three, constitute such fundamental challenges: (1) neutral mutation theory, (2) cladism, and (3) a protest from the perspective of ontogenesis and sometimes also of morphology. The first is not touched on here at all, although it certainly should not be ignored in any comprehensive account of recent theory. That mutations are random even the most fervent selectionist admits: indeed, it is necessary to the structure of a neo-Darwinian, non-Lamarckian account that this be so (see Rensch's essay, for example). But that alleles wholly unconnected with fitness should persist for long periods and in great numbers is a thesis that selectionists find difficult to accept. Yet despite selectionist arguments to the contrary, the view persists (Kimura, 1976; King and Jukes, 1969). [Darwin, it should be noted, in the *Descent of Man*, seems to anticipate even this degree of pluralism. In that late work, he attributes a good deal of evolutionary change to "chemistry"! (Darwin, 1871).] Second, cladism is not only anti-Darwinian. Although it began, with Hennig, as phylogenetic taxonomy in opposition to idealistic morphology, it has by now turned, in its "pattern cladistic" form, against evolution itself, a paradox Maynard Smith refers to in passing (Beatty, 1982; cf. Hull, 1980a). In addition, perhaps one ought to mention a kind of morphogenetic challenge to Darwinism reminiscent of some aspects of Riedl's argument (Wiley and Brooks, 1982; Løvtrup, 1981).

On the whole, however, I find most objections raised largely from within a modified or enriched selectionist approach. This is the case, for example, with Gould and Lewontin (1979), Gould (1980*a,b,c*, 1982, and this volume), Gould and Vrba (1982), Sober and Lewontin (1982). Here, again, however, various lines of argument should be distinguished. Gould and Lewontin's classic paper was subtitled "A Critique of Adaptationism." What is really under attack in that provocative piece is not apparent on its surface; I shall return to that question in a moment. As a whole, however, the "revolt against adaptationism" needs to be distinguished from two other current themes: (1) the controversy about the separability of macro- and microevolution; and (2) the controversy about the units of selection (to which, however, as we shall see, the "adaptationism" argument does bear a close relation). Both these questions are dealt with by Maynard Smith, who is, indeed, one of the most eminent and ingenious defenders of the neo-Darwinian orthodoxy (Maynard Smith in Scudder and Reveal, 1981; cf. Stebbins and Ayala, 1981). Hoffman, as a paleontologist, is also concerned with the former problem and professes skepticism of the new heresy. The vehicle for the separation is the theory of punctuated equilibrium (Eldredge and Gould, 1972; Gould and Eldredge, 1977), which offers a new perspective for reading the fossil record and thereby challenges the gradualist bias of Darwin himself as well as of modern Darwinism. As its originator(s) conceived it, this model was to allow paleontologists to accept the fossil record at face value, rather than having to apologize for the gaps "discovered" in it by a gradualist approach. This seemed to some paleobiologists a vindication of their discipline; to genetically based evolutionists, however, it undercuts precisely one of the great triumphs of the synthesis: its assimilation of macroevolution to its more manageable microevolutionary – that is, population-genetical – base. Yet there is nothing fundamentally antiselectionist about this theory, and it is not, in itself, an attack on but rather a modification of the synthesis (Gould, 1982; cf. Stebbins and Ayala, 1981; Levinton and Simon, 1980).

Even more clearly, the units of selection controversy should be seen as a dispute within the Darwinian tradition. Again, the theory of natural selection is a theory of how adaptations arise. But adaptations of *what*? What is it that is selected? Every least part of every organism – notably every gene? Every trait of every organism? Each organism as a whole? Demes? Populations? Species? G. C. Williams' influential book (1966) arguing against the unpopular and indeed unlikely notion of group selection put the reductionist, gene-oriented answer very well. Still, even in the heyday of the genetically dominated version of the synthesis ("natural selection *is* differential gene frequencies," etc.; see Burian, this book, Chapter 11; and cf. Grene, 1974, chs. 8 and 13), Waddington had warned biologists to remember that although it is genes that differentially survive, it is phenotypes that are selected (Waddington, 1953, 1967, 1969). Since Lewontin's authoritative paper on "The Units of Selection" (Lewontin, 1970; cf. Wright, 1967), and with the work of Wade (1978) and others, however, a more flexible conception of the units of selection has been put forward – most definitively in my view in Sober and Lewontin (Sober and Lewontin, 1982; cf. Brandon, 1982; Richardson, 1982; Lewontin, 1982). Not that this means a global acceptance of group selection and allied concepts. In this volume, Maynard Smith argues against Stanley's concept of species selection for use in any but a very few cases (Stanley, 1975, 1979); and Gould, who had accepted the term, has now renounced its general use (Gould, 1982). Clearly, other things being equal, if a given species speciates faster than another, selection at the species level favors its "daughter species" over those of its competitors. This *is* species selection. But if a species is selected because "it" runs faster, that is just shorthand for the fact that its members (in Hullian terms, its parts!) run faster. In such a case, it is the individual phenotype, not the whole species, that the process of natural selection affects. But neither is it the gene. (Genes don't run, any more than species do.) And the possibility of occasional selection of some larger units remains open. The conceptual changes needed to effect this change in modeling are much more intricate and subtle than I have indicated (see Sober, 1980; Wade, 1978; Wilson, 1980; Wimsatt, 1980, 1981). The point here is simply, first, that these are alterations *in*, not against, what is basically Darwinian evolutionary thought, and second, that this controversy, like the others I have mentioned, should be distinguished from the critique of adaptationism – although, as I shall now proceed to say, it, unlike the others, is closely connected with that critique at least in its "San Marco" version (Gould and Lewontin, 1979).

With those distinctions made, we may look briefly at the direct attack on adaptationism. As I understand it, the "Spandrels of San Marco" was directed chiefly against two kinds of excess or exaggeration in evolutionary theory: a misplaced atomism and an irresponsible teleologism. Both these distortions of Darwinism – let us call them cryptoatomism and pseudoteleology – are especially conspicuous in the arguments of sociobiologists, and it seems reasonable to conjecture that the explosive expansion of that new "discipline" had some connection with the genesis of the "Spandrels" paper.[4] What is important here is to recognize the multidimensionality of biological, and indeed of evolutionary, explanation: that is what is ignored in

---

[4] For a more explicit reference, see Lewontin, 1979; this ought, I believe, to have been a definitive refutation of sociobiological principles.

the simplistic reductionism of purely gene-directed accounts. And that is, of course, one of the themes raised by the units of selection issue. On the other hand, Gould and Lewontin also attack what they consider ill-conceived and ill-tested ethological hypotheses, jerry-built to suit the underlying atomizing account. We may perhaps put the situation in terms of Aristotelian causes, or analogues of these (Grene, 1963; Gotthelf, 1976; Balme, 1980, 1981; Lennox, 1982). Biological process always involves materials hierarchically organized so as to function in a certain way: The parts are constrained by the organizing principle, form, or function that gives them their biological significance and their biological reality (see, e.g., Pattee, 1973; Gould and Vrba, 1982). But reductionistic biologists want to translate form into matter in the sense of least parts: Only biochemistry and cell biology, they hold, are biology at all. Thus they read selection as particulate, affecting always and only genes, not organisms. This is adaptation*ism*, not because it is selectionist, but because it is atomistically so. It is not organisms or populations that are thought to be adapted, but their minute parts (see, e.g., Brandon, 1978, 1982). In Aristotelian terms, formal cause is being suppressed for the sake of its material correlate – and in much the way that Aristotle himself found so inadequate in the case of Democritus. On the other hand, the just-so story aspect of adaptationist explanation, carrying over its atomizing habit to the phenotypic and behavioral level, tells what tales it likes of any and every trait, again, taken on its own. In Aristotelian terms, once more, we may say: All evolutionary biology seeks the moving ("efficient") cause of the phenomena it studies, but moving cause, in the case of living things, is correlated, somehow or other, with the endpoint of the process in question: the *telos*, or final cause. Of course there is only one end in evolution: survival.[5] But to explain why and how a given structure, process, or behavior facilitates survival takes, not only imagination, but detailed biological knowledge and a kind of tact, combining fancy and restraint, in the construction of evolutionary scenarios. Insofar as teleology – or teleonomy – is involved in such explanation, it must be used with care.[6] All evolutionary explanation may well be adaptation*al*; it becomes

---

[5] As Richard Burian has pointed out on reading this introduction, "survival" is a fighting word. Survival of what? In terms of the very general point I am making here, however, I hope the reference is clear: persistence of *kinds* of genes, individuals, behaviors, etc., etc., into another generation, without any concrete specification of special circumstances.

[6] The question of the role of teleology in evolutionary theory is delicate and difficult. See, e.g., "Explanation and Evolution" in Grene, 1974 (ch. 13); Brandon, 1981.

adaptation*ist* when each trait has a "why" fabricated for it in harmony with the overall genes-for-genes' sake theory.[7]

Finally, one more brief terminological comment: Many of the recent discussions of these problems, especially some of Gould's recent papers, emphasize the importance of hierarchy in biological explanation (see also Stebbins, 1969; Lewontin, 1982). Related to these arguments, of course, are others addressed to the problem of levels of organization, or levels of explanation (e.g., Pattee, 1973). Once again, this is a problem our discussions at Bad Homburg did not, in the main, confront explicitly. But I must confess to being puzzled by the use of the concept of hierarchy in hierarchy theory, on the one hand, and in taxonomy and arguments derived from taxonomy, on the other. Ernst Mayr has introduced two distinctions in this context: between "constitutive" and "aggregative" hierarchies (Mayr, 1982*a*, p. 65) and between "exclusive" and "inclusive" hierarchies (*ibid.* pp. 205–207). I am not quite sure how the two distinctions relate to one another; but it is clear that whenever the term "hierarchy" occurs, in any discussion of biological questions, the reader should ask what concept it is that is being employed. What I find to be a chronic confusion about this disturbed me while I was translating Professor Riedl's essay. I submit it as one of many contexts in which our discussion could well be carried forward on some other occasion.

References

Balme, D. 1980. Aristotle's biology was not Essentialist. *Arch. Gesch. Phil.* 62:1–12.

    1981. Aristotle's Natural Teleology. (ms.)

Beatty, J. 1982. Classes and cladists. *Syst. Zool.* 31:25–34.

Boesiger, E. 1980. Evolutionary Biology in France at the Time of the Evolutionary Synthesis. In *The Evolutionary Synthesis*, ed. E. Mayr and W. B. Provine, pp. 309–21. Cambridge, Mass.: Harvard Univ. Press.

Brandon, R. N. 1978. Evolution. *Phil. Sci.* 45:96–109.

    1981. Biological teleology: Questions and answers. *Stud. Hist. Phil. Sci.* 12:91–105.

    1980. A structural description of evolutionary theory. *PSA 1980* 2:427–39.

    1982. The levels of selection. *PSA 1982* 1:315–23.

Brown, H. 1977. *Perception, Theory and Commitment: The New Philosophy of Science.* Chicago: Precedent. Reprinted in paper, 1979. Chicago: Univ. of Chicago Press.

---

[7] Care must be taken also in the application of Aristotelian concepts to modern biological explanation. I must confess to finding Riedl's use of Aristotle in this volume rather confusing – largely because of the ambiguity of "hierarchical" analysis, of which more in the paragraph that follows.

Darwin, C. 1871. *The Descent of Man.* London: Murray.

Eldredge, N., and S. J. Gould, 1972. Punctuated Equilibria: An Alternative to Phyletic Gradualism. In *Models in Paleobiology,* ed. T. J. M. Schopf and J. M. Thomas, pp. 82–115. San Francisco: Freeman, Cooper.

Fleck, L. 1981. On the question of the foundations of medical knowledge. (trl. Trenn.) *J. Med. Phil.* 6:237–256. Original: 1935. Zur Frage der Grundlagen medizinischer Erkenntnis. *Klin. Wochenschr.* 14:1255–9.

Gavaudan, P. 1967. L'Evolution considerée par un botaniste–cytologiste. In *Mathematical Challenges to the Neo-Darwinian Interpretation of Evolution,* ed. P. S. Moorhead and M. M. Kaplan, pp. 129–34. Philadelphia: Wistar Institute.

Ghiselin, M. T. 1974. A radical solution to the species problem. *Syst. Zool.* 23:536–44.

Glick, T. R. 1972. *The Comparative Reception of Darwinism.* Austin: Univ. of Texas Press.

Gotthelf, A. 1976. Aristotle's conception of final causality. *Rev. Met.* 30:226–54.

Gould, S. J. 1980a. The evolutionary biology of constraint. *Daedalus* Spring: 39–52.

1980b. The promise of paleobiology. *Paleobiology* 6:96–118.

1980c. Is a new and general theory of evolution emerging? *Paleobiology* 6:119–30.

1982. Darwinism and the expansion of evolutionary theory. *Science* 216:380–7.

Gould, S. J., and N. Eldredge. 1977. Punctuated equilibria: the tempo and mode of evolution reconsidered. *Paleobiology* 3:115–51.

Gould, S. J., and R. C. Lewontin. 1979. The Spandrels of San Marco and the Panglossian paradigm: a critique of the adaptationist programme. *Proc. R. Soc. Lond. B* 205:581–98.

Gould, S. J., and E. S. Vrba. 1982. Exaptation – A missing term in the science of form. *Paleobiology* 8:4–15.

Grene, M. 1963. *A Portrait of Aristotle.* Chicago: Univ. of Chicago Press.

1974. *The Understanding of Nature.* Dordrecht: Reidel.

1976. Philosophy of medicine: prolegomena to a philosophy of science. *PSA 1976* 2:77–93.

1978a. *Knowledge, Belief and Perception.* New Orleans: Tulane Univ.

1978b. The paradoxes of historicity. *Rev. Met.* 32:15–36.

1981. Changing concepts of Darwinian evolution. *The Monist* 64:195–213.

Hennig, W. 1950. *Grundzüge einer Theorie der Phylogenetischen Systematik.* Berlin: Deutscher Zentralverlag.

1966. *Phylogenetic Systematics.* Trl. D. D. Davis and R. Zangerl. Foreword by D. R. Rosen, G. Nelson, and C. Patterson. Urbana: Univ. of Ill. Press.

Hull, D. L. 1976. Are species really individuals? *Syst. Zool.* 25:174–91.

1980a. Cladism gets sorted out. *Paleobiology* 6:131–6.

1980b. Individuality and selection. *Ann. Rev. Ecol. Syst.* 11:311–32.

Kimura, M. 1976. How genes evolve: a population geneticist's view. *Ann. Génét.* 19:153–68.

King, J. L., and T. H. Jukes. 1969. Non-Darwinian evolution. *Science* 164:788–998.

Kitcher, P. 1982. Genes. *Brit. J. Phil. Sci.* 133:337–59.

Lennox, J. G. 1982. Teleology, Chance and Aristotle's theory of spontaneous generation. *J. Hist. Phil.* 34:1–20.

Levinton, J. S., and C. M. Simon. 1980. A critique of the punctuated equilibria model and implications for the detection of speciation in the fossil record. *Syst. Zool.* 29:130–42.

Lewontin, R. C. 1970. The units of selection. *Ann. Rev. Ecol. Syst.* 1:1–18.

1979. Sociobiology as an adaptationist program. *Beh. Sci.* 24:5–14.

1982. Prospectives, Perspectives and Retrospectives. *Paleobiology* 8:309–13.

Limoges, C. 1980. A Second Glance at Evolutionary Biology in France. In *The Evolutionary Synthesis*, ed. E. Mayr and W. Provine, pp. 322–8. Cambridge, Mass.: Harvard Univ. Press.

Løvtrup, S. 1981. Introduction to Evolutionary Epigenetics. In *Evolution Today*, ed. G. G. E. Scudder and J. L. Reveal, pp. 139–44. Pittsburgh (Carnegie-Mellon): Hunt Institute.

MacIntyre, A. C. 1981. *After Virtue*. Notre Dame: Notre Dame Univ. Press.

Maynard Smith, J. 1981. Evolutionary Games. In *Evolution Today*, ed. G. G. E. Scudder and J. L. Reveal, pp. 1–6. Pittsburgh (Carnegie-Mellon): Hunt Institute.

Mayr, E. 1981. Biological classification: toward a synthesis of opposing methodologies. *Science* 214:510–17.

1982a. *The Growth of Biological Thought: Diversity, Evolution and Inheritance*. Cambridge, Mass.: Harvard Univ. Press.

1982b. Biology is not postage stamp collecting. (Interview) *Science* 216:718–20.

Merleau-Ponty, M. 1968. *Resumés de Cours. Collège de France. 1952–1960*. Paris: Gallimard.

Pantin, C. F. A. 1954. The recognition of species. *Sci. Progr., Lond.* 43(168):578–98.

1968. *The Relations between the Sciences*. Cambridge: Cambridge Univ. Press.

Pattee, H. H., ed. 1973. *Hierarchy Theory: The Challenge of Complex Systems*. New York: Braziller.

Richardson, R. C. 1982. Grades of Organization and the Units of Selection Controversy. *PSA 1982* 1:324–40.

Scudder, G. G. E., and J. L. Reveal. 1981. *Evolution Today*. Pittsburgh (Carnegie-Mellon): Hunt Institute.

Sheppard, P. M. 1958. *Natural Selection and Heredity*. London: Hutchinson.

Simon, H. 1982. Focus on the social sciences, Part II. *Bull. Amer. Acad. Arts & Sci.* 35:26-53.

Sloan, P. 1981. Review of Ruse: *The Darwinian Revolution*. *Phil. Sci.* 48:623–30.

Smith, C. U. M. 1982. Evolution and the problem of mind: Part I. Herbert Spencer. *J. Hist. Biol.* 15:55–88.

Sober, E. 1980. Evolution, Population Thinking and Essentialism. *Phil. Sci.* 47:350–83.

1981. Holism, individualism and the units of selection. *PSA 1980* 2:93–121.

Sober, E., and R. C. Lewontin. 1982. Artifact, cause, and genic selection. *Phil. Sci.* 49:157–80.

Stanley, S. M. 1975. A theory of evolution above the species level. *Proc. Nat. Acad. Sci. USA* 72:647–50.

1979. *Macroevolution. Pattern and Process.* San Francisco: Freeman.

Stebbins, G. L. 1969. *The Basis of Progressive Evolution.* Chapel Hill: Univ. of North Carolina Press.

Stebbins, G. L., and F. J. Ayala. 1981. Is a new evolutionary synthesis necessary? *Science* 213:967–71.

Tuomi, J. 1981. Structure and dynamics of Darwinian evolutionary theory. *Syst. Zool.* 30:22–31.

Turner, J. R. G. 1981. Adaptation and evolution in Heliconius: A defense of Neo-Darwinism. *Ann. Rev. Ecol. Syst.* 12:99–121.

Waddington, C. H. 1953. Epigenetics and evolution. *Symp. Soc. Expt. Biol.* 7:186–99.

1967. The Principle of Archetypes in Evolution. In *Mathematical Challenges to the Neo-Darwinian Interpretation of Evolution,* ed. P. S. Moorhead and M. M. Kaplan, pp. 113–15. Philadelphia: Wistar Institute.

ed. 1969. *Towards a Theoretical Biology, 2.* Edinburgh: Edinburgh Univ. Press.

Wade, M. J. 1978. A critical review of the models of group selection. *Q. Rev. Biol.* 53:101–14.

Wassermann, G. D. 1981. On the nature of the theory of evolution. *Phil. Sci.* 48:416–37.

Wiggins, D. 1980. *Sameness and Substance.* Cambridge, Mass.: Harvard Univ. Press.

Wiley, E. O., and D. R. Brooks. 1982. Victims of history – a nonequilibrium approach to evolution. *Syst. Zool.* 31:1-24.

Williams, G. C. 1966. *Adaptation and Natural Selection.* Princeton: Princeton Univ. Press.

Wilson, D. S. 1980. *The Natural Selection of Populations and Communities.* Menlo Park: Benjamin/Cummings.

Wimsatt, W. C. 1980. Reductionistic Research Strategies and Their Biases in the Units of Selection Controversy. In *Scientific Discovery: Case Studies,* ed. T. Nickles, pp. 213-59. Dordrecht: Reidel.

1981. The units of selection and the structure of the multi-level genome. *PSA 1980* 2:122–83.

Wright, S. 1967. Comments on the Preliminary Working Papers of Eden and Waddington. In *Mathematical Challenges to the Neo-Darwinian Interpretation of Evolution,* ed. P. S. Moorhead and M. M. Kaplan, pp. 117–20. Philadelphia: Wistar Institute.

PART I. THE DEVELOPING SYNTHESIS

PART I. THE DEVELOPING SYNTHESIS

1

Fisher's entrance into evolutionary science: the role of eugenics

BERNARD NORTON

Introduction

The construction of an evolutionary theory is a daring undertaking, involving an attempt to show that millions of individual sequences of biological change, only imperfectly observed via the fossil record, comply with some preferred evolutionary scenario. Under such circumstances there is logical room for rival approaches and for disagreements about the grosser features of the fossil record and about the preferred universal scenario (selectionism, mutationism, etc.).

Historians looking at the rise of evolutionary theory have found rival approaches and disagreement in plenty and have gone on to identify the intellectual and emotional presuppositions that have led the theorists to have such faith in their own products. In so doing, they have brought to the surface levels of motivation that generally remain hidden from view. By such means, it is hoped, the real as opposed to the idealized dynamics of scientific change and progress are to become known.

Hitherto, the bulk of analysis has been focused upon the debates immediately following upon Darwin's work and, especially, upon the biometric–Mendelian debate of the early twentieth century (Mac-Kenzie, 1981, esp. chs. 2–6 and bibliography). The results of these investigations have sometimes been rather alarming. Take, if you will, the case of William Bateson, leader of the Mendelians in the aforementioned debate. Then, if the work of MacKenzie and others is to be trusted – and it seems trustworthy enough – Bateson's insistence that evolutionary change was always discontinuous seems best explained by pointing to his well-attested hatred of utilitarianism. Someone who so passionately hated trade and the nostrums of Samuel Smiles could not abide the gradualist and utilitarian views of traditional Darwinians, who, to Bateson, seemed fit to turn the whole of nature into an exemplification of the most hated of economic philosophies (MacKenzie, 1981, ch. 6).

More recently, under guidance from Professor Provine, historians have turned their attention to the evolutionary works of the twenties and thirties and have learned that rival approaches and disagreements were still the order of the day (Provine, 1971, 1978; Mayr and Provine, 1980). Studies of the works of the Anglo-Saxon "giants" – Wright, Haldane, and Fisher – have shown that "the differences between them are at least as fundamental and instructive as the similarities." Hodge and Provine, for example, have taken an especial interest in the work of Sewall Wright and, naturally, have wondered why he should have vigorously championed ideas of random drift and genic interaction in his work when his colleagues did not. In answer they have detected an interesting range of nonempirical commitments that seem to have been responsible for Wright's scientific direction – notably (a) a belief that nature was efficient and would not depend upon simple mass selection if efficient stockbreeders did not; and (b) a metaphysics of holism and emergentism that led to an interest in interacting levels of hierarchies and in staunch antireductionism. By these factors, Hodge has argued, we can explain Wright's coming to hold a view of evolutionary change in which "superior combinations of genes can turn up in a single individual by statistical luck in the small inbreeding populations, and then spread through the whole species thanks to the partial outbreeding arising from a slight but effective degree of migration linking them together" (Provine, in press).

Interest in Wright is easy to understand, for his works and persona have a very attractive quality, to be contrasted with the mathematical and psychological abrasiveness of Haldane and Fisher (Box, 1978; Werskey, 1978). Of the latter, Fisher has received the greater attention, especially after the publication of a biography by his daughter, which, thankfully, puts truth above filial niceties (Box, 1978).

The Fisher problem

As Fisher's life and works are better understood, a distinctive historical problem – the Fisher problem – has become apparent. This has several interlocking parts. First, there are the personal details: Fisher went to Cambridge as a mathematician, having abandoned biology at school. Yet, later on, he emerges as an evolutionary biologist – not a matter to be explained by market forces![1] Second, there

---

[1] Box (1978) makes clear the difficulties facing the young Fisher when attempting to find employment.

is the matter of his politics: Fisher seems to have been an ardent eugenist, with forceful opinions on such matters as the proper distribution of family allowances among the different classes in society (see, e.g., Fisher, 1932, 1943). Furthermore, this eugenic interest seems not to have been separate from his evolutionary biology. The second half of Fisher's masterwork, the *Genetical Theory of Natural Selection*, was given to eugenic questions – to explaining the collapse of great empires, and so on (Fisher, 1930). The intended integration of the two halves is plain for all to see. Third, there is the very nature of Fisher's evolutionary work: There appears to be no discussion of speciation, for example, and its focus is forever upon natural selection seen as acting upon single-gene effects in very large populations, with no role given to drift or genic interaction. The high spot of his work (as seen by Fisher), the Fundamental Theorem of Natural Selection, asserted that the rate of increase in fitness (actuarially defined) at any time was equal to the genetic variance in fitness. This mathematical masterpiece, only tenuously connected to the appearance of new species on earth, was seen by Fisher as being comparable to the most general laws governing the behavior of gases comprised by large numbers of individual molecules (Fisher, 1930, p. 36; Price, 1972).

To summarize. The "Fisher problem," in crude approximation, is one of understanding why a mathematician should have taken to evolution, why this evolution should have been so seamlessly conjoined with eugenics, and why Fisher should have constructed a preferred evolutionary scenario in which primary importance was given to the action of natural selection upon large Mendelian populations producing ever-increasing levels of fitness in those populations. The elements of a solution to the first two of these matters are at hand, and are confirmed by Joan Fisher Box's biography. It may be argued that the observations that "solve" the first two components of the Fisher problem also solve the third – though such claims have an unsatisfactory vagueness about them, and, as the cliché has it, call for more work.

The solution

The key to understanding Fisher appears to be eugenics. Eugenics was the dog that wagged the tail of population genetics and evolutionary theory, not the other way about. Evidence for this bold assertion has been provided by Alan Cock (in correspondence), Don MacKenzie (1981), and Geoffrey Searle (1976), and was confirmed by the publication of selected contents of the Cambridge University Eugenics Society Minute Book, preserved in the offices of the Eugenics

Society in Ecclestone Square, London (Norton and Pearson, 1976).
From this it appeared that Fisher and other undergraduates had set
up the society in May 1911 with the assistance of such illuminati as
John Maynard Keynes, R. C. Punnett, and Horace Darwin, and that
the Cambridge eugenists shared fully in the wave of eugenic enthusiasm which swept through England in the late-Edwardian period,
largely in response to fears of social unrest at home and imperial decay
abroad. The full extent of Fisher's involvement in the movement is
brought out in his daughter's book, which depicts him as a brilliant
if unsettled undergraduate mathematician with strong idealistic
yearnings to lead a "eugenic life."

Something of his intensity may be garnered from the contents of
a speech of 1914 where he insisted that in any society where members
of small families enjoyed some social advantage over the members
of large ones:

> The qualities of all kinds, physical, mental and moral which
> go to make up what may be called "resultant sterility" tend,
> other things being equal, to rise steadily in the social scale; so
> that in such a society, the highest social strata, containing the
> finest representatives of ability, beauty and taste which the
> nation can provide, will have, apart from individual induce-
> ments, the smallest proportion of descendants; and this dys-
> genic effect of social selection will extend throughout every
> class in which any degree of resultant sterility provides a so-
> cial advantage. [Box, 1978, p. 31]

Under existing social conditions, Fisher felt, members of small families, especially in the middle classes, did have such an advantage.
It followed that the nation could be preserved from genetic erosion,
from dysgenic decay, only by changes in patterns of family-support
payments, encouraged by persons like himself anxious to lead the
eugenic life, which, according to Fisher, involved marrying spouses
of outstanding ability, health, and beauty and having more children
than other people. Fisher, his daughter shows, had the merit of prac-
ticing what he preached, in marked contrast to many other eugenists.
It was this commitment to eugenics, continuing throughout Fisher's
life – which included a spell as Galton Professor of Eugenics at University College London – that led him from mathematics back to biological questions. The route was not an entirely new one, for in
previous years it had been taken by Francis Galton, the father of
eugenics (MacKenzie, 1981, ch. 2), and especially by Karl Pearson,
his greatest prophet and first incumbent of the Galton chair.

Pearson, we now know, was led away from "ordinary" mathematics and into biometrics because it seemed to offer him a way of

putting his abilities into the creation of a truly scientific form of social Darwinism. If one knew the mathematical laws of heredity, Pearson argued, one would be able to control the nature and quality of the *populus*, thereby ensuring the survival, even the dominance, of one's nation in the struggle between races and nations that was always going on. Like Marx, Pearson believed history to be the history of conflicts between social groups with different interests, but whereas Marx saw these social groups as social classes, Pearson saw them as races or nations (Norton, 1978*b*). In this view, the key to history was race improvement. Mathematical evolution – or biometry – was Pearson's version of the key, and it should no longer surprise us that this distinguished statistician should have had a not unkind word for Hitler in his retirement speech in 1934 (Pearson, 1934).

By the time Fisher got to Cambridge and formed the eugenic society there, the bulk of Pearson's work had been done. But, done or not, it was surrounded by controversy because Pearson held that it was inconsistent with, and, in fact, discredited, the rising new science of Mendelian genetics. The reasons for this opposition, now regarded as specious, have been much investigated and do not bear repeating here. Suffice it to say that when Fisher became interested in eugenics, the crucial science of heredity was divided between the two camps of Mendelian and biometrician. The former, led by Bateson, held that heredity was controlled by the transmission of "unit characters" according to Mendelian formulas; the latter, led by Pearson, held that the formulas were of limited applicability and failed completely to deal with the continuous variation so interesting to the eugenist (Norton, 1975). The biometricians, moreover, would have no truck with Mendelian "unit characters" or "factors," but viewed continuous variation as being best dealt with by the use of multiple correlation methods that were in a strong sense "theory-free" and would yield a variety of predictive formulas known generally as the Law of Ancestral Heredity, having the form

$$X_0 = a_1X_1 + a_2X_2 + \cdots + a_nX_n$$

where X_0 was the expected height or intelligence of the offspring, the X_is were the observed heights or intelligences of their fathers, grandfathers, and so forth, and the coefficients a_i were functions of the variances of the different generations in respect of height or intelligence and of the correlation coefficients connecting the different generations in respect to these things (Froggatt and Nevin, 1971). The main point is obvious: Here was a system for dealing with continuous variation that has considerable mathematical sophistication, but seemed at odds with the exciting new science of Mendelism, then being developed at Cambridge under Punnett, Biffen, and others.

Fisher, the Minute Book reveals (Norton and Pearson, 1976), was very concerned about the lack of theoretical harmony in the contemporary study of heredity, and, in an amazing paper given in 1911, extolled the virtues of making a synthesis between Mendelism and biometry. To do this, it would be necessary to show that Mendelism could explain the observed correlations between relatives in respect to continuously varying characters, and that it could explain the Law of Ancestral Heredity in much the same way as the kinetic theory could explain the gas laws. While at Cambridge, Fisher seems to have been unable to complete this task, though it is clear that he took a great deal of interest in the more purely mathematical aspects of biometry. On leaving Cambridge he went on to a variety of unsatisfactory positions, being saved from desolation and despair by a few warm human contacts, notably with Darwin's son, the eugenist Major Leonard Darwin, who found him part-time employment with the Eugenics Education Society and encouraged him in eugenic-related researches (Box, 1978, p. 51). Among the latter, no doubt, was the completion of his scheme for the unification of biometric and Mendelian methods first outlined in the 1911 paper, which was to be published as "The Correlation Between Relatives on the Supposition of Mendelian Inheritance," though not without a struggle (Fisher, 1918b). Records kept at the Royal Society show that the paper, or some version of it, was submitted to the society in 1916, but was withdrawn after being unfavorably refereed by the biometrician Pearson and the Mendelian Punnett (Norton and Pearson, 1976). Only through the financial assistance of Leonard Darwin did the paper ever see the light of day – in the pages of the *Transactions of the Royal Society of Edinburgh*, where it was published in 1918.

This, undoubtedly, was one of Fisher's greatest works, and, even if neglected at the time, is now seen as offering the conclusive proof that Mendelism could, after all, account for the inheritance of continuous variation, for the Law of Ancestral Heredity, and so on. Moreover, it offered a sophisticated theory in which matters like dominance and linkage were taken into account, and that allowed estimates to be made of the proportions of variance in a population that might be ascribed to genetic factors of different sorts (e.g., additive and dominance variance) and those that might be ascribed to environmental factors. Truly, in this paper, Fisher opened a path leading straight to the Jensenist debates of the seventies, concentrated as they were on matters of environment versus heredity (Kamin, 1974). What is particularly interesting is that Fisher's major article was accompanied by a more popular one in the *Eugenics Review* that made clear how it was that Fisher considered that the possibly rather academic business of reconciling biometry and Mendelism gave practical assistance to the advancement of eugenics (Fisher, 1918a; Norton, 1978a). For in this paper "The Causes of Human Variability," Fisher argued that existing methods of establishing the central eugenic thesis of the dominance of nature over nurture were deficient – and that his new method was not. Previously, he argued, eugenists like Pearson had contrasted, say, the value of the parent–son correlation in respect of height with some other correlation coefficient connecting height with some likely environmental factor – for example, level of nutrition. Seeing that the father–son correlation came out much the higher, Pearson would argue for the supremacy of heredity over environment. Fisher could see that this was a very bad argument, and that:

> An advocate of the importance of environment might easily point to a dozen causes to which height or shortness is commonly ascribed, such as regular athletic exercise, or accidental illness in childhood, and it would be difficult to prove without a specially designed investigation for each alleged cause that these do not contribute important proportions of the total. The task of ascertaining the importance of environment in this way is an endless one, since always new environmental causes could be suggested, each more difficult than the last to define, measure and investigate. [Fisher, 1918a, p. 215]

Through the use of this new technique of the analysis of phenotypic variance, Fisher believed that the possibility of appeal to hitherto untested environmental variables could be ruled out, for the new technique automatically considered all environmental considerations. Further evidence that Fisher's first, triumphant entrance into population genetics was shaped and motivated by his eugenics is provided by his touching acknowledgment to Major Leonard Darwin at the end of the paper.

Evidence such as this surely does prove the claim that Fisher's entrance into biological work was consequential upon his espousal of eugenics at Cambridge and, indeed, was shaped by this interest. It is probably futile to speculate upon the causes of this commitment, though it is hard not to follow MacKenzie in seeing eugenics as a movement that advanced the financial and social interests of members of the professional middle class like the Fisher family (MacKenzie, 1976). But, over and above this there seems to have been a set of personal factors leading the young Fisher to embrace a cause to which most of his fellow undergraduates would give only the cold or tepid shoulder. Amateur psychoanalysis is a notorious minefield, but Joan Fisher Box's biography tells the story of a man who had often failed to communicate with his mother, a failure that, in turn, may have been responsible for his development into someone who "grew up without developing a sensitivity to the ordinary human behavior of his fellows" and who was to suffer from insecurity in personal relationships throughout his life (Box, 1978, p. 10). As is often the case, such problems were accompanied by the need for a personal sense of mission, and, in Fisher's case, this seems to have been provided by eugenics. His daughter speaks of his "idealistic nature and eugenic convictions," and it is interesting to note that at one time he seriously considered becoming a subsistence farmer: "As distinct from commercial farming, it was a way of life, not a way of earning money . . . It was the only profession in which a large family was a social advantage" (Box, 1978, p. 38). It would seem, in short, that there was something deeply idealistic about Fisher's eugenics – for, after all, he became a eugenist before his own "proofs" of the supremacy of nature over nurture were provided. Fifty years after leaving Cambridge, his friend the eugenist Stock wrote that Fisher was *"the only man I knew* to practice eugenics" (Box, 1978, p. 32), and so, perhaps, it is unsurprising that he should dedicate his later work, the *Genetical theory of natural selection* to Leonard Darwin "In gratitude for the encouragement, given to the author, during the last fifteen years, by discussing many of the problems dealt with in this book."

So far then, it would seem that Fisher's entrance into serious work in population genetics was brought about and closely shaped by his eugenic interests, and this argument seems to have been generally accepted. But, if this resolution of the first two parts of the Fisher problem is to be presented also as a solution to the third part of the problem, then some further argument is needed. Even the integration of eugenic chapters into the *Genetical Theory* is not, by itself, sufficient to prove that what was distinctive about Fisher's evolutionary work (Hodge, n.d.; MacKenzie, 1981, ch. 8) – his neglect of speciation, diversification, extinction, and related matters and his use of the Fundamental Theorem to buttress his view that evolution occurred mainly, if not exclusively, through the agency of natural selection acting upon Mendelian variation in large interbreeding populations – is the direct consequence of his eugenic orientation. Nor, indeed, is there "proof" in the fact that the work leading up to the *Genetical Theory* was undertaken in the years immediately following the publication of the 1918 paper on the correlation between relatives.

In circumstances such as this, however, conclusive proof is unlikely to be the order of the day. There is little short of an autobiographical note admitting that his whole view of nature was consequent upon opinions formed while he was an undergraduate eugenist that would do the job of *proving* a strong relationship between Fisher's eugenics and his evolutionary biology. Such evidence, for obvious reasons, is unlikely to be forthcoming. It is not surprising, therefore, that a number of authors have advanced the argument by showing, quite simply, Fisher's evolutionary work to be just the sort of work that licensed the sort of eugenics favored by Fisher – one in which those sections of the population with gifts of intellect, beauty, and so forth were encouraged to breed by the payment of money-allowances. Only by these means, Fisher argued, could the collapse of civilization – so often seen in the past – be staved off and the conditions requisite for "permanent civilization" be procured. When one looks at things in this manner, the case for supposing a strong relationship between Fisher's eugenics and his evolutionary theory becomes almost overwhelming. A eugenist is not interested in the origin of species, simply in the best manner of improving the existing one. Such a concern is surely paralleled in Fisher's overriding interest in a theorem like the Fundamental Theorem, which is not about the creation of new species, but about the rate of change of an existing one. Second, the eugenist takes a view of things in which improvement, advancement – call it what you may – can only be brought about by constant, conscious intervention in the reproductive process. Any view of nature that allows important roles to drift or to mutation in the production of population change does not offer a decent basis for claiming that nature sanctions eugenic measures. Indeed, a mutationist like Bateson, who held progress to be the effect of large, randomly occurring discontinuous variations, was unsympathetic – even mocking – to the eugenists. Similarly, one feels that Wright's indifference to eugenics, unusual among his peers, cannot be unrelated to his view of a nature in which progress may come through agencies beyond the eugenists' calculation: "The most effective process either of perfecting an adaptation along a particular line, or of exploiting a major ecological opportunity is thus not, in the long run, the almost deterministic pressure of mass selection acting on recurrent mutations, but rather one of continuous trial and error, made possible by the labile balance among all of the factors" (Wright, 1961).

In short, Fisher's view of evolution (Price, 1972) lacks just those things that a eugenist might be expected to neglect, and, more importantly, it offers a view of nature that is harmonious with the central tenet of eugenism – that progress comes through selection. Observations like these provide evidence for the thesis that Fisher's eugenics – known now to be temporally earlier than his evolutionary work – was an important determinant of the shape of his theory. All in all, there would seem to be good grounds for giving Fisher's eugenic concerns a very important place when trying to understand his life's amazing and wonderful labors in the field of biological theory.

References

Box, J. F. 1978. *R. A. Fisher: Life of a Scientist.* New York: Wiley.

Fisher, R. A. 1918a. The causes of human variability. *Eug. Rev. 10*:213–20.

    1918b. The correlation between relatives on the supposition of Mendelian inheritance. *Trans. Roy. Soc. Edinburgh 52*:399–433.

    1930. *The Genetical Theory of Natural Selection.* Oxford: Clarendon Press.

    1932. Family allowances in the contemporary economic situation. *Eug. Rev. 24*:87–95.

    1943. The birthrate and family allowances. *Agenda 2*:123–33.

Froggatt, P., and N. C. Nevin, 1971. The 'law of ancestral heredity' and the Mendelian Ancestrian controversy in England, 1889–1906, *J. Med. Gen. 8*:1–36.

Hodge, J. n.d. The synthetic theory of evolution: Notes on some historical points. Ms.; copies available from author (Division of History and Philosophy of Science, Philosophy Department, Univ. of Leeds, Leeds, England).

Kamin, L. 1974. *The Science and Politics of IQ.* Potomac, Md.: Erlbaum.

MacKenzie, D. A. 1976. Eugenics in Britain. *Soc. Stud. Sci. 6*:499–532.

    1981. *Statistics in Britain 1865–1930. The Social Construction of Scientific Knowledge.* Edinburgh: Edinburgh Univ. Press.

Mayr, E., and W. B. Provine, 1980. *The Evolutionary Synthesis.* Cambridge: Cambridge Univ. Press.

Norton, B. 1975. Metaphysics and population genetics: Karl Pearson and the background to Fisher's multi-factorial theory of inheritance. *Ann. Sci. 32*:537–53.

    1978a. Fisher and the Neo-Darwinian Synthesis. In *Human Implications of Scientific Advance*, ed. E. G. Forbes, pp. 481–94. *Proc. 15 Int. Cong. Hist. Sci.* Edinburgh: Edinburgh Univ. Press.

    1978b Karl Pearson and statistics: the social origins of scientific innovation. *Soc. Stud. Sci. 8*:3–34.

Norton, B., and E. S. Pearson, 1976. A note on the background to, and refereeing of, R. A. Fisher's 1918 paper "On the correlation between relatives on the supposition of Mendelian inheritance." *Notes and Records of the Roy. Soc. 32*:151–62.

Pearson, K. 1934. Speeches delivered at a dinner held at University College, London, in honour of Professor Karl Pearson, 23 April 1934. Privately printed, copies in the Pearson Archive, University College, London. London WC 1 E 6 BT.

Price, G. R. 1972. Fisher's "Fundamental Theorem" made clear. *Ann. Hum. Gen. 36*:129–40.

Provine, W. B. 1971. *The Origin of Theoretical Population Genetics.* Chicago: Univ. of Chicago Press.

1978. The role of mathematical population genetics in the evolutionary synthesis of the 1930's and 1940's. *Stud. Hist. Biol.* 2:167–92.

In press. *Sewall Wright: Geneticist and Evolutionist.* Chicago: Univ. of Chicago Press.

Searle, G. R. 1976. *Eugenics and Politics in Britain, 1900–1914.* Oxford: Blackwell.

Werskey, P. G. 1978. *The Visible College.* London: Allen Lane.

Wright, S. 1964. Biology and the Philosophy of Science. In *Process and Divinity, the Hartshorne Festschrift,* ed. W. L. Reese and E. Freeman, pp. 101–25. La Salle, Ill.: Open Court.

2

The abandonment of Lamarckian explanations: the case of climatic parallelism of animal characteristics

BERNHARD RENSCH

Introduction

In 1974 I was unable to participate in a conference about the "synthetic theory of evolution," arranged by Ernst Mayr. He, therefore, asked me to write an article about "the historical development of the present synthetic neo-Darwinism in Germany." This article has since appeared in the important book of Mayr and Provine (1980). However, on rereading this, I gained the impression that my comments were relatively brief. It may, therefore, be useful to supplement them by more detailed information concerning my own publications and the relevant literature between 1920 and 1940. I shall quote a number of passages; and as nearly all the relevant literature is published in German, I shall translate into English as literally as possible. I shall not discuss the Lamarckian opinions of paleontologists or the discussions of the common conference of geneticists and paleontologists at Tübingen in 1929 (see Weidenreich, 1929; Plate, 1931) because I have already mentioned this in my article (Rensch, 1980).

Lamarckian explanations between 1920 and 1940

When I was studying biology, chemistry, and philosophy in the University of Halle (S.) between 1920 and 1922, the director of the Zoological Institute, the geneticist Valentin Haecker, suggested that I work on cytological differences during the embryological development of giant, medium-sized, and dwarf races of domestic fowl. My results were to contribute to his research on the developmental analysis of different characteristics, a new field of science that he had called "Phänogenetik" (phenogenetics). His book with that title appeared in 1918 (Haecker, 1918).

In his lectures about genetics and evolution Haecker treated all the relevant problems. We could supplement this by reading his book

Allgemeine Vererbungslehre (General Genetics), the third edition of which had just appeared (Haecker, 1921). We learned that speciation came about by natural selection of hereditary varieties in Weismann's sense. Haecker was convinced that a "heredity of acquired characteristics" must be regarded as being very improbable, because a chain of causal events which phenotypically altered the soma (S) and possibly also genes (G) (S-m-n-o-G) could never be identical with a causal chain leading from the altered genes over embryonic stages back to the characteristic S, but would run G-v-w-x-y-Z. However, he also discussed all possibilities that, in spite of this objection, could lead to such alteration of genes that an induced phenotypical alteration became heritable. In his book (1921) he wrote (p. 154): If one assumes that different "virtual possibilities of individual development" exist, then "a parallel activation (*Parallelaktivierung*) could take place through such latent general potentialities in genetic and somatic cells by means of the altered chemical processes." Such an assumption would perhaps explain the parallelism between the phenotypic and genotypic alteration of wing patterns in certain butterflies. He also believed that it was possible that "constitutional concussion" of genetic and somatic cells could cause a parallel reduction of the resistance against illness. In his book on phenogenetics, Haecker had already written (1918) that a "parallel induction" of somatic and germ cells would be possible only "when characteristics acquired by the parents were already prepared by a disposition, by a virtual potential of the plasma." In such cases, general capabilities of development in somatic and germ cells would become effective.

In *Das Problem der Vererbung erworbener Eigenschaften* (The Problem of the Inheritance of Acquired Characteristics) Semon wrote: "The possibility of an induction of germ cells by external stimuli and corresponding states of germ cells (sensitive period) is not doubted by anybody at present" (1912, p. 177). He explained this by saying that "the new potencies of germ cells are products of stimuli or residuals of excitations, engrams" (p. 175). Strong stimulations by temperature and moisture are the main causes of such engraphic alterations of the germ cells.

Plate (1913) believed that besides speciation by natural selection "inheritance of acquired characteristics" must be assumed to a much higher degree. He was convinced that this had been sufficiently proved by the experiments of Fischer (1901) and Standfuss (1898). Both authors had produced, by means of cold or heat, alterations of the patterns of butterflies that could partly be observed in subsequent generations developed under natural conditions. He also quoted similar results of Tower (1906) with beetles of the genus *Leptinotarsa*. Later on (1931) when Plate discussed the results of the Tübingen conference in 1929, he still defended the possibility of Lamarckian explanations. He pointed out that in long series of generations, characters that became modified phenotypically could become hereditary and he also mentioned that different degrees of genetic stability exist. This article ends with the following sentence: "Hence the present correct opinion means that a biologist and geneticist has to recognize both principles, the Lamarckian as well as the selectionistic" (Plate, 1931, p. 292).

The paleontologist Weidenreich (1921) held the opinion that the evolution of organisms first tried out adaptations to new habitat conditions that then became inherited. He assumed "that no observations exist which justify the conclusion that alterations of the type arise by spontaneous inner alterations of the constitution which cannot be traced back to external effects of some kind" (Weidenreich, 1921, p. 92).

In the leading textbooks of biology that were used between 1920–40, speciation was normally explained by selection of genetical varieties, but the possibility of Lamarckian explanations was not totally denied. In *Allgemeine Biologie* ("General Biology") the cytologist and geneticist Hartmann wrote: "It must be left to future research to decide whether and how far influences of the external world are able to change genes by alteration of the plasma of the germ cells, so that to a certain degree Lamarckian principles become acceptable. Of course, such decisions have to be based on experiments" (Hartmann, 1927, p. 625).

In the tenth edition of the large *Lehrbuch der Zoologie* (Textbook of Zoology) of Claus, Grobben, and Kühn, the part concerning general zoology had been written by the geneticist and developmental physiologist Kühn. He expressed the following opinion: "It is an open question whether a Lamarckian principle which causes direct adaptations plays a role in phylogeny. Among well-known processes only consolidated modifications can be taken into consideration" (Claus, Grobben, and Kühn 1932, p. 382). Such persistent modifications (*Dauermodifikationen*) produced in *Protozoa* and *Drosophila* were attributed to alterations of the protoplasm of germ cells, but they always disappeared after a number of generations (see Hammerling, 1929; Jollos, 1933). But Kühn mentioned: "The most persistent consolidated modification which we know is the increase in heat resistance proportional to size in *Actinophrys sol* effected by the influence of a 35 °C temperature during the period of fertilization and encysting. Such alterations have been maintained for more than four years under normal conditions of cultivation and outlasted eight successive fertilizations" (Claus, Grobben, and Kühn, 1932, p. 377). This case strengthened my own opinion that it should be possible to develop inherited characteristics by means of such persisting modifications.

Also, Stempell expressed a similar opinion in his *Zoologie im Grundriss* (Zoology in Outline): "The assumption that influences of the environment which hit the whole body during long periods (hologenic induction) can change persistent modifications into inherited characteristics is probable. And it is not out of the question that influences which last for shorter periods also at least damage and therefore alter the plasma of germ cells in the sense of indirect somatic induction" (Stempell, 1935, p. 715).

The second volume of the second edition of Hesse and Doflein, *Tierbau und Tierleben* (Structure and Life of Animals), written by the comparative anatomist and ecologist R. Hesse (1943), ends with the following sentence: "As ecology, based on facts of convergence, and paleontology, based on its data, are pressed to assume a direct influence of the environment on the alteration of inherited characteristics, a denial is not in place. We have to do here with the gaps in our knowledge" (Hesse, 1943, p. 820).

In *Grundzüge der allgemeinen Biologie* (Principles of General Biology), Woltereck held the opinion: "that Lamarckism as well as Darwinism has won new credit by the new discoveries of induced mutations and numerous small mutants" (Woltereck, 1932, p. 403). He also assumed an "immanent drive of phylogenetic unfolding" (factor of Göbel), which sometimes led to excessive structures (hypermorphosis). Berg (1926) called such a lawlike drive "nomogenesis." Buchner wrote in *Allgemeine Zoologie* (General Zoology) (1938) that one has to take into consideration all the arguments for Lamarckism and for selectionism: "Then on the one hand the assumption of selection, which increases the quality, will find recognition and on the other hand an organism with all its purposeful adaptations will not be regarded as the result of accidental events but as a being which determines shape and destiny itself and which always works pugnaciously for improvement and reveals a drive of shaping which cannot be measured only by utility" (Buchner, 1938, p. 301). Buchner also mentioned persisting modifications that show the possibility that inherited characteristics might arise. Further on he recognized an endogenous drive for shaping, causing orthogenetic trends in phylogeny.

In the same period, between 1920 and 1940, research on functional anatomy also led to Lamarckian explanations. The anatomist Böker, to whom we are indebted for having analyzed many functional structures in an instructive manner (two vols. 1935, 1937) regarded the capability to react actively to the conditions of the environment by anatomical alterations as "a basic characteristic of living matter." He assumed that such alterations (*Umkonstruktionen*) could become engrained in the germ cells by what he called parallel mnemic effects in the mechanisms of hereditary development (Böker, 1936). He remarked that geneticists do not sufficiently investigate anatomical constructions (1935, p. 20).

Harms believed that his investigations about the adaptations of marine animals to life on land had shown that such evolution proceeded by means of direct influences of the environment, and partly by persisting modifications. He wrote: "The genes and radicals originate through stimuli of the environment that influence germ cells and soma. They are therefore induced and they remain effective after the stimulus of the environment no longer exists" (Harms, 1934, p. 200).

Among paleontologists, Lamarckian explanations of speciation were widely held in the period between 1920–40 (and sometimes even later), although the effect of selection was not denied. In order to explain major phylogenetic steps leading to new types of anatomical construction, several paleontologists supposed an effect of "shaping principles," "autonomous forces of development," an "endogenous drive to unfold," "a creative principle" (Osborn, 1934: "aristogenesis"), an endogenous tendency to "directed phylogenetic development" (orthogenesis), or even a "will to the organism's own free shaping" (Beurlen, 1937). But I will leave out of account a discussion of these ideas, since such concepts only characterized certain phylogenetic processes but did not help to analyze and explain them (see Rensch, 1980). Even in 1952, the paleontologist Peyer still believed that the effects of stimuli could become heritable. His conception was based on the statement that certain chemical compounds release mutations. As all functions of organs are connected with chemical alterations, he argued that it is necessary to weigh (*erwägen*) the possibility that mutations arise that are similar to noninherited phenocopies.

The development of my personal conception between 1920 and 1934

I have tried to make plain that, in the period in question, biologists were normally convinced that natural selection of hereditary varieties was the most important factor of speciation, but that, in some cases, Lamarckian explanations were also justified. It seemed to be particularly significant that many phenotypic and genotypic alterations, both ultimately produced by the same environmental influence, run parallel, and that some persisting modifications have been maintained through many generations without further environmental influence. Jollos had already assumed what we have here are only alterations of the protoplasm of the germ cells. However, I thought that these alterations could also have an effect upon the chromosomes, particularly in the same phase of the division of the nuclei. We often discussed the relation between nucleus and protoplasm in my student years. We learned that genes can probably be regarded as discrete "proteinlike biomolecules." The pleiotropic effect of genes was nearly unknown, although I myself had found that the iridescent structure of the radii of feathers of several birds was only a by-product of the effect of genes that cause the overproduction of melanin (Rensch, 1925a).

When I was a young assistant at the Zoological Museum at Berlin, I was not a pure Lamarckian. I was convinced that the protective coloration of animals came about by natural selection. I carried out experiments in order to prove that the far-reaching similarity between the egg patterns of the European cuckoo (*Cuculus canorus*) and the egg patterns of different host birds came about by the agency of the hosts themselves, which removed a foreign egg from their clutch when its color and pattern differed to some degree. I played the roie of a cuckoo myself. I went in search of thirty-five nests of songbirds and I always painted one egg with black or brownish black spots or points or altered the white ground-color to red. When I did this in clutches of host birds (genera *Sylvia, Lanius, and Emberiza*), the breeding bird removed the painted egg if it contrasted sufficiently with its own eggs, but tolerated the painted egg when it was relatively similar (Rensch, 1924). I supplemented my explanation by referring to the well-known observation that cuckoo eggs found in clutches of the wren (*Troglodytes*) always differ from the eggs of this host, which cannot see the difference in the dark interior of its ball-like nest. I also explained the origin of the instinctive behavior of the newly hatched cuckoo, which removes the eggs and the young host birds from the nest, by assuming processes of selection (Rensch, 1925b).

My research on the histologic structure of the metallic colors of feathers (1925a,c) led to the conclusion that this structure is only caused by increase of melanin in the radii, for albinos have normal radii with hooks and spines. Such an increase is typical for birds in warm and moist regions. Hence, it was an environmental influence that produced the metallic structures. As a similar increase in melanin can be caused experimentally by a warm, moist milieu, I concluded that phenotypic alterations could become genotypic characteristics by the cumulative effect of the same influences in the course of many generations.

During my expedition to the Lesser Sunda-Islands in 1927 (see Rensch, 1930) I was able to prove that the birds of Lombok, Sumbawa, and Flores show a more colorful plumage and also more yellow and green colors than birds in Germany. In both cases the difference was 10 percent. In this case, I also concluded: "that the difference of the coloration between tropical and cooler regions was mainly caused by natural selection and only to a minor degree by direct climatic effects (iridescent colors)" (p. 160).

My Lamarckian explanations were mainly based on my investigations on the climatic parallelism of size and color in geographic races of birds and mammals. Most races of different species follow Bergmann's rule, that is to say, the races that live in cooler regions proved to be larger than races of the same species that live in warmer regions. Neighboring races of different size were often connected by gradual transitions. Races of western Germany are normally a little smaller than races of eastern Germany, Poland, and Scandinavia. I could not imagine that such small size differences could have a selective value. The same holds true for differences of gray and brown colors. Tits like *Parus atricapillus* and *P. palustris* or mammals like hares (*Lepus*) from western Germany are brownish-gray on the back whereas corresponding races of the same species in eastern Germany, Poland, and Scandinavia are more or less gray. In this case, too, it seemed to be improbable that these small but undoubtedly inherited differences should be produced by natural selection, because both color types afford their bearers sufficient protection from birds of prey. I, therefore, assumed a direct influence of the climate that first modified size and color phenotypically and later on, after long chains of generations had been influenced, also altered the corresponding genes. This assumption was supported by experimental observations proving that warm-blooded animals become larger and less brownish when they mature at relatively low temperatures. The same holds for the relative tail-length of mice, which is less in the races of cooler regions (Allen's rule). When Sumner (1923–24) raised mice of the genus *Peromyscus*, the animals became larger but had relatively shorter tails when the temperature was low.

In my first book on speciation (1929), I, therefore, summarized my conception in the following way: "If we assume that geographic races originate by mutations, then the experimentally produced phenotypic varieties would be parallel only incidentally. As it is not only isolated examples that show this parallelism, but we have to do with a general relation, such chance events and the assumption that geographical races came about by mutations do not come into question" (p. 165). Mutations of free-living birds, which had been investigated during this same period by Stresemann (1926), were always characterized by conspicuous differences of color or pattern. I, therefore, concluded: "that inherited geographical races primarily originated as pheno-varieties, which in long chains of generations gradually became hereditary races because of the constant climatic effect" (p. 167). The arguments of Alverdes (1921) concerning "accumulated after effects" endorsed my conception.

I defended my Lamarckian explanations for the last time when I had been invited to report about problems of speciation during the congress of the German Zoological Society in 1933. Meanwhile, the knowledge of these problems had made considerable progress, and I was no longer sure that my conception was sufficiently well founded. I, therefore, introduced my report with a restrictive comment: "The following theoretical discussion seeks not so much to find a solution of the problems as to *exhibit the difficulties* that arise if we assume that geographical races originate by means of undirected mutations" (Rensch, 1933, p. 66). Yet, I added: "The observation that the above mentioned characteristics of size proportions and color of many European birds and mammals are directly determined by climatic factors is a sentence which geneticists as well as systematists must recognize" (Rensch, 1933, p. 67).

Since 1934, I have tried, as far as possible, to explain the climatic parallelism of race characteristics through natural selection. As early as 1934 during the Eighth International Ornithological Congress at Oxford (see Rensch, 1938a), I pointed out that the rules I had recently formulated about the egg number of birds per clutch and about the shape of the wings of songbirds in cooler and warmer regions seem to be the result of natural selection. When a bird species spread into a colder region during the postglacial time, those genetically established varieties that had more pointed wings had an advantage and produced more offspring, because they had to balance the increased losses of individuals during winter. However, such an explanation seemed not to be possible when a species spread from a colder to a warmer region, as was apparently the case with the tit *Parus major.* This species, at present, is also distributed in India and the western Malay Archipelago and has less pointed wings than the European race. And why did this species reduce the number of eggs per clutch to only three to four in India? I emphasized such doubts again in a later publication (1936), in which I discussed more rules of climatic parallelism of the races of a species. But on the other hand, I could show that racial size differences of the tit *Parus atricapillus* run parallel to the isotherms of January in Europe and that the winter minima can, therefore, be regarded as the selecting factor with respect to body size. This assumption could be confirmed by a comparison of the birds of the Canary Islands with the corresponding races of Algiers and Tunis, where the winter is much colder.

Later I realized that it is necessary to take into account the correlations of structures and organs and their different growth gradients. This proved to be particularly important for the understanding of the differences in relative length of tail, ears, and hindfeet of races of mammals and the relative bill-length of birds, both of which depend upon differences in body size (Bergmann's rule). And such relations also led me to a possible understanding why birds that spread into tropical regions become smaller: "Most warm-blooded animals have apparently developed a favorable relation between their body size and their specific physiological constants, including the relations to temperature. Such a favorable relation could be developed in warmer regions by processes of selection which favored a reduction of the body size and, vice versa, relatively greater surface of the body would prevent overheating" (Rensch, 1939a, p. 118).

Finally, I also found an explanation for the fact that the dependence of the size of races of warm-blooded animals runs parallel to experimental modifications that come about by altering the temperature immediately after birth. I had previously failed to take into account a publication of Baldwin that had already appeared in 1902. This author had discussed the possibility that after the spreading of a population into a region with different conditions, the individuals of the first generations might be modified only phenotypically, but that later on corresponding heritable varieties would "step in," because it is more advantageous to be adapted immediately to the altered conditions so that it is not necessary for each individual to alter certain characteristics in order to be successful under the new conditions.

My conviction, which I had held since 1934, that speciation only comes about by altered gene arrangements and natural selection strengthened my opinion that the transspecific development of new structures, new organs, totally new anatomical constructions, and evolutionary progress as well could also be explained by undirected mutations, new gene combinations, processes of isolation, and natural selection. It was particularly important for such explanations to take into account the correlations between structures and their growth-gradients and also the processes of compensation of material in the phase of individual development. This view also allowed an understanding of the so-called explosive development of new species and genera at the beginning of new lines of descent and the slowing down of the evolutionary tempo after the different habitats had been occupied by correspondingly adapted species. I published a relatively brief report immediately before the beginning of the war (1939*b*). A detailed discussion of these problems could only appear later on (1947), when I published *Neuere Probleme der Abstammungslehre* (translated as *Evolution above the Species Level*, 1960).

Summary

During the first three decades of our century the process of speciation was not yet sufficiently understood. Most biologists were convinced that natural selection of heritable varieties was the main factor, but they did not deny that inherited characteristics could also have been caused by direct environmental influences on the genes. The main reason for such an opinion was the idea that the chromosomes in the germ cells could be altered by parallel induction of soma and germ cells. This view was supported by observations of phenocopies of inherited characteristics and also by the experimental production of modifications that lasted during many generations.

These results were decisive for my own view of the particular case of the parallelism of the characteristics of heritable geographic avian races and mammalian races with climatic factors. This had, I believed, to be explained in a Lamarckian sense because natural selection of minor differences of size, proportion, and color shades seemed to be very improbable. We did not know very much of the pleiotropic effects of genes at this period. Later development of genetics and my own experiences with modification of strains of *Drosophila* raised in different temperatures convinced me that selection alone was the guiding principle of the geographical parallelism of characters. Earlier I had already experimentally proved the effect of selection when I investigated the egg mimicry of cuckoos during 1925, when I observed a surplus of bright colors of tropical birds during 1927 and 1930, and when I made correlations between climatic differences and wing shape and egg numbers in 1934.

References

Alverdes, P. 1921. Die Rolle einer kumulierten Nachwirkung in der Stammesgeschichte. *Z. indukt. Abstamm. Vererbungsl.* 27:52–65.

Baldwin, J. M. 1902. *Development and Evolution.* London: Macmillan.

Berg, L. S. 1926. *Nomogenesis or Evolution Determined by Law.* London: Constable & Co.

Beurlen, K. 1937. *Die stammesgeschichtlichen Grundlagen der Abstammungslehre.* Jena: G. Fischer.

Böker, H. 1935, 1937. *Einführung in die vergleichende biologische Anatomie der Wirbeltiere.* 2 vols. Jena: G. Fischer.

1936. Aktives und passives Lebensgeschehen. *Hippokrates 1936, 31*:797–810.

Buchner, P. 1938. *Allgemeine Zoologie.* Leipzig: Quelle and Meyer.

Claus, C., K. Grobben, and A. Kühn. 1932. *Lehrbuch der Zoologie,* 10th ed. Berlin, Vienna: Springer-Verlag.

Fischer, E. 1901. Experimentelle Untersuchungen über die Vererbung erworbener Eigenschaften. *Allgem. Z. Entomol.* 6:49–51, 305–7, 325–7.

Haecker, V. 1918. *Entwicklungsgeschichtliche Eigenschaftsanalyse (Phänogenetik).* Jena: G. Fischer.

1921. *Algemeine Vererbungslehre,* 3rd. ed. Brunswick: Vieweg (ed. 1911).

Hammerling, J. 1929. Dauermodifikationen. In *Handbuch der Vererbungslehre.* Vol. 1. ed. E. Baur and M. Hartmann, Berlin: Bornträger.

Harms, J. W. 1934. *Wandlungen des Artgefüges unter natürlichen und künstlichen Umweltbedingungen.* Tübingen: Heine.

Hartmann, M. 1927. *Allgemeine Biologie. Eine Einführung in die Lehre vom Leben.* Jena: G. Fischer.

Hesse, R. 1943. Das Tier als Glied des Naturganzen. In *Tierbau und Tierleben,* 2nd ed., 2 vols. ed. R. Hesse and F. Doflein, Jena: G. Fischer.

Jollos, V. 1931. Genetik und Evolutionsproblem. *Verh. Deutsch. Zool. Ges.* 252–95.

1933. Die Übereinstimmung der bei *Drosophila melanogaster* nach Hitzeeinwirkung entstehenden Modifikationen und Mutationen. *Naturwiss.* 24:831–83.

Mayr, E., and W. B. Provine. 1980. *The Evolutionary Synthesis. Perspectives of the Unification of Biology.* Cambridge, Mass.: Harvard Univ. Press.

Osborn, H. F. 1934. Aristogenesis the creative principle in the origin of species. *Amer. Natural. 68*:193–235.

Peyer, B. 1952. *Das Problem der Vererbung von Reizwirkungen. Vierteljahresschrift Naturforsch. Ges. Zürich* 97:61–81.

Plate, L. 1913. *Selektionsprinzip und Probleme der Artbildung.* 4th ed. Leipzig, Berlin: Engelmann.

1931. Warum muss der Vererbungsforscher an der Annahme einer Vererbung erworbener Eigenschaften festhalten? *Z. indukt. Abstamm. Vererbungsl. 58*:266–92.

Rensch, B. 1924. Zur Entstehung der Mimikry der Kuckuckseier. *J. Ornithol.* 72:461–72.

1925a. Untersuchungen zur Phylogenese der Schillerstruktur. *J. Ornithol.* 73:127–47.

1925b. Das Problem des Brutparasitismus bei Vögeln. *Sitzungsber. Ges. naturforsch. Freunde Berlin.* Nos. 1–10, 55–69.

1925c. Untersuchungen zur Phylogenese der Schillerstruktur. *J. Ornithol.* 73:127–47.

1929. *Das Prinzip geographischer Rassenkreise und das Problem der Artbildung.* Berlin: Bornträger.

1930. *Eine biologische Reise nach den Kleinen Sunda-Inseln.* Berlin: Bornträger.

1933. Zoologische Systematik und Artbildungsproblem. *Verh. Deutsch. Zool. Ges. 1933*, pp. 19–83.

1936. Studien der klimatischen Parallelität der Merkmalsaus-prägung bei Vogeln und Säugetieren. *Arch. Naturgesch. N. F. 5*:817–86.

1938a. Einwirkung des Klimas bei der Ausprägung von Vogelrassen mit besonderer Berücksichtigung der Flügelform. *Proceedings of the Eighth International Ornithologic Congress, Oxford, 1934*, pp. 285–311.

1938b. Bestehen die Regeln klimatischer Parallelität bei der Merkmalsausprägung von homöothermen Tieren zu Recht? *Arch. Naturgesch. N. F.7*:364–80.

1939a. Klimatische Auslese von Grössenvarianten. *Arch. Naturgesch. N. F. 8*:89–129.

1939b. Typen der Artbildung. *Biol. Rev. 14*, 180–222.

1947. *Neuere Probleme der Abstammungslehre*. Die transspezifische Evolution. Stuttgart: Enke (3rd ed., 1972). [*Evolution above the Species Level*. New York: Columbia Univ. Press, 1960.]

1980. Historical Development of the Present Synthetic Neo-Darwinism in Germany. In *The Evolutionary Synthesis*, ed. E. Mayr and W. B. Provine, pp. 284–303. Cambridge, Mass.: Harvard Univ. Press.

Semon, R. 1912. *Das Problem der Vererbung erworbener Eigenschaften*. Leipzig: Engelmann.

Standfuss, M. 1898. Experimentelle zoologische Studien mit Lepidopteren. *Neue Denkschr. allgem. schweiz. Ges. gesamt. Naturwiss. 36*:3–40.

Stempell, W. 1935. *Zoologie im Grundriss*. Berlin: Borntrager.

Stresemann, E. 1926. Übersicht über die Mutationsstudien 1–XXIV. *J. Ornithol. 74*:377–85, Tables, IV–VIII.

Sumner, F. B. 1923–24. Results of experiments in hybridizing subspecies of Peromyscus. *J. Exp. Zool. 38*:245–92.

Tower, W. L. 1906. An investigation of evolution in chrysomelid beetles of the genus Leptinotarsa. *Carnegie Inst. Publ. 48*.

Weidenreich, F. 1921. *Das Evolutionsproblem und der individuelle Gestaltungsanteil am Entwicklungsgeschehen*. Berlin: Springer.

1929. Vererbungsexperiment und vergleichende Morphologie. *Paläontologische Z. 11*:276–86.

Woltereck, F. 1932. *Grundzüge der allgemeinen Biologie*. Stuttgart: Encke.

3

The development of Wright's theory of evolution: systematics, adaptation, and drift

WILLIAM B. PROVINE

Introduction

Scientists are notoriously ahistorical in their research and writing. The precious little history found in science textbooks at any level is usually "chronological–logical," merely the listing in chronological order of a few major discoveries leading logically to the theories espoused in the textbook. The self-consciously historical approach taken by many of the scientists involved in the current controversies in evolutionary theory is, therefore, unusual.

Ernst Mayr and Stephen Jay Gould have set the historical tone. In addition to being prominent evolutionary biologists, both are influential historians of biology. In *Evolution and the Diversity of Life* (1976), a collection of Mayr's essays, he introduced the section on the history of biology with this comment:

> It is rather remarkable how many current controversies in biology are a continuation of long-standing arguments. The historian of science is not the only one who may profit from investigation of the historical development of these arguments. The working biologist also may benefit greatly from studying the history of such controversies because this often helps him understand the ideological background of those who first originated the controversy, and, what is equally important, it often helps to clarify semantic confusions due to subtle changes in terms. [1976, p. 221]

Both Mayr and Gould, and an increasing number of evolutionary biologists, see current controversies in evolutionary theory as having clear roots in the arguments over Darwin's theory of evolution by natural selection in the last forty years of the nineteenth century, and

I wish to acknowledge the support of the National Science Foundation Grant SES-7926738.

as stemming directly from the period of the "evolutionary synthesis" of the 1930s and 40s (Mayr and Provine, 1980). One consequence of this emphasis upon the historical context of evolutionary theory has been the close interaction of scientists, historians, and philosophers in discussions of the controversies. The intellectual excitement thus generated is visible at conferences (such as the one Mayr organized on the evolutionary synthesis in 1974, the macroevolution meeting at the Field Museum in Chicago in 1980, the Fishbein Center conference on "Persistent Controversies in Evolutionary Theory" in 1981, and, of course, the conference that led to this volume) and also in an emerging host of publications.

The topic of this essay is directly related to the historical analyses of Mayr and Gould. For many years, Mayr has emphasized the contributions of systematists to the evolutionary synthesis of the 1930s and 40s. In particular, he suggested in 1980 that although some systematists like Robson and Richards (1936) emphasized nonadaptive variation between closely related species, "these dissenting voices were, however, very much in the minority. Gradual adaptive variation was clearly in conflict with the mutationist thesis of Bateson and de Vries" (Mayr and Provine, 1980, p. 132). Mayr's historical thesis was that systematists, even the neo-Lamarckians, brought to the synthesis period the view that later clearly dominated – the mechanisms of evolution lead to generally adaptive differences between geographic races and closely related species.

Gould, on the other hand, has stressed (as he does in his essay for this volume) what he calls the "hardening of the synthesis." He thinks that in the early years of the synthesis, up into the early 1940s, evolutionists were more pluralistic about mechanisms of evolution and placed less emphasis upon gradual natural selection and the resulting detailed adaptationism than did the same or younger evolutionists in the 1950s.

Working on the biography of Sewall Wright, I encountered directly these two historical theses of Gould and Mayr. The development of Wright's views on mechanisms of evolution suggests limitations to Mayr's thesis about systematists in the early evolutionary synthesis and illuminates Gould's thesis about the hardening of the synthesis.

Wright's shifting balance theory of evolution

Sewall Wright invented his shifting balance theory of evolution in nature in 1925. Reasoning from the practice of animal breeders, and his own experience in inbreeding guinea pigs, Wright deduced that evolution in nature would proceed most rapidly if natural populations were subdivided into small enough partially isolated subgroups to cause some random drifting of genes, but large enough so that the random drifting did not lead directly to fixation of genes, for this led to degeneration and extinction. The second step of the process was what Wright termed "interdemic" selection, that is, natural selection between the now differentiated small partially isolated subgroups. The mechanism of interdemic selection was selective diffusion from demes with the more advantageous genetic combinations. Viewed in this way, "interdemic selection" was actually individual selection operating within the subdivided population structure. These basic features of Wright's shifting balance theory of evolution in nature have remained the same from 1925 until the present. [For a detailed analysis of the origin and development of Wright's shifting balance theory, see Provine, in press *a* and *b* (ch. 7).]

The question of concern here is the level at which the shifting balance theory of evolution causes adaptive response in natural populations. Beginning no later than 1948, when Wright responded to the challenge R. A. Fisher and E. B. Ford raised to the shifting balance theory (Fisher and Ford, 1947; Wright, 1948), he has maintained consistently that the shifting balance process leads in natural populations to an adaptive response generally at the level of geographical races and certainly by the level of closely related species. Wright's most recent statement (June, 1982) accurately represents his position for the last thirty-five years: "I emphasize here that while I have attributed great importance to random drift in small local populations as providing material for natural selection among interaction systems, I have never attributed importance to nonadaptive differentiation of species" (Wright, 1982, p. 12).

Despite Wright's firmness in presenting this view of his shifting balance process for thirty-five years, his published papers and correspondence from the late 1920s and early 1930s reveal a more pluralistic view of adaptation in relation to mechanisms of evolution. To be sure, he does present early on the view that the shifting balance process can lead to more rapid adaptive advance at the species level than was possible under the mechanism of mass selection (i.e., selection operating upon a large randomly breeding population), so strongly emphasized by R. A. Fisher. In his review of Fisher's *Genetical Theory of Natural Selection* (1930), Wright stated that the consequence of his shifting balance process "would seem to be a rapid differentiation of local strains, in itself nonadaptive, but permitting selective increase or decrease of the numbers in different strains and thus leading to relatively rapid adaptive advance of the species as a whole" (Wright, 1930, p. 355).

Compare, however, this quotation with the following direct quotations from other papers Wright published just before and after:

1. [Assessing the influence of random drift in isolated populations] The nonadaptive nature of the differences which usually seem to characterize local races, subspecies, and even species of the same genus indicates that this factor of isolation is in fact of first importance in the evolutionary origin of such groups, a point on which field naturalists (e.g., Wagner, Gulick, Jordan, Osborn, and Crampton) have long insisted. [1929, pp. 560–1]

2. The actual differences among natural geographical races and subspecies are to a large extent of the nonadaptive sort expected from random drifting apart. [1931, p. 127]

3. Adaptive orthogenetic advances for moderate periods of geological time, a winding course in the long run, nonadaptive branching following isolation as the usual mode of origin of subspecies, species, perhaps even genera, adaptive branching giving rise occasionally to species which may originate new families, orders, etc. . . . are all in harmony with this interpretation. [1931, p. 153]

4. [Under the shifting balance theory] complete isolation of a portion of a species should result relatively rapidly in specific differentiation, and one that is not necessarily adaptive. The effective intergroup competition leading to adaptive advance may be between species rather than races. Such isolation is doubtless usually geographic in character at the outset but may be clinched by the development of hybrid sterility. [1932, p. 363]

5. That evolution involves nonadaptive differentiation to a large extent at the subspecies and even the species level is indicated by the kinds of differences by which such groups are actually distinguished by systematists. It is only at the subfamily and family levels that clear-cut adaptive differences become the rule. [Robson, 1928; Jacot, 1932] The principal evolutionary mechanism in the origin of species must then be an essentially nonadaptive one. [1932, pp. 363–4]

6. Subdivision into numerous local races whose differences are largely nonadaptive has been recorded in other organisms wherever a sufficiently detailed study has been made. [1932, pp. 364–5]

The biologist reading Wright's papers in the years 1929–32 would almost certainly have concluded that Wright believed nonadaptive random drift was the primary, but not exclusive, mechanism of the origin of geographical races, subspecies, species, and perhaps genera. The historical questions now emerge clearly: What led Wright to adopt the views on nonadaptive differentiation he expressed in the years 1929–32, and why did he change his emphasis before 1948? The answers to these questions will aid in the evaluation of the historical theses of Gould and Mayr.

In 1925, when Wright first conceived of and wrote down his shifting balance theory of evolution, his personal store of knowledge from direct observation of natural populations was meager. As a young boy he had carefully observed and identified birds, but of the science of systematics he knew nothing. In his thesis research at Harvard (1912–15), his experimental work was on guinea pigs. He continued research on guinea pigs and breeding in domestic animals while at the USDA in Washington, D.C. (1916–25), yet in all these years he never saw a guinea pig in the wild. Thus his knowledge of natural populations in 1925 was drawn from the literature, not experience. In the late 1920s he collected spiders at the Indiana Dunes, but the collection was sporadic, and by 1930 he gave up this work on natural populations for lack of time. When, therefore, Wright published his major early works on evolutionary theory (1929, 1930, 1931, 1932), his knowledge of systematics and natural populations was necessarily drawn from the experts.

Luckily for the historian, Wright carefully listed in his early evolutionary papers the experts upon whom he relied. These were Moritz Wagner, John T. Gulick, David Starr Jordan, Henry Fairfield Osborn, Henry Edward Crampton, G. C. Robson, Arthur Paul Jacot, Alexander Ruthven, Vernon L. Kellogg, Wilfred Osgood, Alfred C. Kinsey, Bernhard Rensch, Johannes Schmidt, David Thompson, and F. B. Sumner. Before turning directly to these systematists, however, a brief look at the nineteenth-century setting of the relationship between mechanisms of evolution and adaptation is essential.

The nineteenth-century background

Charles Darwin in *The Origin of Species* (6th ed., 1872) states at the end of the introduction, "I am convinced that natural selection has been the most important, but not the exclusive, means of modification" (p. 4). This famous statement was taken more seriously by Darwin's contemporaries and evolutionists in the late nineteenth and early twentieth centuries than after the evolutionary synthesis of the 1930s and 1940s. In the *Origin*, Darwin proposed at least seven distinct mechanisms of evolution, gradual natural selection operating upon small heritable individual differences being, of course, the most important. Next to natural selection, Darwin thought the use and disuse of parts was the most effective mechanism of adaptive evolution. Family selection, as in the cases of altruistic social behavior or neuter castes, was a third and far more restricted mechanism of adaptive evolution.

In addition to these three mechanisms of adaptive evolution, Darwin proposed (and here is the surprise to most neo-Darwinians) four mechanisms of nonadaptive evolution. These were (1) sexual selection; (2) directed variations, when, according to Darwin, certain rather strongly marked variations simply spread over a population in the absence of selection (1872, p. 72); (3) correlated variation – a maladaptive or nonadaptive character was correlated with another of adaptive value such that their combination was positively adaptive; and (4) spontaneous variations that simply appeared spontaneously and were passed on by heredity (the appearance of a nectarine on a peach tree was one of Darwin's examples). To this list of nonadaptive mechanisms of evolution found in the *Origin* should be added the effects of geographical isolation upon small populations, a concept forcefully advanced by the German naturalist Moritz Wagner (1868). Later in his life, Darwin believed that nonadaptive evolution was particularly likely in small isolated populations, as he wrote to Wagner in 1876 (F. Darwin, 1887, Vol. III, p. 159). The sixth edition of the *Origin* exhibits clearly Darwin's belief that nonadaptive evolution was widespread in natural populations (Darwin, 1872, pp. 35, 157–8, 175–6).

Why would Darwin, who advanced the idea of natural selection as the primary mechanism of evolution, document so carefully the existence of nonadaptive characters that did not evolve by natural selection? The answer is that Darwin's allegiance was not only to the idea of natural selection, but to the concept of evolution by descent with its implications for the problem of classification.

Darwin believed simply that classification should reveal community of descent. The true system of classification is based upon the evolutionary tree. The problem is that taxonomists could not use for purposes of classification those distinctive characters most clearly adapting the organism to its environment:

> It might have been thought . . . that those parts of the structure which determined the habits of life, and the general place of each being in the economy of nature, would be of very high importance in classification. Nothing can be more false. No one regards the external similarity of a mouse to a shrew, of a dugong to a whale, of a whale to a fish, as of any importance. [Darwin, 1872, p. 365]
> On the view of characters being of real importance for classification, only insofar as they reveal descent, we can clearly understand why analogical or adaptive characters, although of

the utmost importance to the welfare of the being, are almost valueless to the systematist. For animals, belonging to two most distinct lines of descent, may have become adapted to similar conditions, and thus have assumed a close external resemblance; but such resemblances will not reveal – will rather tend to conceal their blood-relationship. [Darwin, 1872, p. 374]

With adaptive characters useless for classification by descent, the value of the nonadaptive characters becomes obvious as does Darwin's reason for documenting them in a book extolling the adaptive effect of natural selection. In moving from one taxonomic level to another, a common nonadaptive character was the surest sign of common descent and natural classification. So Darwin was actually pleased that natural selection did not determine all characters distinguishing a species.

Some nineteenth-century neo-Darwinians were unwilling to grant so prominent a role to mechanisms of nonadaptive evolution and emphasized more strongly than Darwin the role of natural selection and its consequent adaptationist view. Alfred Russel Wallace, for example, argued that characters used as taxonomic markers were, in fact, adaptive, and had evolved by natural selection. When Darwin's student and friend G. J. Romanes argued in his essay "Physiological Selection" that natural selection could not be the mechanism of speciation, because the features distinguishing allied species were generally nonadaptive (Romanes, 1886), Wallace countered with strong support for the pervasive power of natural selection (Wallace, 1889, pp. 141–2). Other neo-Darwinians in England followed Wallace's strong selectionist, adaptationist point of view. Prominent among them were E. Ray Lankester, Raphael Meldola, and E. B. Poulton.

Opposed to these neo-Darwinians were both taxonomists and experimental biologists who argued that the differences between geographical races, closely allied species, and often genera were nonadaptational. William Bateson, Hugo de Vries, and Thomas Hunt Morgan were influential adherents of this view in the late nineteenth and early twentieth century, carrying with them most of the early Mendelians. These biologists, rejecting natural selection as the mechanism of microevolution (up to the species level), and as the mechanism of speciation, advocated the then popular mutation theory of evolution.

From this very brief analysis of the nineteenth-century background to the question of mechanisms of evolution, adaptation, and systematics (for a more detailed account see Provine, in press *a*), one point is central for this essay. Whether the differences between geographical races and allied species were adaptational or not was a hotly disputed question, the answer to which was inevitably tied to conceptions of the mechanisms of microevolution and speciation.

Experts upon whom Wright relied

In citing his list of experts on differences between geographical races and species Wright did not claim in 1932 that all emphasized the prevalence of nonadaptive differences. But he did suggest that all their data could be interpreted in that way: "Many of these authors [referring to Gulick, Crampton, Jordan, Ruthven, Kellogg, Osgood, Kinsey, and Rensch] insist on the nonadaptive character of most of the differences among local races. Others attribute all differences to the environment, but this seems to be more an expression of faith than a view based on tangible evidence" (Wright, 1932, p. 364). It is worth examining systematically just what the experts cited by Wright actually said.

1. Moritz Wagner (1813–1887)

Wright's knowledge of Wagner's work came primarily from the discussion of it in Kellogg's *Darwinism Today* (1907, pp. 234–237). Wagner argued that geographical isolation was the necessary antecedent to speciation, and that natural selection played a subordinate or often insignificant role. He pointed out that small isolated populations were most likely to diverge clearly from their parent populations, a point accepted by Darwin. The apparently nonadaptational differences Wagner observed between geographical races and species were, to him, evidence for geographical isolation rather than natural selection as being the primary mechanism for their origins.

2. John T. Gulick (1832–1923)

Gulick agreed with Wagner that geographical isolation was the key factor in the evolution of races and species:

> I believe that no process of natural selection, or of sexual selection, or of any other form of selection, can transform one species into two or more species without the prevention of free crossing between the branches that are thus transformed. Isolation is, I believe, an essential factor in the production and maintenance of divergent types, whether they be varieties or species; and any theory that fails to consider the causes and effects of isolation is an insufficient explanation of divergent evolution. [Gulick, 1905, pp. 6–7]

Gulick did field research on the Hawaiian land snails Achatinellidae.

He discovered that the many geographical races and allied species differed primarily by characters for which he could see no possible use, among them color-banding patterns and right- or left-handed coiling. His explanation for this nonadaptive evolution was that a natural population possessed much heritable variation, and any isolated portion of the population was unlikely to represent the remaining portion of the population. This initial divergence led to greater divergence over time.

3. David Starr Jordan (1851–1931)

Jordan, one of the most prominent evolutionists in the United States, was known throughout the world for his taxonomic work on fishes. In his writings on evolution Jordan constantly emphasized what he called "the survival of the existing" as a mechanism alternative to survival of the fittest in the origin of species, and pointed out the implications of this mechanism for systematics:

> The process of natural selection has been summed up in the phrase "survival of the fittest." This, however, tells only part of the story. "Survival of the existing" in many cases covers more of the truth. For in hosts of cases the survival of characters rests not on any special usefulness or fitness, but on the fact that individuals possessing these characters have inhabited or invaded a certain area. The principle of utility explains survivals among competing structures. It rarely accounts for qualities associated with geographic distribution.
>
> The nature of the animals that first colonize a district must determine what the future fauna shall be. From their specific characters, which are neither useful nor harmful, will be derived, for the most part, the specific characters of their successors.
>
> It is not essential to the meadow lark that he should have a black blotch on the breast or the outer tailfeather white. Yet all meadow larks have these characters just as all shore larks have the tiny plume behind the ear. Those characters of the present stock, which may be harmful in the new relations, will be eliminated by natural selection. Those especially helpful will be intensified and modified, but the great body of characters, the marks by which we know the species, will be neither helpful nor hurtful. These will be meaningless streaks and spots, variation in size of parts, peculiar relations of scales or hair or feathers, little matters which can neither help nor hurt, but which have all the persistence heredity can give. [Jordan, 1898, p. 218]

4. Vernon L. Kellogg (1867–1937)

Kellogg was an entomologist whose specialties were bird lice and silkworms. His review of the status of evolutionary theory in 1907, *Darwinism Today*, was an even-handed and influential book. Sewall Wright, for example, read the book carefully while in college. Kellogg taught at Stanford with Jordan and wrote three books with him, the most famous being *Evolution and Animal Life* (1907). Discussing the influence of geographical isolation, they examined the differences between the various species of orioles in North America, and concluded:

> Not one of these varied traits is clearly related to any principle of utility. Adaptation is evident enough, but each species is as well fitted for its life as any other, and no transposition or change of the distinctive specific characters or any set of them would in any conceivable degree reduce this adaptation. No one can say that any one of the actual distinctive characters or any combination of them enables their predecessors to survive in larger numbers than would otherwise be the case. [Jordan and Kellogg, 1907, p. 129]

After discussing similar cases of nonadaptive specific differentiation in honeybees and ladybird beetles, they summarized the implications of isolation:

> In these characters, there is, therefore, no rigorous choice due to natural selection. Such specific characters, without individual utility, may be classed as indifferent, so far as natural selection is concerned, and the great mass of specific characters actually used in systematic classification are thus indifferent . . . Adaptation is presumably the work of natural selection; the division of forms into species is the result of existence under new and diverse conditions. [Jordan and Kellogg, 1907, p. 130]

5. Henry Edward Crampton (1875–1956)

Beginning in 1899, very early in his career, Crampton initiated a series of selection experiments on larvae of silkworm moths. He used biometrical methods modeled after those used by H. C. Bumpus (1899) to measure selective death rate in the English sparrow and by W. F. R. Weldon (1895) in the shore crab. Reporting on his initial results in *Biometrika* (Crampton, 1904), Crampton found selective elimination based upon certain characters in the larval stages. One difficulty was that the characters used were of no apparent use in the larval stages, but Crampton was, at this time, oriented toward a selectionist interpretation of evolutionary change in nature, a view he clearly ex-

pressed in his paper "On a General Theory of Adaptation and Selection" (Crampton, 1905).

In 1906, however, Crampton began a major series of studies on the
variation, geographical distribution, and evolution of the Polynesian
lung snails of the genus *Partula*. Working primarily on the islands of
Tahiti and Moorea, Crampton found that the differences between the
geographical races or closely related species of *Partula* could be correlated with no detectable environmental differences. Differentiation
followed upon isolation, but not through natural selection. Crampton's results, therefore, agreed perfectly with those reached earlier
by Gulick. Two of the three monographs from this work, with their
beautiful color plates of the snails, were widely known and cited by
systematists (Crampton, 1916, 1932), and he also published a summary article in *The American Naturalist* (1925), this being the one cited
by Wright in 1932.

6. *Wilfred Osgood (1875–1947) and F. B. Sumner (1874–1945)*

Osgood made his reputation as a systematist by studying the pro-
fusion of species and subspecies of the deer mouse, genus *Peromyscus*.
His 1909 monograph on *Peromyscus* was much cited in the systematics
literature through the 1940s and even later (Osgood, 1909). In general,
Osgood held an adaptationist interpretation of the differences between geographical races and between species.

F. B. Sumner, however, initiated in 1913 a far more careful experimental and quantitative analysis of a few of the species and geographical races of *Peromyscus*. Until 1925 a neo-Lamarckian who (along
with such major figures as Bernhard Rensch and Ernst Mayr) believed
in the direct action of the environment in producing the differences
between geographical races, Sumner challenged Osgood's adaptationist outlook. Differences in intensity of pigment, for example, appeared to be correlated with humidity rather than the color of the
background. Thus Sumner argued that the differences between geographical races did not result from natural selection and were nonadaptive. Because he enjoyed a well-deserved reputation as an exceedingly careful experimentalist, his reinterpretation of the adaptive
value of the differences between geographical races carried much
weight with both systematists and geneticists. Sewall Wright read all
of Sumner's early papers on *Peromyscus*, and although (like most geneticists) he rejected Sumner's neo-Lamarckism, he accepted Sumner's conclusions about nonadaptive evolution in *Peromyscus*.

After 1925, Sumner began to reject neo-Lamarckism and adopted
the Mendelian multiple-factor theory as the source of differences between geographical races of *Peromyscus* (Provine, 1979). In some cases

of racial differences, Sumner changed to a straight selectionist interpretation. But in Sumner's 1932 summary paper on the *Peromyscus*
work, he still left considerable room for nonadaptive mechanisms of
speciation (Sumner, 1932, pp. 56–86). Later Mayr would view this
paper as adaptationist, but when Wright cited the paper in 1932, it
was because he saw Sumner's work as indicating nonadaptive differentiation at the lowest taxonomic levels.

7. Johannes Schmidt (1877–1933)

Schmidt was a well-known Danish marine biologist who, for a middle
period in his career, concentrated upon morphological differences
between geographical populations of fish. His best-known work was
upon the viviparous blenny (*Zoarces viviparus*). Examining some
25,000 specimens from the coast of northern Europe, Schmidt discovered that local populations from fjord to fjord differed in the number of vertebrae and other characters. Heincke, having earlier found
such geographical variation in the herring (Heincke, 1898), had suggested that the differences were correlated with differences in salinity.
Schmidt, although having no mechanism to account for observed racial differences, argued that neither selective forces nor direct action
of the environment could account for the differences he found:

> Our investigations thus by no means support the hypothesis
> that the racial characters are determined exclusively by envi-
> ronment. On the contrary, they seem rather to indicate that
> differences of environment are not sufficient to explain the
> structural differences between the races, and that the impor-
> tance of the salinity, especially, has doubtless been greatly
> over-estimated. [Schmidt, 1918, p. 116].

8. G. C. Robson (1888–1945) and O. W. Richards (b. 1901)

More than any other taxonomists in the early synthesis period, Robson and Richards promoted the concept of nonadaptive differentiation at the level of geographical races and species. In his 1932 paper,
Wright cited Robson's book *The Species Problem* (1928), which appeared
in the prestigious Oliver and Boyd "Biological Monographs" series.
I suspect that Wright had not read the earlier Richards and Robson
paper, "The Species Problem and Evolution," which had appeared
in *Nature* two years earlier (Richards and Robson, 1926). This paper
made an eloquent plea for evolutionists to give up the charade of
inventing adaptationist stories to account for differences between geographical races and allied species. These apparently nonadaptive differences were, they argued, just what they appeared. In his 1928
book, Robson not only provided evidence of nonadaptive differen-

tiation, but also a very careful critique of selection experiments purporting to exhibit natural selection in action. Natural selection, Robson argued, could not be the mechanism leading to the ubiquitous nonadaptive differentiation found at the species level in nature. In 1936, Robson and Richards published *The Variation of Animals in Nature*, which was perhaps the best-known general book on zoological taxonomy in English until the appearance of Julian Huxley's edited book *The New Systematics* (1940).

The discussions of mechanisms of evolution in these works by Robson and Richards are far too complex to summarize here, but the evidence and arguments presented indicated clearly that differentiation at the level of geographical races and species was primarily nonadaptive rather than adaptive.

9. A. C. Kinsey (1894–1956), D. H. Thompson (b. 1897), and A. P. Jacot (b. 1890)

In 1932, all three of these taxonomists had recently published works indicating nonadaptive differentiation at lower taxonomic levels. Kinsey (1930) had published the first part of his work on the taxonomy of the gall wasp, genus *Cynips*. One of the basic taxonomic characters used by Kinsey was wing-length:

> There seems no basis for believing the shortened wings or any of the concomitant variations of any adaptive value to any of these insects. The short wings are not confined to warmer or colder climates, and long- and short-winged forms of various species are active at the same season in the same localities. The field data suggest nothing as to the survival value of these outstandingly basic modifications of structure. [Kinsey, 1930, p. 34]

Thompson had studied variation in fishes in Illinois as a function of distance both within a single stream and between streams. He found that even great environmental differences along the length of a single stream led to far less racial differentiation than found in the same species in different streams, but in apparently similar environments. In general terms, Thompson explained his view:

> Every species that has been studied genetically has shown heritable variations of greater or less degree. Furthermore, these heritable variations are continually arising *de novo* by changes in the germ plasm due to causes entirely unrelated to ordinary variations in the physical environment, or at least not as adaptive responses to them. While most of these heritable variations are disadvantageous to the animal and are rap-

idly eliminated, many of them are of indifferent selective
value and may persist. [Thompson, 1931, p. 278]

Finally, in the July–August 1932 issue of the *American Naturalist*
(Wright delivered his 1932 paper in August 1932), Jacot cited Robson's
1928 book as evidence that no proof existed for the argument that
natural selection caused interspecific differences. If evidence for the
adaptationist argument did exist, Jacot asked, "after so much observation and experimentation, should it not be plainly evident?" (Jacot,
1932, p. 351).

10. H. F. Osborn (1857–1935), B. Rensch (b. 1900), and A. Ruthven (1882–1971)

To be sure, these three of the sixteen systematists cited by Wright
did argue for a primarily adaptive view of evolutionary change. One
can easily see, however, how Wright could have evaluated their views
as being less compelling than the authorities previously discussed.
Ruthven's study of speciation in garter snakes used dwarfing and
scutellation for taxonomic markers. Scutellation was highly correlated
with degree of dwarfing. Ruthven admitted that the amount of dwarfing "does not seem to be directly associated with the nature of the
environment," but then concluded that the dwarfing was "associated
in some way with the environment" in an adaptive manner (Ruthven,
1908, p. 193). But by what authority did Ruthven make this surmise?
The answer was W. L. Tower's 1906 monograph on the inheritance
of acquired characters in Colorado potato beetles, work later revealed
to have been faked (Tower, 1906; Weinstein, 1980).

Although Osborn and Rensch both emphasized adaptive evolution,
each left considerable room for nonadaptive differentiation. In his
1927 summary, Osborn had divided speciation into gradual adaptive
and mutational nonadaptive. His concluding statement was that:

> Speciation is a normal and continuous process; it governs the
> greater part of the origin of species; it is apparently always
> adaptive. Mutation is an abnormal and irregular mode of ori-
> gin, which while not infrequently occurring in nature is not
> essentially an adaptive process; it is, rather, a disturbance of
> the regular course of speciation. [p. 42]

Rensch's 1929 book had a substantial section devoted to nonadaptive
differentiation, which he considered to be of two types: (1) characters
that arose apparently randomly; and (2) characters exhibiting geographic change by a gradual succession of steps, but apparently in
the absence of selection. Most of the cases cited by Rensch (e.g.,
Allen's rule, Gloger's rule, etc.) for adaptive change were very general

and were more applicable to the genus level than to the level of geographical races or closely allied species.

Wright drew the obvious conclusion from the writings of his experts in systematics. Having no independent first-hand knowledge of the careful and long-term study of natural populations required for research in systematics, he accepted from those who had studied natural populations that geographical races and closely allied species most frequently differed by nonadaptive characters.

I am not arguing that the systematists upon whom Wright relied were justified in concluding that evolution at the lower taxonomic levels was primarily nonadaptive. Indeed, these systematists suggested a confusing plethora of mechanisms to account for nonadaptive evolution, from geographical isolation to direct action of the environment to random mutation to sampling effects in small populations derived from larger ones. But I can say that Wright accepted their conclusion about nonadaptive differentiation in geographical races and closely allied species.

Furthermore, many evolutionists were influenced by these same systematists in the same way as Wright. J. B. S. Haldane, for example, argued in *The Causes of Evolution* (1932) that natural selection was the primary mechanism of evolution in nature, yet added this caveat:

> But when we have pushed our analysis as far as possible, there is no doubt that innumerable characters show no sign of possessing selective value, and, moreover, these are exactly the characters which enable a taxonomist to distinguish one species from another. This has led many able zoologists and botanists to give up Darwinism. [Haldane, 1932, pp. 113–4]

When the ecologist Charles Elton published his first book, *Animal Ecology* (1927), it appeared in a series edited by Julian Huxley and had a glowing introduction by him. In the book, Elton, who had worked with Richards as well as Huxley, summarized the Richards and Robson article in *Nature* (1926) with approval:

> The gist of their conclusions is that very closely allied species practically never differ in characters which can by any stretch of the imagination be called adaptive. If natural selection exercises any important influence upon the divergence of species, we should expect to find that the characters separating species would in many cases be of obvious survival value. But the odd thing is that although the characters which distinguish genera or distantly allied species from one another are often obviously adaptive, those separating closely allied species are nearly always quite trivial and apparently meaningless. [Elton, 1927, p. 184]

Huxley himself, certainly a champion of Darwin's idea of evolution by natural selection, accepted the argument of Richards and Robson that many of the differences between closely allied species were non-adaptive. Indeed, Huxley's work on allometry in the 1920s and early 1930s (summarized in Huxley, 1932) provided explicit proof of correlation that could lead to the evolution of nonadaptive characters.

The most telling proof of the seriousness with which the arguments of Richards and Robson were taken comes from E. B. Ford. R. A. Fisher and E. B. Ford were perhaps the two most influential and vocal adherents of a purely selectionist view of the evolutionary process during the 1930s. Fisher's theoretical view rejected the significance of all mechanisms of evolution with the sole exception of natural selection (Fisher, 1930, pp. 20–1). Ford's background and research on natural populations led him also to become a staunch selectionist (Ford, 1980). In 1931 he published the first edition of his widely read book *Mendelism and Evolution*, which by 1960 had gone through a total of seven editions. After arguing that the effects of genes were probably multiple, Ford stated:

> This consideration may throw some light on the nature of the characters which separate local races and closely allied species. That these are sometimes entirely non-adaptive has been demonstrated, we believe successfully, by Richards and Robson (1926). It is evident that certain genes which either initially or ultimately have beneficial effects may at the same time produce characters of a non-adaptive type, which will therefore be established with them. Such characters may sometimes serve most easily to distinguish different races or species; indeed, they may be the only ones ordinarily available, when the advantages with which they are associated are of a physiological nature. Further, it may happen that the chain of reactions which a gene sets going is of advantage, while the end-product to which this gives rise, say a character in a juvenile or the adult stage, is of no adaptive significance . . .
>
> J. S. Huxley has pointed out another way in which non-adaptive specific differences may arise. For he has shown that changes in absolute body-size, in themselves probably adaptive, may automatically lead to disproportionate growth in a variety of structures, such as horns and antlers in Mammalia and the appendages in Arthropoda. The effects so produced may be very striking, but, as they are the inevitable result of alteration in size, they can rarely have an adaptive significance.

It is not perhaps always recognized how complete has been the demonstration provided by the above authors that the characters available to systematists for the separation of allied species may be of a wholly non-adaptive kind. [Ford, 1931, pp. 78–9)]

Of course, Ford goes on to argue that one was not justified in assuming that the nonadaptive characters were produced by a process of nonadaptive evolutionary change – quite the contrary. The nonadaptive characters used by systematists were simply the necessary by-products of adaptive evolution. Ford's statements about nonadaptive differences between local races and allied species quoted above remained unchanged through the fifth edition of 1949.

Other examples can easily be cited but are unnecessary. Sewall Wright and many other evolutionists accepted in the early 1930s the dominant view of systematists that geographical races and closely allied species differed primarily by nonadaptive characters by which such races and species were most easily distinguished.

Wright's random genetic drift

As mentioned earlier, the systematists who agreed that geographical races and closely allied species differed by primarily nonadaptive characters proposed many mechanisms to account for such apparently nonadaptive evolution. The two favored explanations were correlation of parts (Huxley's allometry) and sampling from larger populations (Gulick's favorite explanation). Elton, who had studied systematically the subject of fluctuating population sizes, argued strongly for sampling effect as the cause of nonadaptive differentiation:

Many animals periodically undergo rapid increase with practically no checks at all. In fact, the struggle for existence sometimes tends to disappear almost entirely. During the expansion in numbers from a minimum, almost every animal survives, or at any rate a very high proportion of them do so, and an immeasurably larger number survives than when the population remains constant. If therefore a heritable variation were to occur in the small nucleus of animals left at a minimum of numbers, it would spread very quickly and automatically, so that a very large proportion of numbers of individuals would possess it when the species had regained its normal numbers. In this way it would be possible for non-adaptive (indifferent) characters to spread in the population, and we should have a partial explanation of the puzzling facts about

closely allied species, and of the existence of so many apparently non-adaptive characters in animals. [Elton, 1927, p. 187]

The major problem with Elton's scenario was that even very small
samples of most natural populations were not very different from the
parent populations – nor was it clear why a small isolated population
would diverge ever farther in nonadaptive characters from its parent
population. The problem with the allometry argument was that it
could be proved or disproved only with great difficulty, if at all. In
the light of this dilemma in explaining the origin of nonadaptive differentiation, the impact of Wright's theory of random genetic drift
upon systematists becomes more understandable.

The six quotations from Wright's published papers in the years
1929–32 given earlier show clearly that Wright was suggesting random genetic drift as the most plausible mechanism for nonadaptive
differences at the lower taxonomic levels. Wright's concept of random
drift was unquestionably a conceptual advance over the mechanisms
offered previously by systematists. Wagner, Gulick, D. S. Jordan,
Kellogg, Elton, and many others had emphasized the importance of
geographical isolation or fluctuation of population size in producing
nonadaptive differentiation by isolation of a nonrepresentative sample of the parent population. Yet most of these same systematists
were uneasy about the power of such sampling effects to produce
the observed results. Wright's concept of random genetic drift, a statistical consequence of enforced inbreeding in a small population, was
precisely the required mechanism to produce the observed differentiation. Even in the absence of initial nonrepresentative sampling
in the isolated population, random drift could cause increased differentiation generation after generation, as long as the population
remained small enough – up to the limit of total homozygosity, when,
of course, differentiation must cease.

Wright's contention in 1929–32 that random genetic drift could explain nonadaptive differentiation as observed by systematists did not,
in fact, fit well with his general view of evolution. He originally saw
random drift as valuable in creating novel genic interaction systems
in highly inbred domestic populations, thus enabling the breeders to
select the most desirable combinations. Random genetic drift was,
therefore, crucial in producing variation, not in providing an evolutionary mechanism as an alternative to selection. Indeed, as Wright
had observed in his experimental populations of highly inbred strains
of guinea pigs, the effect of random drift in a population of very small
effective size was nearly always deleterious to the fitness of the population. Very small isolated populations were, according to Wright
in 1931 and 1932, doomed to extinction.

Wright's shifting balance theory of evolution emphasized large but subdivided populations as the key to rapid adaptive evolution in nature. There the balance between random drift, selection, mutation, migration, and other factors produced the optimum conditions for adaptive advance. From the earliest attempt Wright made to devise a theory of evolution in nature, he stressed the proper *balance* of all the variables. Why then did Wright in the years 1929–32 emphasize so strongly the role of random drift in the evolution of geographical races and species, when, at the same time, his general shifting balance theory of evolution in nature did not require an imbalance favoring random drift as the primary variable at the racial and species level?

The answer has already been provided. Wright thought from what he read in the systematics literature that the differences between geographical races and allied species were primarily nonadaptive, and so he interpreted the balance in his balance theory to favor random drift at the lower taxonomic levels. Wright's theory of evolution would actually have fitted better with a more adaptive view of evolution than systematists provided in the early 1930s.

Even though Wright's connection of systematists' observations and his concept of random genetic drift may not have harmonized well with his own shifting balance theory of evolution in nature, the connection certainly solved for systematists and evolutionists, in general, the problem of the unknown primary mechanism of the evolution of nonadaptive differences between related races and species. After 1932, Wright's name and the concept of random drift were invoked constantly whenever the evolution of nonadaptive characters came under discussion. Proof for this assertion may be found in almost any book on systematics and evolutionary theory from the mid-1930s up into the 1950s. I will accordingly give only a few examples here from major books.

Theodosius Dobzhansky's *Genetics and the Origin of Species* (1937; 2nd ed., 1942; 3rd ed., 1951) was perhaps the central book of the evolutionary synthesis period, and was very widely read by students in genetics, systematics, and evolutionary theory. Dobzhansky had begun his career as a systematist of the ladybird beetles (Coccinellidae), had written on the biological species concept (Dobzhansky, 1935), and had tried to synthesize systematics and population genetics in his book (for more background on Dobzhansky, see Provine, 1981). Dobzhansky was greatly influenced by Wright's shifting balance theory of evolution, as can easily be seen in Chapter VI, "Selection," in the 1937 book. At the end of the chapter, Dobzhansky lauded Wright's thesis that differentiation of a population into numerous semi-isolated colonies was the most favorable for progressive evolution:

Indeed, a number of considerations speak in its favor. The present writer is impressed with the fact that this scheme is best able to explain the old and familiar observation that races and species frequently differ in characteristics to which it is very hard to ascribe any adaptive value. Since in a semi-isolated colony of a species the fixation or loss of genes is to a certain degree independent of their adaptive values (owing to the restriction of the population size), a colony may become different from others simultaneously in several characters. One or a few of the latter may be adaptive, and may enable the population to conquer new territories or ecological stations. The rest of the characters may spread concomitantly with the adaptive ones. For example, the chromosome structures that are so variable in *Drosophila pseudoobscura* can hardly be regarded as anything other than neutral characters, although some of them have become racial characteristics in subdivisions of the species population. [pp. 190–1]

Next to Dobzhansky's *Genetics and the Origin of Species*, Julian Huxley's *Evolution: The Modern Synthesis* (1942) was probably the most influential book of the synthesis period. Huxley had been much taken by Wright's concept of random genetic drift in relation to systematics, and cited Wright over and over again in the introduction to *The New Systematics* (1940) and in *Evolution: The Modern Synthesis*.

The proof given by Wright, that non-adaptive differentiation will occur in small populations owing to "drift," or the chance fixation of some new mutation or recombination, is one of the most important results of mathematical analysis applied to the facts of neo-mendelism. It gives accident as well as adaptation a place in evolution, and at one stroke explains many facts which puzzled earlier selectionists, notably the much greater degree of divergence shown by island than mainland forms, by forms in isolated lakes than in continuous river systems. [1942, pp. 199–200]

Non-adaptive or accidental differentiation may occur where isolated groups are small. This "drift," which we have also called the Sewall Wright phenomenon, is perhaps the most important of recent taxonomic discoveries. It was deduced mathematically from neo-mendelian premises, and has been empirically confirmed both in general and in detail. [p. 260]

The major problem, according to Huxley, had not been finding the causes of adaptive differentiation in species with wide geographic distribution; that was a relatively easy task. Instead:

On the contrary, the major biological problem has been that of accounting for that fraction of the divergence which is not adaptive, and this would now appear to have been settled in principle, as due to the Sewall Wright phenomenon of drift. [1940, p. 265]

Huxley's name for random genetic drift, the "Sewall Wright effect," accurately reflects the general tendency to associate Wright with it.

Ernst Mayr's *Systematics and the Origin of Species* appeared in 1942, the same year as *Evolution: The Modern Synthesis.* In a section on population size and variability, introducing his concept of "founder effect," Mayr began with this statement:

Naturalists have known for a long time that island populations tend to have aberrant characteristics. Wright (1931, 1932, and elsewhere) found the theoretical basis for this by showing that in small populations the accidental elimination of genes may be a more successful process than selection. [Mayr, 1942, p. 234]

In general, of course, Mayr emphasized the adaptive aspects of evolution in *Systematics and the Origin of Species.* After 1942, he attributed less importance to random drift as an explanatory factor in differences between island populations.

David Lack's change of view on the adaptive value of specific and subspecific differences in Darwin's finches (Geospizinae) is particularly instructive (Kottler, personal communication; Gould, this volume). With Julian Huxley's backing, Lack went to the Galapagos Islands in the fall of 1938, studying the finches there for six months. He spent April–August 1939 at the California Academy of Sciences, and several weeks with Ernst Mayr at his home in New Jersey before returning to England in September 1939. There he analyzed his data on Darwin's finches and by June 1940 had completed the final draft of his monograph on them. The California Academy of Sciences, because of the war, was unable to publish the manuscript until May 1945 (Lack, 1945). Before the monograph appeared, Lack had completed by 1944 his popular work *Darwin's Finches*, which appeared only after a delay of three years (Lack, 1947). In the 1945 monograph, he argued that the specific and subspecific differences were mostly nonadaptive, and in the 1947 book that they were mostly adaptive. In autobiographical remarks written several years before his death and published in an obituary, Lack explained his dramatic shift in view:

Since so knowledgeable a man as Arthur Cain asked me, in the mid-1950s, why I postulated that various subspecific differences in the finches are non-adaptive, it is worth stressing that, before my book, almost all subspecific differences in animals were regarded as non-adaptive (hence the importance of Sewall Wright's theory of genetic drift). The main exceptions were size differences, such as those in accordance with Bergmann's rule, and the coloration of certain desert larks. Still more were the differences between closely related species regarded as non-adaptive, except for the specific recognition characters which reduce hybridisation. *The Variation of Animals in Nature* by G. C. Robson and O. W. Richards in 1936 fairly reflected current opinion on these problems. I reached my conclusion, that most subspecific and specific differences in Darwin's finches are adaptive, and that ecological isolation is essential for the persistence of new species, only when reconsidering my observations five years after I was in the Galapagos. [Lack 1973, p. 429]

Lack's comment clearly supports my thesis about the way systematists generally understood the significance of Wright's theory of evolution in nature, but I am left puzzled on one point. About the weeks Lack spent visiting Mayr in August 1939, Lack said that he "learned much about systematics and speciation from him" (Lack 1973, p. 428), and Mayr said, "we discussed again and again the problem of speciation on islands and elsewhere" (Lack 1973, p. 432). Mayr has had, since the 1940s, a reputation for advocating and defending an adaptationist view of evolution, yet after all the detailed discussion with Mayr, Lack immediately thereafter wrote about the nonadaptive quality of most of the specific and subspecific differences in Darwin's finches. Mayr has, at this time, no specific recollection of having discussed the adaptation question with Lack in 1939 (personal communication, 1/10/83). I suspect that the adaptation question was not burning in 1939, and that Mayr's 1939 views on island speciation were probably mirrored in the quote above from *Systematics and the Origin of Species*.

And finally, from late in the synthesis period, comes this comment from Ledyard Stebbins' well-known *Variation and Evolution in Plants* (1950):

Population size is important chiefly in connection with the effects of random fluctuations in gene frequency. Numerous workers (cf. Dobzhansky, 1941, pp. 161–5) have shown that while in infinitely large populations gene frequencies tend to remain constant except for the effects of mutation and selection pressure, in populations of finite size there is a gradual reduction in variability owing to chance fluctuations in gene frequency and random fixation of individual alleles. This phe-

nomenon of "random fixation," "drift," or the "Sewall
Wright effect" is undoubtedly the chief source of differences
between populations, races, and species in nonadaptive characteristics. [Stebbins, 1950, p. 145]

Although neither Mayr nor Stebbins believed nonadaptive evolution
was widespread in nature, they believed that, to the extent it did
occur, Wright's random drift was the primary explanatory factor.

These few quotations could easily be increased by a factor of ten
from other books and articles published between 1935–50. Wright
was believed to have supplied the crucial missing factor for explaining
nonadaptive evolution, in general, and the nonadaptive differences,
in particular, between geographical races and closely allied species.

None of these quotations, it should be clearly understood, accurately reflects Wright's shifting balance theory of evolution in nature.
Wright had denied repeatedly that random drift in small isolated populations led to anything but extinction. Yet it was true that Wright
himself had clearly suggested the connection of random drift with
observed nonadaptive differentiation in nature.

The historical consequences were ironic. Systematists in the 1940s
and 1950s, led strongly by Ernst Mayr, E. B. Ford, and David Lack
(after he had given up his earlier ideas about nonadaptive differentiation in Darwin's finches), guided systematists to a more adaptationist interpretation of differences between geographical races and
allied species. The three most frequently cited examples of nonadaptive differentiation by random drift, the inversions mentioned by Dobzhansky above, banding- and coiling-patterns in snails, and blood
groups, were all found to be subject to substantial or even enormous
selection pressures. As this shift toward adaptationism occurred, systematists had no further need for their conception of Wright's random
drift and minimized the importance of the concept in their interpretation of the evolutionary process. At the same time, of course, Wright
had no further reason to jerk the balance in his evolutionary theory
around to fit what systematists had said about nonadaptive racial and
species differences in nature. He could now emphasize the view his
shifting balance theory had incorporated from the very beginning –
that random drift served the important function of providing novel
genetic interaction systems upon which natural selection could act to
yield a more rapid process of adaptive evolution than could occur
under mass selection alone. Thus, just as Wright was providing what
many prominent evolutionists *now* judge to be a very robust and
useful view of the evolutionary process, many evolutionists came to
dismiss Wright's evolutionary theory as mere random drift, unimportant in the adaptive world of species in nature.

By the time Wright became involved in the dispute over *Panaxia dominula* with Fisher and Ford in 1947–51, he had redressed the balance in his shifting balance theory of evolution to lessen the role of random drift. Wright, unfortunately, contributed to the confusion by maintaining that he had never emphasized random drift in small populations as an important mechanism of evolution. This was true, but he neglected to say that he had, in fact, changed his mind about random drift in relation to differences between geographical races and allied species.

A good measure of Wright's adaptationist views at the time he retired from the University of Chicago in 1955 to take a position at the University of Wisconsin comes from the perceptions of Motoo Kimura, who studied with Wright at Madison in 1955–6 and who was again with him in Madison in 1961. Kimura came to work with Wright wishing to extend the quantitative analysis of the effects of random drift. Although Kimura's analyses indicated to him that random drift should be a substantial factor in evolution in nature, he could not himself, at that time, accept this view because his mentor was so clearly and deeply adaptationist in his outlook. Indeed, Kimura's perception of the leading evolutionists was that they were all adaptationist, from Wright, Dobzhansky, Mayr, Crow, and Stebbins right on down the line. Even in 1961, Kimura, though ever more convinced from his quantitative analysis of the apparent importance of random drift, could not bring himself to really believe in this nonadaptationist process. The depth of adaptationist thinking in the early 1960s is thus apparent from Kimura's perceptions of Wright's views at that time, and from the strongly negative reaction of such evolutionists as Dobzhansky to Kimura's shift to belief in the "neutralist" theory (drawn from personal conversation with Kimura, June 22, 1982).

Conclusions

On the basis of evidence presented in this essay, I agree with Mayr's thesis that the systematists had a strong influence in the evolutionary synthesis. Their influence was, however, at least in the United States and England, not in the adaptationist mold suggested by Mayr. Instead, systematists contributed primarily a nonadaptationist view of the differences between geographical races and allied species. This view was strongly reflected in Wright's initial presentation of his shifting balance theory of evolution in nature, and can be seen in the work of other evolutionary theorists, particularly Julian Huxley. Systematists did play a prominent, if not the crucial, role in shaping the more selectionist and adaptationist view found later in the period of the evolutionary synthesis. Mayr's own influence was important here.

Many systematists, like Mayr, already held adaptationist views early in the synthesis period, but their voices did not dominate until into the 1940s.

The development of Wright's views in the period 1929–48 exemplifies nicely Gould's thesis about the "hardening" of the evolutionary synthesis toward a more adaptationist view. Kimura's example also supports Gould's thesis. A corollary of Gould's thesis is that the selectionism/adaptationism of the later evolutionary synthesis went too far and that current evolutionary theory should provide a more balanced view, including nonadaptationist mechanisms of speciation.

The evidence presented here also helps to resolve the puzzling problem of why Wright has insisted, over and over again, that his shifting balance theory of evolution in nature was misinterpreted by Fisher, Ford, Huxley, and most of the other major evolutionists. What happened is that these evolutionists understood perfectly Wright's attempt in 1929–32 to connect random genetic drift with observed nonadaptive differentiation in nature. But they did *not* understand well the adaptive aspects of the shifting balance theory and thought Wright's theory was damaged by the discovery in the 1940s and 1950s of a greater range of adaptive evolution. To Wright, the move toward a more adaptationist interpretation of evolution in nature enhanced his theory rather than diminished it. Part of the problem resulted because Wright did not clarify his change of views from the early 1930s to the late 1940s.

Finally, the development of Wright's views as seen in this essay helps elucidate the complex relationship between theoretical population genetics and field research in systematics during the evolutionary synthesis. In addition to suggestions I have made elsewhere (Provine, 1978), the evidence here indicates a close interactive relationship. Wright's early published formulation of his shifting balance theory was much influenced by the observations and interpretations of systematists working in the field. In turn, his connection of random drift and nonadaptive differentiation at the racial and species levels influenced many field systematists, until they took a more thoroughgoing adaptationist view. This interdependence of theory and field research, or of theorists and field researchers (as in the cases of Fisher and Ford or Wright and Dobzhansky), helped mold the evolutionary synthesis itself, and continues to stimulate evolutionary biology today.

References

Bumpus, H. C. 1899. The elimination of the unfit as illustrated by the introduced sparrow, *Passer domesticus*. *Biological Lectures from the Marine Biology Laboratory, 1898*, pp. 209–26. Boston: Ginn.

Crampton, H. E. 1904. Experimental and statistical studies upon Lepidoptera. I. Variation and elimination in *Philosamia cynthia*. *Biometrika* 3:115–30.

————. 1905. On a general theory of adaptation and selection. *J. Exp. Zool.* 2:425–30.

————. 1916. *Studies on the Variation, Distribution, and Evolution of the Genus Partula. The Species Inhabiting Tahiti.* Publ. No. 228, Carnegie Institution of Washington, Washington, D.C.

————. 1925. Contemporaneous organic differentiation in the species of *Partula* living in Moorea, Society Islands. *Amer. Natural.* 59:5–35.

————. 1932. *Studies on the Variation, Distribution, and Evolution of the Genus Partula. The Species Inhabiting Moorea.* Publ. No. 410, Carnegie Institution of Washington, Washington, D.C.

Darwin, Charles. 1872. *The Origin of Species*, 6th ed. London: John Murray.

Darwin, Francis, ed. 1887. *The Life and Letters of Charles Darwin.* 3 vols. London: John Murray.

Dobzhansky, Theodosius. 1935. A critique of the species concept in biology. *Phil. Sci.* 2:344–55.

————. 1937. *Genetics and the Origin of Species.* New York: Columbia Univ. Press. [2d ed. 1941, 3rd ed. 1951]

Elton, Charles. 1927. *Animal Ecology.* London: Macmillan.

Fisher, R. A. 1930. *The Genetical Theory of Natural Selection.* Oxford: Oxford Univ. Press.

Ford, E. B. 1947. The spread of a gene in natural conditions in a colony of the moth *Panaxia dominula* L. *Heredity* 1:143–74.

————. 1931. *Mendelism and Evolution.* London: Methuen.

————. 1980. Some recollections pertaining to the evolutionary synthesis. In *The Evolutionary Synthesis,* ed. E. Mayr and W. B. Provine, pp. 334–42. Cambridge: Harvard Univ. Press.

Gulick, John T. 1905. *Evolution, Racial and Habitudinal.* Publ. No. 25, Carnegie Institution of Washington, Washington, D.C.

Haldane, J. B. S. 1932. *The Causes of Evolution.* London: Longmans, Green.

Heincke, E. 1898. Naturgeschichte des Herings. *Abhandlungen des Deutschen Seefischereivereins* 2:1–223.

Huxley, Julian S. 1932. *Problems of Relative Growth.* London: Methuen.

————. ed. 1940. *The New Systematics.* Oxford: Oxford Univ. Press.

————. 1942. *Evolution: The Modern Synthesis.* London: Allen & Unwin.

Jacot, Arthur Paul. 1932. The status of the species and the genus. *Amer. Natural.* 66:346–64.

Jordan, David Starr. 1898. *Footnotes to Evolution.* New York: D. Appleton.

Jordan, David Starr, and Vernon L. Kellogg. 1907. *Evolution and Animal Life.* New York: D. Appleton.

Kellogg, Vernon L. 1907. *Darwinism Today.* New York: Holt.

Kinsey, A. C. 1930. *The Gall Wasp Genus Cynips: A Study in the Origin of Species.* Indiana Univ. Studies 16, Nos. 84, 85, 86.

Lack, David L. 1945. *The Galapagos Finches (Geospizinae): A Study in Variation.* Occasional Papers XXI. San Francisco: California Academy of Sciences.

1947. *Darwin's Finches.* Cambridge: Cambridge Univ. Press.

1973. David L. Lack. Obituary. *Ibis 115*:421–41.

Mayr, Ernst. 1942. *Systematics and the Origin of Species.* New York: Columbia Univ. Press.

1976. *Evolution and the Diversity of Life.* Cambridge, Mass.: Harvard Univ. Press.

Mayr, E., and William B. Provine, eds. 1980. *The Evolutionary Synthesis.* Cambridge, Mass.: Harvard Univ. Press.

Osborn, H. F. 1927. The origin of species, V: Speciation and mutation. *Amer. Natural. 61*:5–42.

Osgood, Wilfred. 1909. *Revision of Mice of the American Genus Peromyscus.* U.S. Dept. of Agriculture, Biological Survey.

Provine, William B. 1978. The role of mathematical population geneticists in the evolutionary synthesis of the 1930s and 1940s. *Studies in the History of Biology 2*:167–92.

1979. Francis B. Sumner and the evolutionary synthesis. *Studies Hist. Biol. 3*:211–40.

1981. Origins of the Genetics of Natural Populations series. In *Dobzhansky's Genetics of Natural Populations I–XLIII*, ed. R. C. Lewontin, John A. Moore, W. B. Provine, and Bruce Wallace, pp. 1–83. New York: Columbia Univ. Press.

In press *a.* Adaptation and Mechanisms of Evolution after Darwin: a Study in Persistent Controversies. In *The Darwinian Heritage,* ed. D. Kohn. Wellington: Nova Pacifica.

In press *b. Sewall Wright – Geneticist and Evolutionist.* Chicago: Univ. of Chicago Press.

Rensch, Bernhard. 1929. *Das Prinzip geographischer Rassenkreise und das Problem der Artbildung.* Berlin: Bornträger.

Richards, O. W., and G. C. Robson. 1926. The species problem and evolution. *Nature 117*:345–7, 382–4.

Robson, G. C. 1928. *The Species Problem.* London: Oliver and Boyd.

Robson, G. C. and O. W. Richards. 1936. *The Variation of Animals in Nature.* London: Longmans, Green.

Romanes, G. J. 1886. Physiological selection; an additional suggestion on the origin of species. *J. Linnean Soc. London 19*:337–411.

Ruthven, Alexander G. 1908. *Variations and Genetic Relationships of the Garter-snakes.* Smithsonian Institution: U.S. National Museum, Bulletin 61.

Schmidt, Johannes. 1918. Racial studies in fishes. I. Statistical investigations with *Zoarces viviparus L. J. Genet. 7*:107–18.

Stebbins, G. L. 1950. *Variation and Evolution in Plants.* New York: Columbia Univ. Press.

Sumner, F. B. 1932. Genetic, distributional, and evolutionary studies of the subspecies of deer mice (*Peromyscus*). *Bibliog. Genet. 9*:1–106.

Thompson, David H. 1931. Variation in fishes as a function of distance. *Trans. Ill. State Acad. Sci. 23*:276–81.

Tower, W. L. 1906. *In Investigation of Evolution in Chrysomelid Beetles of the*

Genus Leptinotarsa. Publ. No. 48, Carnegie Institution of Washington, Washington, D.C.

Wagner, Moritz. 1868. *Die Darwinsche Theorie und das Migrationsgesetz der Organismen*. Leipzig: Duncker and Humbolt.

Wallace, A. R. 1889. *Darwinism*. London: Macmillan.

Weinstein, Alexander. 1980. A note on W. L. Tower's *Leptinotarsa* work. In *The Evolutionary Synthesis*, ed. E. Mayr and W. B. Provine, pp. 352–3. Cambridge: Harvard Univ. Press.

Weldon, W. F. R. 1895. Attempt to measure the death-rate due to the selective destruction of *Carcinus moenas* with respect to a particular dimension. *Proc. Roy. Soc. Lond.* 58:360–79.

Wright, Sewall. 1929. The evolution of dominance. Comment on Dr. Fisher's reply. *Amer. Natural.* 63:557–61.

　　1930. The genetical theory of natural selection. *J. Hered.* 21:349–56.

　　1931. Evolution in Mendelian populations. *Genetics* 16:97–159.

　　1932. The roles of mutation, inbreeding, crossbreeding, and selection. *Proceedings of the Sixth International Congress of Genetics*. Menasha, Wis.: Brooklyn Botanic Garden, Vol. 1, pp. 356–66.

　　1948. On the roles of directed and random changes in gene frequency in the genetics of populations. *Evolution* 2:279–94.

　　1982. The shifting balance theory and macroevolution. *Ann. Rev. Genet.* 16:1–19.

4

The hardening of the modern synthesis

STEPHEN JAY GOULD

Pluralistic and hard versions

In 1937, just as Dobzhansky published the book that later generations would laud as the foundation of the modern synthesis, the *American Naturalist* published a symposium on "supra-specific variation in nature and in classification." Alfred C. Kinsey, who later became one of America's most controversial intellectuals for his study of basic behaviors in another sort of WASP,[1] led off the symposium with a summary of his extensive work on a family of gall wasps, the Cynipidae.

In his article, Kinsey strongly advocated the central theme of the developing synthesis: Evolution at all scales, particularly macroevolution, could be explained by the genetic mechanisms observed in laboratories and local populations. He first complained that some geneticists and naturalists were still impeding a synthesis with their insistence upon causal separation of levels: "Just as some of the geneticists have insisted that the laboratory genetics may explain the nature and origin of Mendelian races, but not of natural species, so others indicate that the qualities of higher categories must be explained on bases other than those involved in species" (1937, p. 208). He then defended the central postulate:

> On the contrary, recent taxonomic studies and such experimental analyses as have been made of natural species indicate that Mendelian genetics provides all the hereditary mecha-

---

[1] Apologies to readers unfamiliar with American slang, but I couldn't resist the allusion. Yes, Kinsey the celebrated wasp taxonomist is the same man who later published the famous "Kinsey Reports" on sexual behavior in the human male and female – the first dispassionate studies, based on sufficiently large samples, of what people really do do. WASP is a colloquial American acronym for "white Anglo-Saxon Protestant," still, despite all our ethnicity, the largest group in our country.

nism necessary for the evolution of species as well as for laboratory races; and I shall undertake to show that the same genetics is all that is involved in the origin and development of the characteristics of higher categories. [1937, p. 208]

Yet in his major monograph of 1930, the first "Kinsey Report" if you will, Kinsey defended a position that would put him outside the synthesis by modern definitions (and he continued to hold these views in 1937). In his section on "mutation" in the origin of *Cynips* species, he contrasted two points of view: the first, supported by "laboratory studies" held that "changes in genes occur . . . as mutations which are complete as soon as they have occurred"; and the second, defined as the "neo-Darwinian conception of 'fluctuating variations' which, by being accumulated over many generations and bent in a given direction by the force of natural selection, would gradually give rise to the characters of new species" (1930, p. 25). Kinsey was not merely talking about the origin of new features by sudden mutation, but of new species. Later in the paragraph, he said so explicitly and again contrasted his position with Darwinism:

> Under this latter [neo-Darwinian] interpretation there may be incipient species of the sort conceived by many taxonomists in their definitions of varieties and subspecies; under the genetic interpretation the first mutant individual embodies all that the new species will contain, and is the new species as soon as it is given an opportunity to perpetuate its mutant characters through a population of individuals. [1930, p. 25]

Let me emphasize that the "mutations" Kinsey defended for the origin of species were generally of small – though discontinuous – morphological effect, entirely comparable to (indeed homologous with) standard laboratory mutations. Kinsey did not ally himself with the true de Vriesian saltationists who refuted the central syntheticist postulate by contrasting large-scale mutations that make species all at once with small-scale variation that differentiates local populations. Instead, Kinsey supported a unified genetics for all levels. His "macro" mutants were relatively small, not discombobulations of entire genetic systems à la Goldschmidt's "systemic mutants." They did not constitute a different genetics acting above the level of processes that continually occur within populations (the antisyntheticist claim). Rather, they represented the standard kind of change that, according to Kinsey, occurred within populations *and* also made new species.

Kinsey then explicitly defended the "genetic" view for speciation in *Cynips* (1930, p. 34) and promptly committed a second, and consequent, apostasy from later conceptions of the synthesis: He denied that most of the morphological expressions of species-forming mutations had any adaptive significance:

> There seems no basis for believing the shortened wings or any of the concomitant variations of any adaptive value to any of these insects. The short wings are not confined to warmer or colder climates, and long- and short-winged forms of various species are active at the same season in the same localities. The field data suggest nothing as to the survival value of these outstandingly basic modifications of structure. [1930, p. 34]

This brief synopsis of Kinsey's views raises a dilemma: How could a man who seemed so much "in" the synthesis (as later defined) by his defense of genetic continuity in causation, be so far out of it in his explicitly antiadaptationist views and his defense of discontinuous origin (however small the steps) for new species. Consider, for example, Mayr's definition (1963, p. 586; but see also 1980, p. 1, to appreciate his long advocacy of this account): "The proponents of the synthetic theory maintain that all evolution is due to the accumulation of small genetic changes, guided by natural selection, and that transpecific evolution is nothing but an extrapolation and magnification of the events that take place within populations and species."

This dilemma can be resolved (ironically, since professional students of evolution must face it) by recognizing a property of science that monolithic, textbookish interpretations ignore at their peril: theories evolve, too. I shall argue that the synthetic theory underwent a major change in intent between its formulation in the 1930s and its hardening in the late 1940s. Most biologists are unaware of this change because we only read the last editions of famous works (Dobzhansky, 1951; and Simpson, 1953 for example) and ignore as superseded historical curiosities the original presentations (Dobzhansky, 1937; and Simpson, 1944, in these cases). Kinsey's work lay within the original boundaries for its defense of ordinary genetics as the source of all evolutionary change, while it clearly fell beyond the pale of the later, hardened version in its non-Darwinian approach to the roles of adaptation and natural selection in speciation.

The change involved intent even more than content. During the 1930s, the main concern of the primary builders (for a movement as yet unnamed) was methodological. They knew the great Darwinian hope of the late nineteenth century: The mechanism of heredity, when elucidated, would vindicate Darwin's views on small-scale variation and the creative action of natural selection working in the cumulative mode. They remembered the crushing of this hope during the early twentieth century when Mendelism, in its first evolutionary incarnation, contributed yet another anti-Darwinian theory in the form of de Vriesian saltationism. They rued the resulting anarchy of incommensurate, competing theories that had driven so many great naturalists to the despair of nearly total agnosticism, as in Bateson's famous declaration: "Less and less was heard in genetical circles and now the topic is dropped. When students of other sciences ask us what is now currently believed about the origin of species we have no clear answer to give. Faith has given place to agnosticism" (1922, p. 57).

As genetics had sown confusion, so too (they hoped) might it eventually breed concord by setting a unified causal base for all evolutionary change. Advances in genetics had established the ground for such a hope. De Vries' systemic saltations had proven to be, quite literally, a primrose path; a genetics of smaller-scale change might now prove sufficient for all levels of evolution, from the lowest (by observation and experiment) to the highest (by extrapolation). Fisher had proven that continuous variation could have a Mendelian underpinning. Morgan had abandoned his commitment to saltational change and had, late in life, accepted the idea that micromutations underlay evolutionary change. The time was ripe for causal unity.

Yet, just as geneticists were weeding out their confusion and settling upon a consensus about how heredity worked, many evolutionists refused to recognize the unifying potential of this emerging agreement, and continued to insist upon a separation between small-scale (and, by implication, unimportant) evolution produced by these known genetic mechanisms and a larger-scale evolution produced by other genetic processes as yet unknown. Kinsey, as a synthesist of the first period, expressed his frustration at this defeatist argument in ridiculing (see quote on p. 71) those evolutionists who liked Mendelism for lab stocks but not for natural populations, or for change within population but not for transspecific evolution.

The original version of the synthesis, therefore, sought to remove this idea of a separate but unknown genetics for large-scale changes and to render all of evolution by *known* genetic mechanisms that could be studied directly in field and laboratory. It was primarily a plea for knowability and operationalism, for a usable and workable evolutionary unity. It did not attempt to crown any particular cause of change, but to insist that all permissible causes be based on known Mendelian mechanisms. In particular, it did not insist that adaptive, cumulative natural selection must underlie nearly all, or even most, change – though many synthesists personally favored this view. Any theory of change would be admitted, so long as its causal base lay in known Mendelian genetics. In the 1930s, for example, genetic drift was often granted a predominant role in phenotypic change, not only at the level of demes, but also in the origin of many species. (Dobzhansky's 1937 book may favor selection slightly, but it comes far closer to asserting this role for drift than his later work would lead one to expect.) Such views about the relative importance of drift (and consequently diminished role for selection) were admitted to the original synthesis because they invoked only known and small-scale genetic change. Likewise, the original version admitted Kinsey's view that the primary phenotypic changes of speciation were generally nonadaptive. For Kinsey vigorously supported the key methodological claim of the original synthesis: Small-scale genetic changes within natural populations are the stuff of all evolution. I have called this original version "pluralistic" because it admitted a range of theories about evolutionary change, Darwinian and otherwise, and insisted only that explanations at all levels be based upon known genetic causes operating within populations and laboratory stocks.

This pluralistic version was slowly and subtly altered, primarily during the 1940s (and perhaps with the 1947 Princeton conference as a focal point), as the intent of explanation by known genetics shifted to the content of one particular theory – neo-Darwinism and its insistence that cumulative natural selection leading to adaptation be granted pride of place as *the* mechanism of evolutionary change. The synthesis hardened by elevating one theory to prominence among the several that supported the primary methodological claim of the original version – and eventually (particularly among many second-string votaries) by insisting to the point of dogma and ridicule that selection and adaptation were just about everything. The hard version had no more room for Kinsey, who was off doing something else by then anyway.

In the rest of this essay, I shall illustrate this shift by discussing the increasing commitment to adaptationist explanations by three key figures: Dobzhansky, who was certainly *primus inter pares* of the whole movement; Simpson, who brought the most disparate field of paleontology under the aegis of the synthesis; and Wright, the great geneticist who did so much to create the original version and then largely stood outside its later hardening (his own shift to adaptationism is especially interesting in this light). If this subject strikes some readers as a bit arcane or unimportant, I can only assert an insider's view, admittedly partisan (Gould and Lewontin, 1979; Gould and Vrba, 1982): The rather strict Darwinism of the hard version established a research program that has directed (and in some ways restricted) the field for 30 years. Adaptationist commitments extend from work on the origin of life to sociobiology. I am also painfully aware that the "triumph of adaptationism" is an exceedingly complex historical subject. I do not wish to fall into the typological traps for history that all evolutionists are trained to avoid for nature. Variation abounded at all times, and it was as unconstrained as human intellects are broad. I speak of no lock step affecting all evolutionists together as an imposed essence upon a field of variation, but only of general tendencies that, like evolutionary trends generated by species selection within clades, often sneak up on you and get established while virtually no one notices what is happening.

Three examples of increasing commitment to adaptation[2]

1. Increasing emphasis on selection and adaptation between the first (1937) and last (1951) edition of Dobzhansky's *Genetics and the Origin of Species*.

Dobzhansky's original probe (1937) toward synthesis was more a methodological claim for knowability than a strong substantive advocacy of any particular genetic argument – though his general Darwinian preferences are clear enough. It held, contrary to his own mentor Filipchenko in Russia or H. F. Osborn in America, that the methods of experimental genetics can provide enough principles to encompass evolution at all levels, but it did not play favorites among the admitted set of legitimate principles. It did not, in particular, proclaim the pervasive power of natural selection leading to adaptation as a mechanism of evolutionary change. Macroevolution, it stressed, is not a thing apart, unknowable in principle from experimental work on laboratory and natural populations and requiring different (and perhaps unfathomable) modes of genetic change. Dobzhansky writes cautiously, emphasizing the hope for complete knowability based on microevolutionary genetics:

> Experience seems to show, however, that there is no way toward an understanding of the mechanisms of macroevolutionary changes, which require time on a geological scale, other than through a full comprehension of the microevolutionary processes observable within the span of a human lifetime and often controlled by man's will. For this reason we are compelled at the present level of knowledge reluctantly to put a

---

[2] I have written the Dobzhansky and Simpson examples in other contexts before (Gould, 1980 and 1982a) and much of the material in this discussion is taken from my previous work. William Provine knows so much more about Sewall Wright than anyone else does that I have based my discussion upon his excellent analysis in the manuscript cited in the references.

sign of equality between the mechanisms of macro- and microevolution, and, proceeding on this assumption, to push our investigations as far ahead as this working hypothesis will permit. [1937, p. 12]

Some inkling of the chaotic and depressed state of evolutionary theory before the synthesis can be glimpsed in a simple list of previously popular arguments that Dobzhansky regarded as sufficiently important to refute – claims that denied his hope for synthesis by suggesting that Mendelian processes observed in the laboratory do not represent the genetic style of "important" evolutionary change in nature. Dobzhansky rebuts explicitly the following arguments: Continuous variation in nature is non-Mendelian and different in kind from discrete mutational variation in laboratory stocks (p. 57); Mendelian variation can only underlie differences between taxa of low rank (races to genera), while higher taxa owe their distinctions to another (and unknown) genetic process (p. 68); chromosomal changes are always destructive and can only lead to degeneration of stocks (p. 83); differences between taxa of low rank are directly induced by the environment and have no genetic or evolutionary basis (p. 146); Johannsen's experiments on pure lines showed the ineffectiveness of natural selection as a mechanism of evolutionary change (p. 150); selection is too slow in large populations to render evolution, even in geological time (p. 178); genetic principles cannot account for the origin of reproductive isolation (p. 255).

Dobzhansky's fifth chapter, on "variation in natural populations," stresses the pluralism of the early synthesis. Observable genetic phenomena are the source of *all* evolution; we find continuity from studies in the laboratory, to variation within natural populations, to formation of races and species:

It is now clear that gene mutations and structural and numerical chromosomal changes are the principal sources of variation. Studies of these phenomena have been of necessity confined mainly to the laboratory and to organisms that are satisfactory as laboratory objects. Nevertheless, there can be no reasonable doubt that the same agencies have supplied the materials for the actual historical process of evolution. This is attested by the fact that the organic diversity existing in nature, the differences between individuals, races, and species, are experimentally resolvable into genic and chromosomal changes that arise in the laboratory. [1937, p. 118]

But what forces shape and preserve this variation in nature? Dobzhansky stresses natural selection (p. 120) as he does throughout the book, but he does not grant it the dominant role that later "hard" versions of the synthesis would confer upon it. He emphasizes genetic drift (which he calls "scattering of the variability") as a fundamental process in nature, not as an odd phenomenon occurring in populations too small to have an evolutionary legacy. He argues that local races can form without the influence of natural selection, and supports Crampton's (1916, 1932) interpretation of the nonadaptive and indeterminate character of substantial racial differentiation in the Pacific land snail *Partula*. He emphasizes that evolutionary dynamics depend, in large measure, upon the size of populations *because* selection is not always in control (and we, therefore, need information about numbers of individuals and their mobility in order to assess the effects of drift, migration, and isolation). He coins the term "microgeographic race" and argues that most are nonadaptive and genetically based, contrary to many naturalists who regard them as adaptive and nongenetic.

Genetics and the Origin of Species went through three editions (1937, 1941, and 1951). As in the various versions of Darwin's *Origin*, the differences are not trivial or cosmetic, but represent a major change in emphasis – a change that set the research program for most of evolutionary biology throughout the 1960s and 1970s. As the synthesis developed, the adaptationist program grew in influence and prestige, and other modes of evolutionary change were neglected, or redefined as locally operative but unimportant in the overall picture.

Dobzhansky's third edition (1951) clearly reflects this hardening. He still insists, of course, that not all change is adaptive. He attributes the frequency of some traits to equilibrium between opposed mutation rates (p. 156) and doubts the adaptive nature of racial variation in blood group frequencies. He asserts the importance of genetic drift (pp. 165, 176) and does not accept as proof of panselectionism one of the centerpieces of the adaptationist program – Cain's work on frequencies of banding morphs in the British land snail *Cepaea* (p. 170).

But inserted passages and shifting coverage have, as their common focus, Dobzhansky's increasing faith in the scope and power of natural selection and in the adaptive nature of most evolutionary change. He deletes the two chapters that contained most material on nonadaptive or nonselected change (polyploidy and chromosomal changes, though he includes their material, in reduced form, within other chapters). He adds a new chapter on "adaptive polymorphism" (pp. 108–34). He argues that anagenesis, or "progressive" evolution, works only through the optimizing, winnowing agency of selection based on competitive deaths; species adapting by increased fecundity in unpredictably fluctuating environments do not contribute to anagenesis (p. 283).

But the most remarkable addition occurs right at the beginning. I label it remarkable because I doubt that Dobzhansky really believed what he literally said; I feel sure he would have modified his words had anyone pointed out that he had allowed a fascination for adaptationism to displace the oldest of evolutionary truths.

Dobzhansky poses the key question of why morphological space is so "clumped" – why a cluster of so many cats, another of dogs, a third of bears, and so much unoccupied morphological space between? He begins by transferring Wright's model of the "adaptive landscape" to a hierarchical level where Wright did not intend it to apply. In so doing, Dobzhansky subtly shifts the meaning of the model from an explanation for nonoptimality (with important aspects of nonadaptation) to an adaptationist argument based on best solutions. Wright devised the model to explain differentiation among demes *within* a species. He proposed it to justify a fundamentally nonadaptationist claim: If a "best solution" exists for a species' phenotype (the highest peak in the landscape), why then do not all demes evolve it? When the model is "upgraded" to encompass differences *between* species within a clade, then it becomes a framework for strict adaptationism. Each peak is now the optimal form for a single species (not the nonoptimal form for some demes within a species), and related peaks represent a set of best solutions for different adaptations of separate evolutionary entities within the clade.

Dobzhansky then attempts to solve the problem of clumping with an adaptationist argument based upon the organization of ecological space into preexisting optimal "places" where good design may find a successful home. If evolution has produced a cluster of cats, this is because an "adaptive range," studded with adjacent peaks, exists in the economy of nature, waiting, if you will, for creatures to discover and exploit it.

> The enormous diversity or organisms may be envisaged as correlated with the immense variety of environments and of ecological niches which exist on earth. But the variety of ecological niches is not only immense, it is also discontinuous . . .
>
> The adaptive peaks and valleys are not interspersed at random. "Adjacent" adaptive peaks are arranged in groups, which may be likened to mountain ranges in which the separate pinnacles are divided by relatively shallow notches. Thus, the ecological niche occupied by the species "lion" is relatively much closer to those occupied by tiger, puma, and

leopard than to those occupied by wolf, coyote, and jackal. The feline adaptive peaks form a group different from the group of the canine "peaks." But the feline, canine, ursine, musteline, and certain other groups of peaks form together the adaptive "range" of carnivores, which is separated by deep adaptive valleys from the "ranges" of rodents, bats, ungulates, primates, and others. In turn, these "ranges" are again members of the adaptive system of mammals, which are ecologically and biologically segregated, as a group, from the adaptive systems of birds, reptiles, etc. The hierarchic nature of the biological classification reflects the objectively ascertainable discontinuity of adaptive niches, in other words, the discontinuity of ways and means by which organisms that inhabit the world derive their livelihood from the environment. [pp. 9–10]

Thus, Dobzhansky renders the hierarchical structure of taxonomy as a fitting of clades into ecological spaces. Discontinuity is not so much a function of history as a reflection of adaptive topography. But this cannot be, for surely the cluster of cats exists primarily as a result of homology and historical constraint. All felines are alike because they arose from a common ancestor shared with no other clade. That ancestor was well adapted, and all its descendants may be. But the cluster and the gap reflect history, not the current organization of ecological topography. All feline species have inherited the unique cat *Bauplan*, and cannot deviate far from it as they adapt, each in its own particular (yet superficial) way. Genealogy, not current adaptation, is the primary source of clumped distribution in morphological space.

2. The shift in G. G. Simpson's explanation of "quantum evolution" from drift and nonadaptation (1944) to an exemplar of strict adaptation (1953).

Although Simpson, probably more than Dobzhansky, personally favored selectionist arguments in the first version of his seminal work (1944), he was equally pluralistic and nonrigid. Indeed, he developed an explicitly nonadaptationist theory to resolve his problem, and he considered this theory of "quantum evolution" as the crowning achievement of his book.

Like Dobzhansky, Simpson viewed consistency of all evolutionary change with principles of modern genetics as his primary claim. The major challenge to unity and consistency arose from the famous "gaps" or discontinuities in the fossil record – the appearance of new *Baupläne* without fossil intermediates. He wrote:

The most important difference of opinion, at present, is between those who believe that discontinuity arises from intensification or combination of the differentiating processes already effective within a potentially or really continuous population and those who maintain that some essentially different factors are involved. This is related to the old but still vital problem of micro-evolution as opposed to macro-evolution . . . If the two proved to be basically different, the innumerable studies of micro-evolution would become relatively unimportant and would have minor value to the study of evolution as a whole. [1944, p. 97]

To explain discontinuities, Simpson relied, in part, upon the classical argument of an imperfect fossil record, but he concluded that such an outstanding regularity could not be entirely artificial. He also recognized that his favored process of gradualistic Darwinian selection in the phyletic mode would not suffice, and he, therefore, framed the hypothesis of quantum evolution. He was clearly quite pleased with this formulation, for he ended the book with a twelve-page defense of quantum evolution, calling it "perhaps the most important outcome of this investigation, but also the most controversial and hypothetical" (p. 206). Faced with the prospect of abandoning strict selection in the gradual, phyletic mode, he framed a hypothesis that stuck rigidly to his more important goal – to render macroevolution by genetical models operating within species and amenable to study by neontologists. Thus, he focused upon the one major phenomenon in the literature of population genetics that granted control of direction to a phenomenon other than selection – Sewall Wright's genetic drift.

He envisaged the major transition as occurring within small populations (where drift might be effective and preservation in the fossil record virtually inconceivable). He chose the phrase "quantum evolution" because he envisioned the process as an "all-or-none reaction" (p. 199) propelling a small population across an "inadaptive phase" from one stable adaptive peak to another. Because selection could not initiate this departure from the ancestral peak, he called upon drift to carry the population into its unstable intermediary position, where it must either die, retreat, or be drawn rapidly by selection to a new stable position. Simpson felt that with quantum evolution he had carried his consistency argument to completion by showing that the genetical models of neontology could encompass the most resistant and mysterious of all evolutionary events. Quantum evolution, he wrote, "is believed to be the dominant and most essential process in the origin of taxonomic units of relatively high rank, such as families, orders, and classes. It is believed to include circumstances that explain the mystery that hovers over the origins of such major groups" (p. 206). Simpson could, therefore, conclude: "The materials for evolution and the factors inducing and directing it are also believed to be the same at all levels and to differ in mega-evolution only in combination and in intensity" (p. 124).

When pressured for a new edition of *Tempo and Mode*, Simpson realized that too much had happened in the intervening ten years to permit a reissue or even a simple revision. The field that he pioneered had stabilized and flourished: "It was [in the late 1930s] to me a new and exciting idea to try to apply population genetics to interpretation of the fossil record and conversely to check the broader validity of genetical theory and to extend its field by means of the fossil record. The idea is now a commonplace" (1953, p. ix). Thus, Simpson followed the outline of *Tempo and Mode*, but wrote a new book more than double the length of its ancestor – *The Major Features of Evolution*, published in 1953.

The two books differ in many ways, most notably by Simpson's increasing confidence in selection within phyletic lineages as the only important cause of change. Consider this addition to the 1953 book, a speculative comment on trends in titanothere horns and too rapid a dismissal (in my view) of the venerable argument that incipient stages of useful structures may have no evident function themselves:

> This long seemed an extremely forceful argument, but now it can be dismissed with little serious discussion. If a trend is advantageous at any point, even its earliest stages have *some* advantage; thus if an animal butts others with its head, as titanotheres surely did, the slightest thickening as presage of later horns already reduced danger of fractures by however small an amount. [pp. 270-1]

But the most dramatic difference between the two books lies in his demotion to insignificance of the concept that was once his delight and greatest pride – quantum evolution. It had embodied the pluralism of his original approach – reliance on a *range* of genetical models. For he had advocated genetic drift to propel very small populations off adaptive peaks into an ultimately untenable inadaptive phase. And he had explicitly christened quantum evolution as a mode different in *kind*, not only in rate, from phyletic transformation within lineages. But now, as the adaptationist program of the synthesis hardened, Simpson decided that genetic drift could not trigger any major evolutionary event: "Genetic drift is certainly not involved in all or in most origins of higher categories, even of very high categories such as classes or phyla" (p. 355).

In an "intermediate stage" – his presentation to the Princeton conference – Simpson (1949, p. 224) had emphasized the dominance of selection in quantum evolution, but had not denied other factors. But by 1953, he had completed his own transition. In 1953, quantum evolution merits only four pages in an enlarged final chapter on modes of evolution. More importantly, it has now become what Simpson explicitly denied before – merely a name for phyletic evolution when it proceeds at its most rapid rates, a style of evolution differing only in degree from the leisurely, gradual transformation of populations. Quantum evolution, he now writes, "is not a different sort of evolution from phyletic evolution, or even a distinctly different element of the total phylogenetic pattern. It is a special, more or less extreme and limiting case of phyletic evolution" (p. 389). On page 385, quantum evolution is listed as one among the four styles of phyletic evolution – and all four are characterized by "the continuous maintenance of adaptation." The bold hypothesis of an absolutely inadaptive phase has been replaced by the semantic notion of a relatively inadaptive phase (an intermediary stage inferior in design to either ancestral or descendant *Bauplan*). But *relative* inadaptation is no threat to the adaptive paradigm, for it matters little that an intermediate is not so well suited for its environment as its ancestor was for a different habitat (because they are not in competition). And it matters even less that the intermediate is not so well designed as its descendent will be (for there is even less opportunity for competition here!). In short, relatively inadaptive populations are fully adaptive to their own environments in their own time – and quantum evolution moves comfortably under the umbrella of the adaptationist program. Simpson even suggests that quantum evolution may be more rigidly controlled by selection than other modes of evolution (though he still invokes inadaptation for the initial trigger): "Indeed the relatively rapid change in such a shift is more rigidly adaptive than are slower phases of phyletic change, for the direction and the rate of change result from strong selection pressure once the threshold is crossed" (p. 391).

3. Sewall Wright's early change on the role of adaptation in the formation of species.

Sewall Wright, when interviewed today (as both W. Provine and I can attest), complains bitterly that his views on the evolutionary role of genetic drift, once called the "Sewall Wright effect," have been consistently misrepresented. Since genetic drift is incontestably about stochastic change in gene frequencies via sampling error alone, one might assume (as many have, given the lamentable "scholarly" tradition of pronouncement without reading) that Wright had advocated a radically non-Darwinian approach to evolutionary change by demoting selection and adaptation in favor of accident. Not so, Wright insists. He argues, with evident justice apparent to anyone who studies his works of the past thirty years, that his theory of "shifting balance," which does specify an important role for genetic drift, is strongly adaptationist – but that adaptation arises at a level higher than the traditional Darwinian focus on individuals.

In brief, Wright believes that most evolutionary change within species occurs by the differential success of some demes over others (rather than by strictly individual selection within a panmictic population). Wright views this "interdemic selection" as adaptive and argues that his theory lies within Darwinian traditions because it identifies (higher-level) selection as the cause of evolutionary change, and adaptation as its result.

But an appeal to higher-level or demic selection poses another problem: How does a species get divided into demes in the first place? In other words, what process supplies the raw material (differences among demes) that higher-level selection must utilize to produce adaptive change? Wright invokes genetic drift to resolve this problem, and argues that drift acts as a generator of higher-level variability among demes, not as an agent of evolutionary change.

Consider the founding deme of a new species. It occupies one adaptive peak of the potential landscape. Other peaks are available for habitation, but how can they be populated? Other peaks cannot be reached by selection alone, because organisms descending the slopes of the original peak – a prerequisite to any climbing of a new slope – will be eliminated by selection, even though adjacent peaks may be higher than the original habitation. Wright invokes genetic drift as a mechanism for permitting small demes to descend slopes and cross adaptive valleys. When a small deme has, by accident, entered a valley and approached another slope, selection can then draw it up to the new peak. In this way, many peaks can be inhabited and the raw material for interdemic selection can be generated.

Wright's theory is rooted in the conceptually unfamiliar notion of hierarchy (see Gould, 1982b). Indeed, Wright told me that he originally thought of naming his shifting balance theory the "two-level" theory for its emphasis on the concerted interaction of ordinary Darwinian selection upon organisms and the higher-level process of interdemic selection. In this context, we can understand the major source of the misrepresentation that Wright so bitterly deplores: Few evolutionists are used to thinking in terms of hierarchically super-posed and interacting levels of selection, for the strict Darwinian tra-dition advocates nearly complete explanation by selection acting upon organisms within a population. If evolutionists tend to translate everything they hear into their favored terms of organisms within populations, then what else can genetic drift be but a mechanism of evolutionary *change*, for it indisputably alters gene frequencies within demes – and this *is* the stuff (by extension) of all evolution. Genetic drift, in this limited view, must, therefore, be a non-Darwinian agent and opposed to selection. Hierarchical thinking is a prerequisite to understanding Wright's true intent, for drift can then act (as he be-lieves) as an agent of variation, supplying raw material in the form of differentiated demes to the higher-level process of interdemic se-lection.

But we must also acknowledge a different and historical (or on-togenetic) reason for the confusion. Wright's later shifting balance theory is adaptationist as he claims and does invoke drift only as a source of variation. But Wright, though in many ways estranged from the developing synthesis, followed the same trend toward increas-ingly exclusive emphasis upon adaptation in evolutionary change. The version of shifting balance that he has advocated for 30 years did not arise by sudden creation, complete in its final form. It had an-tecedents of differing tenor in Wright's earlier work, and these arti-cles, written during the pluralistic phase of the synthesis, granted a much greater role to nonadaptation in evolutionary change. In short, Wright did sometimes invoke drift as a non-Darwinian agent of change in articles written during the early pluralistic phase of the synthesis.

William Provine, who is writing a complete scientific biography of Wright, has catalogued Wright's ambiguities and multiple intents during the crucial period 1929–32. The later selectionist view is al-ready in the wings, but several key passages also advocate the non-adaptationist role for drift that Wright would later reject. Wright wrote in 1931 (p. 158) that shifting balance "originates new species differing for the most part in nonadaptive respects." In the following year, he stated (1932):

> That evolution involves nonadaptive differentiation to a large
> extent at the subspecies and even the species level is indi-
> cated by the kinds of differences by which such groups are
> actually distinguished by systematists. It is only at the
> subfamily and family levels that clear-cut adaptive differences

become the rule. The principal evolutionary mechanism in the origin of species must then be an essentially nonadaptive one. [pp. 363–4]

Provine concludes "The careful reader in 1932 would almost certainly conclude that Wright believed nonadaptive random drift was a primary mechanism in the origin of races, subspecies, species, and perhaps genera. Wright's more recent view that the shifting balance theory should lead to adaptive responses at least by the subspecies level is found nowhere in the 1931 and 1932 papers" (ms., p. 58).

Was hardening a parochial American phenomenon?

The community of American evolutionists, although large enough, was itself sufficiently focused and hierarchical that the changed opinion of a few key leaders might have swayed an entire field for no particular good reason beyond authority. We must, therefore, ask whether the hardening of the synthesis was simply a parochial American phenomenon, traceable to one or a few key people, and without general significance.

England provides a good test for the generality of hardening because adaptationist traditions have been so strong in Darwin's own land (though Darwin himself was a pluralist). Perhaps the synthesis was "hard" in Britain from the first, and all the change I have documented merely represents a few recalcitrant Americans finally falling into line. After all, England has produced a long line of hyperselectionists, from Wallace in Darwin's own day, through the purity of R. A. Fisher's adaptationism, to the convictions of E. B. Ford and the *Cepaea* school launched by A. J. Cain.

Yet we can demonstrate that hardening also affected the work of several prominent British evolutionists. Consider, for example, Julian Huxley who, after all, gave the synthetic theory its name (1942). Huxley, to be sure, and as a good Englishman (but unlike his famous grandfather), was always more committed to adaptation than, say, the early Dobzhansky. Yet his early views had the pluralistic cast so characteristic of the 1930s and early 1940s, and he undertook what many biologists regard as his most important research in a largely nonadaptationist context. Throughout the 1920s, and culminating in his book of 1932, Huxley developed and largely founded the study of allometric growth expressed by power functions. Though this theme could have been developed in a strictly adaptationist context (see next paragraph), Huxley took a radically different approach and used the allometric insight primarily to argue that major features of an organism's phenotype are probably not adaptations *in se*, but complex nonadaptive consequences of inherited growth tendencies, when selection works on one feature and "drags along" a large set of correlated traits.

I say this with personal diffidence because I am, in some measure, responsible for the reinterpretation of this work along hard adaptationist lines (Gould, 1966). In one of my first papers, written with all the fervor of a graduate student asked to compose a review article, I argued that virtually all allometries should be recast as adaptations because parameters of power functions are as subject to selection as the morphology of phenotypes. I well remember, in my unquestioning adaptationist commitment of the time, feeling that Huxley had somehow "betrayed the cause" in departing from the straight and narrow of strict Darwinism when an adaptationist interpretation for the same phenomena was available. My general contention about selection on growth rates was not wrong, but I now rather suspect that Huxley's original perspective is the more important theme for allometricians.

Huxley continued his active pluralism into the book that gave the synthesis its name (Huxley, 1942). He emphasized adaptation and the revival of Darwinism, but held throughout that differences among taxa of lower rank were largely nonadaptive and that Wright's genetic drift supplied an explanation. Provine (ms.) has counted the references and documented that Huxley gives about equal time to Fisher's strict adaptationism and to the nonadaptationist interpretation of Wright's genetic drift.

Yet, by the 1950s, Huxley had adopted the hard version of strict adaptationism. At the major symposium of the 1959 Darwin centennial – a three volume paean to the hard version (Tax, 1960) – Huxley presented the keynote address (Huxley, 1960) and defended panselectionism with an explicit renunciation of his former view that nonadaptation regulates small-scale diversification. He quotes some lines of Darwin (slightly out of context I think), and then comments: "The first sentence refers to small-scale processes and makes intelligible the omnipresence of detailed adaptation" (1960, p. 11), thus recalling Weismann's famous line, the rallying cry of strict Darwinism at the turn of the century, about the omnipotence (*Allmacht*) of selection.

Hardening is also reflected through the several versions of what may be the most celebrated empirical study of the time – David Lack's work on the Galapagos finches. (I thank Malcolm Kottler for calling my attention to this example and for documenting it in a long letter of May 4, 1981.) Lack published his first monograph on the finches in 1945, though a first draft had been written just before war broke out in 1939. In this work, Lack strongly supported the common view that small scale differences were largely nonadaptive. Kottler writes:

> In this first monograph, Lack took the position that all sub-specific differences and almost all differences between closely-related species were non-adaptive. The major exceptions in the latter case were the beak differences. But at the time Lack explained them entirely in terms of reproductive isolation. He explicitly rejected the view that closely-related coexisting species had to be ecologically isolated and denied that their beak differences were correlated with significant dietary differences.

In his major work of 1947 (*Darwin's Finches*), Lack continued to defend nonadaptation for many small-scale differences, but now advanced an adaptive interpretation for several others. He also introduced the famous, and now well-established, claim that beak differences are adaptive responses to diet and act to prevent competition.

In 1960, as the Darwin centennial marked the heyday of the hard version, Harper Torchbooks reissued Lack's 1947 monograph. In the major point of a one-page preface, Lack explicitly renounced his non-adaptationist view, citing the general change in concept that I have labeled as the hardening of the synthesis:

> The reader may therefore be reminded that this text was completed in 1944 and that, in the interval, views on species-formation have advanced. In particular, it was generally believed when I wrote the book that, in animals, nearly all of the differences between subspecies in the same genus, were without adaptive significance. I therefore specified the only exceptions then known and reviewed the various ideas as to how non-adaptive differences might have been evolved. Sixteen years later, it is generally believed that all, or almost all, subspecific and specific differences are adaptive, a change of view which the present book may have helped to bring about. Hence it now seems probable that at least most of the seemingly non-adaptive differences in Darwin's finches would, if more were known, prove to be adaptive.

Conclusion

After all this documentation, I fear that the conclusion will seem a bit flat. I now arrive at the point where I should be giving a conclusive and erudite explanation of why the synthesis hardened. Yet truly, I do not know. Some guidelines and factors seem clear enough. First of all, the community of evolutionary biologists is sufficiently small and sufficiently stratified – a few lead and many follow, as in most human activities – that we need not seek some deep and general scientific or societal trend to render a change in opinion of so many evolutionists of different nations. A change by a few key people, themselves in close contact and with mutual influence, might trigger a general response.

Empirical aspects certainly influenced the hardening. Once the pluralistic version had reemphasized classical Darwinism as a respectable alternative, the search for actual measures of selection and adaptation in nature intensified and succeeded. The British panselectionist school, headed by Ford and Cain (see Ford, 1963, for example), presented many examples from butterflies and snails. More importantly, Dobzhansky, the key figure in the transition, discovered that his favorite example of potential nonadaptation needed to be reinterpreted in selectionist terms. In 1937, he had attributed differences in inversion frequencies within natural populations of *Drosophila* to genetic drift, but he then discovered (see 1951) that these frequencies fluctuate in a regular and repeatable way from season to season, and decided (with evident justice) that they must be adaptive. Still, we surely cannot attribute such a major change as the hardening of the synthesis entirely to induction from a few empirical cases. After all, no pluralist had doubted that selection regulated many examples, so the elegant display of a few should not have established a generality. After all, the arch antisynthesist Richard Goldschmidt argued forcefully (1940, see Gould, 1982c) that most changes within populations had a selectionist base (that could not be extrapolated to explain higher-order evolution).

More importantly, most evolutionists brought up wholly within the generation of the synthesis do not know that an "official" (though admittedly not exclusive) viewpoint among taxonomists during the early 1930s identified as nonadaptive the great majority of phenotypic differences separating subspecies and species. (Kinsey's position, cited in the first section, reflects this professional consensus; the highly influential book of Robson and Richards, 1936, advanced this view as its primary contention.) Deference to experts is an important phenomenon in science – all the more so in a diffuse and largely nonexperimental field like evolutionary biology, where no one can have direct experience in all subdisciplines, where simple laboratory repetition cannot validate many conclusions, and where deference to expert opinion, therefore, plays a larger role than usual. One cannot read Wright and Huxley, in particular, during their pluralistic period without getting the definite impression that they personally rooted for a stricter form of Darwinism, but hesitated to advance it because taxonomists proclaimed so certainly that the required frequency of adaptation did not exist at low levels in nature.

As the theoretical basis for neo-Darwinism grew, a new generation of systematists (Mayr, 1942, in particular) asserted a much wider role for selection and adaptation, and the earlier view of Robson and Richards eventually faded. One might give two different interpretations to such a scenario. The "heroic" version holds that the prevalence of adaptation is an empirical truth whose recognition had been impeded by an inadequate theoretical base. The supply of proper theory opened the eyes of systematists to an empirical reality that, in turn, reinforced the theory. The more "cynical" version holds that we still do not have a proper assessment of the relative importance of adaptation in small-scale differences. Since the world is so full of a number of things, cases of both adaptation and nonadaptation abound, and enough examples exist for an impressive catalogue by partisans of either viewpoint. In this light, historical trends in a science might reflect little more than mutual reinforcement based on flimsy foundations – in this case, an assertion of neo-Darwinian theory might lead taxonomists to emphasize the cases of adaptation that manifestly exist, and theorists might then feel sufficiently strengthened to assert a harder version on the false assumption that its empirical base had been independently and securely established.

Questions of relative frequency are among the hardest to resolve in science, and natural history abounds with them. No crucial experiment is possible, nor can we (given the multitude of potential cases) establish an adequate relative frequency by simple enumerative induction (What, after all, is a "random" sample in a world of protean taxonomic diversity?) We still do not know the relative frequency of adaptation in small-scale differences. The adaptationist context of current theory leads us to focus upon established examples, but how often do we find any discussion, or even acknowledgment, of the hundreds of other unexplained differences that separate most taxa?

Finally, a methodological point in conclusion. Darwin's assertion of evolution was an event of such unrivaled importance in the history of science and human society that we tend to view it as a watershed for all concepts in biology – as though everything should be discussed primarily in terms of before or after evolutionary theory. Such a perspective is inadequate, for many traditions of thought persist through evolutionary theory, emerging with a reinterpretation of causality to be sure, but intact nonetheless. In particular, certain "national styles" persisted from the eighteenth century, through Darwin's era, and into our own time. Views on adaptation provide a good example.

I have said nothing about German biology because it has generally held a view of adaptation outside the scope of this essay (Rensch and other synthesists notwithstanding). Adaptation is seen as real but superficial, a kind of jiggling and minor adjustment within a *Bauplan* evolved by some other mechanism – not, in any case, a general mechanism (by extrapolation) for evolutionary change at higher levels. This viewpoint is firm in the nonevolutionist transcendental morphology of Goethe and many of the *Naturphilosophen*. It affects the entire "laws of form" tradition, underlies the pre-Darwinian evolutionism of Etienne Geoffroy de Saint-Hilaire, and persists to our time among such German evolutionists as Remane, Schindewolf, and Goldschmidt.

The adaptationist tradition, on the other hand, has been an English pastime for at least two centuries. If continental thinkers glorified God in nature by inferring the character of his thought from the laws of form linking his created species, or incarnated ideas (as Agassiz maintained), then Englishmen searched for him in the intricate adaptation of form and function to environment – the tradition of natural theology and Paley's watchmaker. Darwin approached evolution in a quintessentially English context – by assuming that adaptation represented the main problem to be solved and by turning the traditional solution on its head. Few continental thinkers could have accepted such a perspective, since adaptation, in their view, was prevalent but superficial. The centrality of adaptation among English-speaking evolutionists in our own times, and the hardening of the synthesis itself, owes much to this continuity in national style that transcends the simple introduction of evolutionary theory itself. One might say that adaptation is nature's truth, and that we had to overthrow the ancient laws-of-form tradition to see it. But one might also say that twentieth-century panselectionism is more a modern incarnation of an old tradition than a proven way of nature. It may now be impeding a more pluralistic account that the early synthesists groped toward and then let slip away. The only honest answer at the moment is that we do not know.

References

Bateson, W. 1922. Evolutionary faith and modern doubts. *Science* 55:55–61.

Crampton, H. E. 1916. *Studies on the Variation, Distribution, and Evolution of the Genus Partula, the Species Inhabiting Tahiti.* Publ. No. 228. Washington, D.C.: Carnegie Inst. of Washington.

1932. Studies on the Variation, Distribution, and Evolution of the Genus Partula. The Species Inhabiting Moorea. Publ. No. 410. Washington, D.C.: Carnegie Inst. of Washington.

Dobzhansky, Th. 1937. *Genetics and the Origin of Species*. New York: Columbia Univ. Press.

　　1951. *Genetics and the Origin of Species*, 3rd ed. New York: Columbia Univ. Press.

Ford, E. B. 1963. *Ecological Genetics*. London: Methuen.

Goldschmidt, R. 1940. *The Material Basis of Evolution*. New Haven: Yale Univ. Press.

Gould, S. J. 1966. Allometry and size in ontogeny and phylogeny. *Biol. Rev.* 41:587–640.

　　1980. G. G. Simpson, Paleontology and the Modern Synthesis. In *The Evolutionary Synthesis*, ed. E. Mayr and W. B. Provine, pp. 153–72. Cambridge, Mass.: Harvard Univ. Press.

　　1982a. Introduction to T. Dobzhansky, *Genetics and the Origin of Species*, pp. xvii–xli. The Columbia Classics in Evolution Series, ed. N. Eldridge and S. J. Gould. New York: Columbia Univ. Press.

　　1982b. Darwinism and the expansion of evolutionary theory. *Science* 216:380–7.

　　1982c. *The Uses of Heresy*: An Introduction to Richard Goldschmidt's "The Material Basis of Evolution," pp. xiii–xlii. New Haven and London: Yale Univ. Press.

Gould, S. J., and R. C. Lewontin. 1979. The spandrels of San Marco and the Panglossian paradigm: a critique of the adaptationist programme. *Proc. Roy. Soc. Lond. B* 205:581–98.

Gould, S. J., and E. S. Vrba. 1982. Exaptation – a missing term in the science of form. *Paleobiology* 8(1):4–15.

Huxley, J. 1932. *Problems of Relative Growth*. London: MacVeagh.

　　1942. *Evolution, the Modern Synthesis*. London, Allen & Unwin.

　　1960. The Emergence of Darwinism. In *Evolution after Darwin*, Vol. 1, ed. S. Tax, pp. 1–21. Chicago: Univ. of Chicago Press.

Kinsey, A. C. 1930. The gall wasp genus *Cynips*, a study in the origin of species. *Indiana Univ. Studies*, Vol. 16.

　　1937. Supra-specific variation in nature and in classification, from the viewpoint of zoology. *Amer. Natural.* 71:206–22.

Lack, D. 1945. The Galapagos finches (Geospizinae). *Occasional Papers of the California Academy of Sciences*, No. 21.

　　1947. *Darwin's Finches*. Cambridge: Cambridge Univ. Press.

　　1960. *Darwin's Finches*, 1960 reprint with preface. New York: Harper Torchbooks.

Mayr, E. 1942. *Systematics and the Origin of Species*. New York: Columbia Univ. Press.

　　1963. *Animal Species and Evolution*. Cambridge, Mass: Belknap Press of Harvard Univ.

　　1980. Some Thoughts on the History of the Evolutionary Synthesis. In *The Evolutionary Synthesis*, ed. E. Mayr and W. B. Provine, pp. 1–48. Cambridge, Mass.: Harvard Univ. Press.

Provine, W. B. Ms., "Adaptation and mechanisms of evolution after Darwin:

a study in persistent controversies." Paper presented at conference "Persistent controversies in evolutionary theory," Univ. of Chicago, March 1982.

Robson, G. C., and O. W. Richards. 1936. *The Variation of Animals in Nature.* London: Longmans, Green.

Simpson, G. G. 1944. *Tempo and Mode in Evolution.* New York: Columbia Univ. Press.

1949. Rates on Evolution in Animals. In *Genetics, Paleontology and Evolution,* ed. G. L. Jepsen, E. Mayr, and G. G. Simpson, pp. 205–28. Princeton: Princeton Univ. Press.

1953. *The Major Features of Evolution.* New York: Columbia Univ. Press.

Tax, S. 1960. *Evolution after Darwin.* Chicago, Univ. of Chicago Press, 3 vols.

Wright, S. 1931. Evolution in Mendelian populations. *Genetics* 16:97–159.

1932. The Roles of Mutation, Inbreeding, Crossbreeding, and Selection. *Proceedings of the Sixth International Congress of Genetics,* Vol. 1 pp. 356–66.

PART II. MIMETIC THEORY:
ITS RELATION TO THE HISTORY OF
EVOLUTIONARY BIOLOGY

5

Mimicry: views of naturalists and ecologists before the modern synthesis

WILLIAM C. KIMLER

Perennially a testing ground for theory, considerations of mimetic coloration mirror, in a general way, the development of evolutionary biology since 1859 and publication of *On the Origin of Species*. The first important independent application of the theory of descent with modification by natural selection was that of mimetic butterflies, by Henry Walter Bates in 1861. His work was seen not only as application of theory to the question of species and speciation, but also as a confirmation of Darwin's theoretical mechanisms: heritable variation, struggle for existence, natural selection, and divergence. Darwin delighted in this case, brought forward by Bates, as another biological phenomenon made sensible only under the umbrella of evolutionary notions. The phenomenon has rested there ever since; controversial and disputed, but a literally colorful exemplar for hypothesized evolutionary mechanisms.

Bates, a noted taxonomist and natural history collector, developed his theory of mimetic resemblance (Bates, 1862) from the perspective of the problem of speciation, for mimetic series, indeed, seemed to show in the laboratory of nature the very process of the origin of species. But as evolution, or the transmutation of species, itself became acceptable to growing numbers of naturalists, other considerations under the general heading of evolution came into prominence. Alfred Russel Wallace approached the question of animal coloration and mimicry from his customary viewpoint, an ecologically oriented interest in biogeography and adaptation (Wallace, 1867). For him, mimicry offered not only a chance to see speciation occurring but also a case that would demonstrate the importance of evolutionary theory for the interpretation of any ecological pattern, and, hence, for the explanation of natural phenomena in general. Following Bates's formulation, mimicry theory developed in this direction of increasing ecological sophistication and the categorization of ecological patterns.

Wallace, who was typical of the naturalists of this period, felt no great need for genetics, for a science of the sources of variation or of the fundamental patterns of heredity. The broader ecological problems of animal coloration were what attracted naturalists, including Fritz Müller, Raphael Meldola, and E. B. Poulton, over the remaining years of the nineteenth century. They extended the number of known cases, analyzed the components of these cases into ecological influences (perception by predators, geographic trends, direct influence of environment), and, in general, accepted as dogma the explanation derived from natural selection theory.

But mimicry was also used by rival theorists, especially by those more concerned with the mechanisms of heredity. It became an example and even a specially telling case for the Lamarckian orthogenesis of E. D. Cope and the more physiological orthogenesis of T. Eimer. At the same time it could serve as a major example within A. Weismann's combination of germ-plasm heredity and selection theory. In short, however, through the general decline of Darwinism within biology during the late 1800s, natural selection theory lost its exclusive claim to being the explanation of mimetic phenomena. The gradualistic approach to the mimicked pattern seemed particularly vulnerable to all the rival theorists, and for William Bateson and other newly ascendant mutationists and Mendelians, that gradualism finally appeared fatal to the explanation of mimicry by natural selection. The ecological developments by Wallace, Müller, and Poulton seemed irrelevant in the face of mutationist genetics, as R. C. Punnett argued in his 1915 book *Mimicry in Butterflies*. That book illustrates the depth and strength of antiselectionism during this low point for Darwinism.

Nor were physiologists leaning toward orthogenesis and geneticists favoring discontinuity alone in rejecting Darwinism around the turn of the century. The efficacy of natural selection was denied within the ranks of field naturalists as well, a principal disputant on the mimicry argument being the American ornithologist W. L. McAtee (1912). Though never persuasive to all, strong antiselectionist arguments simmered and flared until the 1940s. It is difficult to characterize by field the views held in the first forty years of this century; the best picture must display the multiplicity of views. Biologists held to various admixtures of selectionism, Lamarckism, and directed variation; the notions of continuous variation, discontinuous variation, mutation, and genetic expression were similarly nonuniform.

Thus it is difficult to talk of the contributions made by various fields toward the eventual modern synthetic theory. Conceptions certainly changed within the disciplines, but individual workers still held to bewildering combinations of modern and older bits of theory. To date, the histories have concentrated on the geneticists. Mayr (1980*a*) made a strong attempt to reorient views about the synthesis with his dichotomy between naturalist–systematists and geneticists, emphasizing contributions of the naturalist–systematists. Understandably, Mayr paid most attention to the taxonomists (Mayr, 1980*b*, p. 123), and even though he discussed the variety of views within the discipline, he also presented them as a camp, split from geneticists. But naturalist and systematist are not the same thing, and many ecologically minded naturalists were not involved in taxonomic concerns in this presynthesis period. A focus on the species problem or on the nature of variability does not completely cover the concerns of naturalists. What I think we can isolate is a strain of thought that is primarily ecological in outlook, and mimicry can serve as an interesting case in the changing fortunes of this viewpoint. Darwin's and Wallace's evolution was conceived principally from an ecological perspective, which was focused on geographic distribution and adaptation to environment by gradual steps, and in this perspective we can perhaps see what it was that declined and revived, that was called "Darwinism," and, as such, contributed to the "Darwinian" nature of the modern synthesis.

What is meant by an ecological perspective in evolution? Any evolutionary theory deals with the nature of variability, of speciation, of adaptation; the ecologist stresses the distribution and abundance of organisms and emphasizes adaptations, both resulting from the evolutionary process. In this view, in which evolution is understood as a dynamic process, it is the selective conditions that contain the most interest. Genetic variability is just raw material, and speciation a result of less concern than the adaptive structure and function of organisms. In this view the study of mimicry naturally demands an answer to the question of how adaptation comes to be. *Fitness* of variations rather than *origin* of variations is seen as the answer to the question: "How did it arise?"

The British entomologist Edward B. Poulton may be taken as illustrative of this heritage. A working period from roughly 1880 to 1930 makes him a bridge from the nineteenth-century naturalists, especially as he self-consciously (and self-righteously) took up, at the turn of the century, the task of defending the Darwinian "speculations" and the importance of natural history. He was a firm supporter of the continuity and gradualism of evolution, with interests in ecology and in complex adaptations. Adaptations were explained by Darwin's argument – if small variations always exist, are heritable, and affect fitness, then selection leads to adaptedness. For Poulton, this was no simple faith. He had been among the first to attempt field experiments demonstrating natural selection, most importantly in his experiments on the protective nature of butterfly pupa coloration. But despite this, Poulton was an easy target for the genetic experimentalists, for he defended explicitly teleological and utilitarian reasoning in the explanation of adaptations, while downplaying the importance of the new Mendelian genetics for this purpose.

Poulton saw William Bateson and the other enthusiastic Mendelians ignoring and depreciating the lines of investigation "inspired" by Darwin and Wallace. In particular, Bateson's writings were a "hindrance in the attempt to solve the problem of Evolution," "injurious" because of their prejudice, narrowness, dogmatism, and the exaggerated importance ascribed to them (Poulton, 1908, p. xiii). "The investigation of heredity by experimental methods," Bateson had declared, "offers the sole chance of progress with the fundamental problems of evolution" (Poulton, 1908, p. xvi). This approach was just too restrictive for Poulton. He preferred to use the systematists' work, the geographic studies of varieties, which lead to a view of continuous evolution. As Darwin had pointed out, evolution is seen in geographic distribution, and fifty years of collections and taxonomic work since 1859 had led most systematists to take the data on distribution and species change at range borders as demonstrating continuity.

But what of the evidence for discontinuity, for De Vriesian mutation? Poulton simply denied its place in nature, ignoring the "only important evidence . . . of certain Oenothera" (1908, p. xix) as being too restricted, and even problematic within that genus in nature. He saw the whole mutation theory as one more case of the faulty but perennial argument that "before a thing can be selected it must be." This was the usual argument against gradualism and the creative power of selection, and to counter it Poulton cited the example of protective resemblance and striking mimicries: "It is as unlikely that a key could be made to fit a complicated lock by a number of chance blows upon a blank piece of metal, as that the elaborate pattern on the wings of a butterfly should have been reproduced on those of its mimic by Mutation" (1908, p. xxiii). He had turned what had originally been an objection to Darwin – that "blind chance" could not direct evolution – into a support for Darwinism! Of course he was able to do that because, in fact, evolution by natural selection is not random; selection is the directive force that preserves advantageous variations. Thus, Mendelism was too limited for Poulton in its explanatory power because single-step mutational speciations could not explain the very complexity of adaptation that fascinated the naturalist.

Speciation itself seemed to have been of little concern to Poulton. He was an entomologist famous, not for taxonomic work, but for insect ecology, and especially for his grand organizing system for mimicry theory. But De Vries's *Mutationstheorie* was really talking about speciation: What was its relation to adaptation? Here, too, Poulton found the theory inadequate; he followed Darwin in holding that "divergence," resulting simply from the passage of time, allowed merging of the two processes. He also, like Darwin, admitted the difficulty of deciding when a variety became a species. Diagnostic characters alone, when arranged into the variational series that any comprehensive collection would reveal, were not sufficient to yield unambiguous species. Criteria of free interbreeding and of common descent were needed as well. The idea of the "interbreeding community" (Poulton, 1903, p. 68) gave meaning to the character series and made sense of transition. The geographic subspecies (= incipient species) were connected in interbreeding until by "the development of pronounced preferential mating or by the accumulated incidental effects of isolation prolonged beyond a certain point, [the subspecies broke] up into distinct and separate species" (Poulton, 1903, p. 76). Because the geographic variations were themselves established by selection, speciation was just a result of continuous adaptive change, with final divergence sufficient to inhibit interbreeding.

There is a distinctly postsynthesis cast to these 1904 remarks of Poulton's, and yet before the 1930s this was increasingly *not* the systematists' position. They held too firmly to the nonadaptive nature of species taxonomic characters, as summarized in such a standard work as G. C. Robson's *The Species Problem* (1928). Whether or not the mechanism of speciation was seen to be adaptive, it was widely believed that useful characters for the taxonomist were frequently rather incidental features of morphology. The naturalists who thought otherwise were the followers, like Poulton, of A. R. Wallace's ultrautilitarian line; it was ecology and not taxonomy that stressed the adaptive nature of all the features of organisms. Indeed, the stimulating new spate of series-collecting that informed turn of the century systematists supported the belief in the nonadaptiveness of taxonomic characters, as stress, for example, by Bateson in his *Materials for the Study of Variation* (1894) and revealed by Crampton in his famous *Partula* snail monograph of 1916.

However, the museum-specimen series held another lesson. The morphological variation revealed in an extensive collection did, indeed, show many small differences, arrangeable into continuous series. Poulton could then use that continuity, together with his gradual adaptationist mechanism, to argue against any discontinuous process of evolution. Mutation theory was under fire particularly for its inadequacy on the problem of gradual trends of change.

Yet, despite such naturalists' claims that mutation theorists simply did not understand the museum evidence on continuity of variation, it could be asked, who among them had demonstrated speciation? After all, De Vries was justifiably delighted with his *Oenothera* results: A simple series of novel experiments had revealed not just a little information on change, but an actual case of speciation. After years of defending evolution, even though it might be unobservable in action, biologists had an actual, well-attested, experimentally repeatable speciation event. Of course it was persuasive.

The difference of approach to speciation theory was as crucial here as were species concepts, whether populational or typological. Mutationists had experimental results that allowed them to dissociate speciation from adaptation, and their speciation event overcame such standard problems with Darwinism as insufficient geological time for slow evolution, uselessness of incipient stages, and the nonadaptive nature of some taxonomic characters. The Darwinian emphasis on range of variation seen within a species meant that the definition of any species was broader than one "type"; to the mutationists, it did not necessarily follow that these variants became incipient species. Moreover, since nature exhibited much more variation than it did speciation, it seemed to follow that something was keeping the species constant as well as variable. For Darwin, that uniformity was produced by blending (in sexual reproduction) and by ecological constraints (the demand for fitting a "place" in the economy of nature). But Mendelism introduced its own mechanism for constancy, with results that had little to do with evolutionary change. Under a simple interpretation, Mendel seemed to have shown just why a species should produce not illimitable variation, but a few persistently recurring variants. To effect change, the constant hereditary material of the germ cells had itself to be altered; thus Mendelism was quickly tied to mutation theory in the explanation of speciation. Those who believed that De Vries had produced good species saw the rival view as an expression of faith that Darwin's "divergence" existed and eventually produced good species, too.

However, it was not only the geographic variation displayed in collections that supported a gradualist, adaptational speciation theory; the mimicry case made that variation *useful*. It countered Bateson's celebrated denial of continuous adaptive variation in nature by pointing to what seemed an obvious adaptation, and it linked that adaptation to speciation. Darwinians like Poulton still relied for their convincing argument upon that originally presented by Bates. His concerns had been taxonomic, and his 1862 paper was intended as a demonstration of the process of speciation; as its title clearly stated, the paper was ostensibly about classification of the Heliconiid butterflies. Bates used quite an eclectic approach to the problem of the family's taxonomy. Argument over the standard of wing-venation was followed by discussion of behaviors and ranges; he used habits and ecology and an evolutionary biogeography to separate a Danaoid from an Acraeoid group within a family that others had lumped together. With mimicry previously unrecognized, appearances had, understandably, led to confusion in ascribing affinities. Geographic trends in Brazil were striking, for the diversity and abundance of species was remarkably high. Especially striking to Bates, given that the country was fairly uniform, were the limited ranges of the species. Many had "the appearance of being geographical varieties" (Bates, 1862, p. 500), especially as the differences often were only of color or pattern. But the differences, he reported, are "well-defined, so ordinarily common to all the individuals concerned, and there is so generally an absence of connecting links, that they are held on all hands to be good and true species" (1862, p. 500). Even more compelling for Bates, where they occurred together there were, with one important exception, no hybrid forms.

That exceptional kind of species, polymorphic with local varieties, gave him the key to an explanation of the origin of the rest. The great variability presented "all the different grades between simple individual differences and well-marked local varieties or races, which latter cannot be distinguished from true species, when two or more of them are found coexisting in the same locality without intercrossing" (Bates, 1862, p. 501). To Bates it was evident that the now distinct species had arisen from local varieties: Varieties were, indeed, incipient species. His best examples connected two distinct species "by an unbroken series of varieties" that because of their geographic separation could not be hybrid progeny. The intermediate forms appeared to be "remnants of the steps of modification." So, in the example of the polymorphic *Mechanitis Polymnia*,

> We see here the manufacture, as it were, in process. The species is widely disseminated and variable. The external conditions in certain localities are more favourable to one or more of the varieties there existing than to the others; those favoured ones, therefore, prevail over the others. We find, in this most instructive case, all the stages of the process, from the commencement of the formation of a local variety (var. *Egaensis*) to the perfect segregation of one (var. *Lysimnia*) considered by all authors as a true species. [Bates, 1862, p. 501]

By analogy, the connections of the local varieties to their ancestors through individuals showing all the "shades of variation" were the same as the relations of affinity within a group of good species. The connection through variation in the forms of a few exemplary species was the only way to discover forms to be varieties and not true species.

Bates made the mimetic case his exemplar for this theory of speciation. He used the variation also seen in the mimics of the South American Heliconiids to demonstrate again how geographic variation is linked to incipient speciation, and the adaptive nature of mimicry was proof that the mechanism responsible for this process was natural selection. Thus, explanation of the demonstrable case was extrapolated to all cases.

Interestingly, Bates gave no reason for the existence of the local varieties in the models, although he did not discount the possibility of direct influence of habitat itself as a sufficient cause for variability. But the polymorphic case did at least show, from the geographic data, that "the local conditions favoured the increase of one or more varieties in a district at the expense of the others" (Bates, 1862, p. 511). He noted a tendency for the insects to mate with "none but their own counterparts" (Bates, 1862, p. 501) as probably accelerating species formation from the varieties. With the mimetic species there was a similar segregation of local forms, and additionally the direction for selection was known in each locality – that is, the appearance of the model. Especially remarkable were the parallel changes of mimic as models varied geographically, although no parallelism was observed in the variability itself of local forms. The cause of geographic trends of mimicry was not simply the mechanisms of variation. Also, the coexistent polymorphism of mimics in some locales and the departure from the facies of closely related nonmimics made "direct action of physical conditions" (Bates, 1862, p. 500) unlikely to be the cause of local mimic varieties. They were not hybrids nor were they sports. All in all, the inference from distributional data was that only natural selection explained the preservation of variation in the direction of the model until close mimicry was achieved.

As in many a Darwinian case, the argument was indirect, and it suffered from lack of knowledge of the laws of variation and heredity; yet it did not become any less convincing with the coming of genetic information. The ecological picture explained an adaptation functionally, and the Darwinian ecologist of 1900 knew from observation, from museum taxonomists, and even from the literature on variation in domestic breeds that there would be variation available. From that it was a short step to operating on the assumption that, for any adaptive scenario one might devise, sufficient variability would be present. At least what Bates did use of the laws of variation – variability is not directed nor is it uniform between species or regions – was not denied by later genetics. Genetic theory did not have to provide much to the argument, because it was not given much place in it.

In continuing the line initiated by Bates, Poulton realized, however, that Mendelism was yielding dramatic new data on some of the very problems of variation still faced by the naturalist. Breeding experiments with subspecies or incipient species could reveal possible scenarios of relationship. But more importantly, Mendelism confirmed the particulate nature of heredity and so removed the problem of the supposed swamping of a new trait on intercrossing. Far from eliminating the need for natural selection in an evolutionary explanation, that very fact made gradualism easier. By the same token, the new (Mendelian) trait would be prevented "from penetrating the mass of the species" (Poulton, 1908) unless it possessed some selective advantage. Further, the Mendelians had already admitted that their traits might be small effects and not the large changes of De Vries's theory. Thus Poulton, in 1908, saw a possible synthesis based on accommodating Mendelian patterns of heredity into his Darwinian ecology.

This naturalist was, then, far from unaware of the new genetics; in fact, he insisted that the divergence between Darwinism and Mendelism had been exaggerated and was really slight. He muted the revolutionary quality of Mendelism by emphasizing Weismann's reasoning from particulate heredity and he cleverly pointed to Bateson's own statement that "when the unit of segregation is small, something mistakably like continuous evolution must surely exist" (Poulton, 1908, p xxxviii). Like Darwin, Poulton simply rejected the notion that large variations were commonly important. But although he was well aware of, and sensitive to, the genetic information, Poulton did not really use it. He considered the arguments contrary to Darwinism to be based on hyperbole and on extraordinary cases, and he also saw enough agreement between the two approaches to presume that the genetic nature of the continuous variations he needed could be discovered.

Moreover, Poulton was not himself going to use information from genetics, for that was not where he would find meaning in evolution. As E. B. Ford wrote in a memoir, "Poulton was not really anti-Mendelian. He thought Mendelism interesting and, at a superficial level, important" (Ford, 1980, p. 337). He was prepared to accept Mendelian results as they applied to mimicry, but he did not believe "the fundamental properties of organisms were determined on Mendelian lines" (Ford, 1980, p. 337). He expected further research in genetics to reveal more, but still he held that selection was the necessary director of change. The question for the evolutionary naturalist was how organisms had become what they were, and for Poulton the answer was to be found in the designing (selective) forces. Although Ford, a geneticist, also wrote that Poulton "knew absolutely no genetics at all" (Ford, 1980, p. 337), Poulton was not ignorant of controversy over mechanisms of heredity; but ultimately it was the purposive quality of the adaptation that interested him. Certainly the evolution of adaptation was tied to the nature of variation, but more crucial in the origin of adaptation were the necessary ecological conditions for selection to work on any variation that happened to exist. Few of the geneticists were experienced naturalists like Poulton, with such a strong interest in ecology. This interest was reflected in his methodological arguments, explicitly allowing – demanding – teleological or utilitarian reasoning in the study of adaptation.

No wonder a laboratory experimental geneticist could claim that natural selection had simply replaced the Creator in design arguments. Not Bateson but his colleague, R. C. Punnett, produced the Mendelian answer to the staunch Darwinian view of mimicry and its mechanism of origin. Punnett began his *Mimicry in Butterflies* (1915) with a sarcastic attack on the ease with which Darwinian naturalists identified adaptations, claiming that: "Probably there is no structure or habit for which it is impossible to devise some use, and the pursuit has doubtless provided many of its devotees with a pleasurable and often fascinating exercise of the imagination" (Punnett, 1915, p. 6). In the case of mimetic resemblances, he saw that a commitment to belief in their adaptive nature was essential if selection was to have "anything to do with either their origin or with their survival" (Punnett, 1915, p. 61). Punnett, however, saw two possibilities for the evolution of mimicry: "(a) that the resemblance has been brought about through the gradual accumulation of very numerous small variations in the right direction through the operation of natural selection; and (b) that the mimetic form came into being as a sudden sport or mutation, and that natural selection is responsible merely for its survival and the elimination of the less favoured form from which it sprang" (Punnett, 1915, pp. 61–2). Now this second possibility, of course, was just that "chance" that Poulton found inconceivable. Punnett had theoretical objections, though, to the first mechanism and empirical support for the second. He was claiming that genetics did deny the likelihood of the Darwinian scenario and that it did have to direct theory rather than simply to blend in as comfortable background support.

The difficulty in supposing that selection accumulates small variations in the right direction lies in the lack of adaptive value of that original variation if model and would-be mimic are originally markedly different in pattern, as, for example, in the Ithomiines and Pierines in South America. From the first, the common hypothesis had been that the ancestral appearances of model and mimic may not have been as divergent as the present differences between the model and the nonmimetic relatives of the mimic – which presumably resemble the ancestral form of the mimic – would lead one to suppose. Presumably, the model species itself had originally looked somewhat like the relatives of the mimic and subsequently had acquired a greater distinctiveness. More difficult are mimicry rings, in which there are several mimics and more than one model. For the origin of a complicated series system like this, a simple mimicry compounded in difficulty by the simultaneous mimicking of any one model by several species or genera, Punnett doubted the natural selection explanation completely. The number of presently dissimilar species required to have an ancestral form like that of the model seemed improbable. Worse, in many rings the nonmimetic males of sexually dimorphic mimic species presumably showed the ancestral condition, and the males of the different dimorphic species all looked different. A special hypothesis of ancestral similarity between model and potential mimic broke down when it had to apply to more than one potential mimic. In such cases, a coevolution of model and mimic toward greater distinctiveness from a common ancestral appearance – a hypothesis that preserved adaptive gradualism in the acquisition of mimicry – became extremely improbable. The key doubt remained, then, whether slight variations around the originally dissimilar forms could have had any utility.

But these rings involved Muellerian mimicry as well, and perhaps there was a greater plausibility there of the accumulation of small variations through natural selection. Punnett countered this with G. A. K. Marshall's arithmetic argument against the Muellerian approach by model and mimic to an intermediate pattern. Marshall thought he had demonstrated (Marshall, 1908), using hypothetical numerical examples, that predatory destruction of variants could not bring about a resemblance of a more numerous to a less numerous species; nor could there be an approach when the numbers were roughly equal. In other words, the requirements for Muellerian mimicry were, as for the Batesian, that the less numerous species take on the pattern of the more numerous; coevolution to an intermediate pattern would not occur. Thus the same difficulties concerning ancestral distinctiveness made the Muellerian case no more likely than the Batesian to have come about through a long series of slightly advantageous variants.

In Punnett's version of the faulty "older view" (Punnett, 1915, pp. 152-3), this gradual accumulation, working "as if upon a plastic organism," molded the more and more perfected adaptation. Admittedly this hypothesis had been attractive in its simplicity, but great additional problems arose from the new genetic information. In the experimental breeding of polymorphic species, the crossing of the final form of mimic with the "original" had revealed no intermediate steps. Here, with modern genetic experiments, was one of the most carefully considered and studied cases of mimicry, the perfect case study to test theory. This case was *Papilio polytes.*

Papilio polytes is a common swallowtail butterfly of Ceylon and India. The female is polymorphic, and the three distinct forms, f. *cyrus,* f. *polytes,* and f. *romulus,* resemble, respectively, (1) the males of their own species, thus remaining nonmimetic; (2) individuals of the species *Pachlioptera aristolochiae;* and (3) *Pachlioptera hector.* Poulton accepted A. R. Wallace's original explanation (Wallace, 1869) that *P. polytes* is palatable and the nonmimetic form has the original coloration. The mimetic female morphs diverged from this toward mimicry of two distasteful models, the *Pachlioptera* species. J. C. F. Fryer of Ceylon had bred these butterflies for two years and reported his results in 1913. The immediately striking result (Fryer, 1913) had been that no intermediate forms resulted from the crosses. Although the mimicry is sex-limited, the males can transmit the mimetic factors to their female offspring. The nonmimetic form is recessive to the mimics, so that the differences producing the mimicry seem to result from a single hereditary factor. The difference between the two mimetic forms is produced by one additional factor, in effect converting f. *polytes* into f. *romulus.* That factor alone, however, fails to produce either mimetic pattern. Out of this extremely complex case of mimicry, comes an explanation in simple Mendelian terms. Since Poulton thought that the mimicry had evolved progressively to f. *polytes* and then to f. *romulus,* Punnett reasoned that crossing f. *cyrus,* the nonmimetic start of the series, with *romulus* ought to give "our series of hypothetical intermediates, or at any rate some of them" (Punnett, 1915, p. 91). This did not result, and the "Bates expectation" of geographic multiple variants failed.

Thus Fryer and Punnett both believed the polymorphism was to be explained in terms of only two independently assorting genes. Mendelism told Punnett that such discontinuous inheritance was regular and normal. He drew on the work of Bateson and co-workers on flower color in the sweet pea (*Lathyrus*) (Bateson, Saunders, and

Punnett, 1906); they had also had the idea (Punnett, 1915, p. 91) that independently assorting genes affect the same character, with discontinuous phenotypes. This distinguishes their 1905 idea from E. M. East's work on multiple factors, because that produced continuous variability of the phenotype (East, 1910). As in the peas, Punnett argued, the forms arose as sudden mutations and not by the gradual accumulation of slight differences. Critically, he assumed that the present phenotypic difference and its associated single gene difference entailed the evolutionary course. The present phenotype was the same as when the mutation first appeared and had not been modified. Relying on genetics, he saw cause to deny the Bates expectation of intermediates as the typical geography of variants.

Punnett did not stop with genetical criticisms. Standard theory owed to Wallace the enumeration of five conditions that mimicry fulfills. Of these, Punnett accepted only (1) that the imitation is external or superficial and not internal or structural. For his test case, he denied (2) that mimics and models occupied the same localities throughout their ranges. Moreover, claiming the mimics to be faster fliers than the models contradicted the condition (3) that mimics be more defenseless. Nor, (4) were the mimics always less numerous, although they usually were. Last, the condition (5) that mimics differ from the normal appearance of their taxonomic group had little meaning, because a survey of the Lepidoptera and of his own genetical results led Punnett to a conclusion that mimicry "runs in certain groups" (Punnett, 1915, p. 145). Since four of the conditions are not absolutely fulfilled, as he thought the theory of selected cumulative resemblance required, Punnett, therefore, doubted the natural selection explanation for all cases. Drawing upon the dissident natural history literature for more criticisms of mimicry theory, he also expressed doubts about the usual ecological story. He could find no certain evidence for distastefulness of the butterfly model. He doubted that bird color vision possessed the necessary discriminatory ability to allow precise mimicry. Moreover, he even doubted that birds were predatory enemies of butterflies. The net result is the absence of the selective agent required by Darwinian mimicry theory.

Thus, selection had nothing to do with the formation of mimicry, building on progressive variation. But Punnett reserved a limited role for selection. If the new form had some advantage, it would be conserved. Punnett turned to H. T. J. Norton for calculations of hypothetical rates of change in the frequency of the new mutant. They were surprisingly fast for a dominant allele with even slight selective advantage, and Punnett judged that evolution, "so far as it consists of the supplanting of one form by another, may be a very much more rapid process than has hitherto been suspected" (Punnett, 1915, p. 96). This was a geneticist's mathematical argument for the power of selection (though not of gradualistic selection) in 1915! But, paradoxically, for *Papilio polytes*, population estimates indicated to Punnett a historically stable equilibrium among forms and so there could not be significant selection going on with respect to mimetic resemblance. That an adaptive evolution *could* go rapidly he illustrated by reference to the melanic form of the peppered moth – soon to become a major neo-Darwinian example! Punnett would grant a role to selection, then, but not as the source of adaptive differences between these forms of *P. polytes*; thus, "as may quite possibly happen," there could be selectively neutral differences. Instead of the selectionist view that all characters are useful, the new view was that "the new character that differentiates one variety from another arises suddenly as a sport or mutation" (Punnett, 1915, p. 3).

Here then are reasons from genetics for agreement with systematists claiming nonadaptive character differences between species. Here, too, is reason from genetics to refrain from using a Darwinian geographical argument to deduce an evolutionary mechanism. Similar variation in different mimetic cases and the way the Mendelian alleles controlled color in *P. polytes* led Punnett to see (nonadaptive) regularities of variation. This put the direction of evolution more firmly under the control of the variation itself. The species problem was then, as he stated in his 1911 book *Mendelism*, "not one that can be resolved by the study of morphology or of systematics. It is a problem in physiology" (Punnett, 1911, p. 152). Punnett had taken an ecologically explained phenomenon, set up the problem as that of gradualism versus discontinuity, and thrown the solution into the arms of the new, experimental genetics. Note, however, that this geneticist was also using ecological doubts in order to emphasize his directing mechanism based on genetics.

If the camps are split, moreover, it is not neatly between geneticist and naturalist. Punnett was certainly making the grandest claims possible from a genetical point of view, but his criticisms of Darwinism accord with common statements by taxonomists about adaptiveness and by naturalists skeptical of the protectiveness of mimicry. This may be a case of making selective use of the literature, but it is not a case of ignorance.

Poulton's attitude and style, as already mentioned, were similar. He heard the geneticist's claims, but for him the answer of interest lay in ecology. His published response to *Mimicry in Butterflies* is fascinating in this regard. He reviewed it for *Nature*, spending five of his seven paragraphs caviling at the "principal feature of the book"

(Poulton, 1916, p. 237), its illustrations. What of Punnett's ecological and taxonomic objections? The review leaves the clear implication that an author who cannot properly identify or label his illustrated species is not to be trusted in the complex field of mimicry. There is no mention of breeding experiments and the troublesome Mendelian evidence. Criticism of the text is limited to a passing shot at the "numerous errors contained in the text" (Poulton, 1916, p. 238) and a closing general criticism to the effect that butterfly mimicry is not alone a general enough view of mimicry from which to derive conclusions for general evolutionary theory. Although that seems odd after all of Poulton's trumpeting of the value of mimicry for theory, it is actually a sly but devastating argument. Poulton is really pointing out that mimicry *between* insect orders, or even between plant and animal, also occurs, and in such cases using genetic uniformity of variation to explain the mimicry is absurd. The complexity of mimetic relations, like the complexity of most adaptations, can be explained not by "internalist" drives or rules, but only by "externalist" factors of environment and ecology, that is, by selection pressures.

Externalist causes might include Lamarckism in all its variants, but, in one of his concurrences with Mendelians, Poulton rejected these. Lamarckism had its influential supporters before the 1920s, and animal coloration was certainly an important example there, too; but Poulton was himself influential in steering the mimicry debate to two contenders: selection versus internalist causes. Geneticists fell easily into the trap of thinking that physiological mechanisms explained more than they did, and Poulton viewed any idea that evolution was driven by mutation pressure as one more overextended physiological claim. It would look too much like Eimer's orthogenesis or other directed variation theories of the past. None of those internalist causes could really explain multiple mimics of one model (as Punnett thought he had done), nor mimicry across different orders, nor why a particular form was the model, nor why it was only superficial changes that occurred in the production of the mimic. And on the side of selection, it alone could fit consistently with *all* the mimetic cases. They showed that mimicry was a special case in a broad class of protective coloration, that different mimics could proceed by separate paths to copy the appearance of some one model, that the models possessed protection of some kind, and that whereas the distribution of mimic with its model coincided, there was no matching of mimetic series with series displaying taxonomic affinity. Thus the explanation of mimicry was like Darwinism itself: It drew its convincing power from its consistency and from its ability to account for many different bits of information. A feature of an organism could not be explained simply by showing that the feature followed the rules of Mendelian heredity; of course heredity would have a determinable mechanism, but that would not suffice to show how the different cases originated. The solution to the meaning of a feature, that is, the key to its evolution, lay in fitting the feature into the whole ecology of the organism in question.

Complexity and coadaptation crop up here again. Constructing the adaptive scenario is satisfying because it seems to work through using the complexity rather than by denying it. Interconnections are made for nature, fulfilling Darwin's expression about "how complex and unexpected are the checks and relations between organic beings" (Darwin, 1859, p. 71). Like so much of this classical naturalist's style, Poulton's own ecological answer to criticisms such as Punnett's harks back to the world of Darwin and Bates. As I have argued elsewhere (Kimler, 1982), the world in which Darwin's natural selection operates is characterized by a dynamic but sensitive balance of components. Since the selectionist view admits change, this is not the static, God-given balance drawn from the tenets of natural theology. The Darwinian balance emphasizes the importance of the web of relations between components and the ecological effects of their interaction in attaining the end results. The original evolutionary theory is certainly an ecological rather than genetical theory of natural selection. The general sensitivity of the ecological balance that is required for the control of the evolutionary dynamic by natural selection is found in high levels of competition and predation, which act on the fruits of Malthusian fecundity. It is the constraint of these selection pressures that gives the utilitarian argument its meaning and power. In the case of potential prey, for instance, it is this constraint that makes any defense mechanism advantageous. In a world of model forms already able to deter predatory threat, mimicry immediately makes sense as a way to gain protection. Mimicry theory also provides an immediately sensible explanation in a world where coloration is mismatched to taxonomic affinity, where the ecology of predation rather than genetics is the connecting link between similar appearances.

In these circumstances, all that is necessary is to show these ecological conditions to be fulfilled. The special conditions of mimicry had been argued for anecdotally in the earliest accounts, and even in the twentieth century most new cases were presented with the same kind of field observations and accompanying reasonableness of argument. Mimicry examples tended to draw qualitatively on their connection to other cases of adaptive coloration. Interestingly, although, even in the 1800s, mimicry was one of the areas in which naturalists did attempt to adopt experimental and quantitative methods, mimicry theory was still subject to sharp methodological attack because of its storylike nature. Perhaps this was because the experiments dealt with the behavior of "typical" predators, such as Thomas Belt's pet monkey and domestic ducks in Nicaragua, whereas the theories all extrapolated results from such artificial conditions to avian predators in the wild. Naive experimentalism was as bad as none. At least this was the view of W. L. McAtee, the influential ornithologist of the Biological Survey division of the United States Department of Agriculture. McAtee had made extensive surveys of the food of common birds in order to assess quantitatively their economic impact for agriculture. He entered the mimicry debate in 1912 with a direct attack on the utility of mimetic coloration by concluding from bird gut contents that all insect prey were taken solely in proportion to their abundance (McAtee, 1912). In a situation of generally not very high predation on butterflies, mimicry seemed to confer no protection.

As "naturalists" became "ecologists" through quantification of their subject matter, such criticisms as McAtee's carried enough weight to counterbalance the support given by biogeographical patterns. Those details of mimicry were consistent with Darwinism, but an extensive investigation had failed to confirm the more basic belief in the prevalence of high predation pressure. In short, the necessary selective agent did not exist. The most difficult problem for a broad science like ecology is always how to find empirical evidence to certify its broad picture of nature: What kind of data are needed to confirm the idea that competition and predation are the major selective forces in nature? In the case of mimicry, the classic example in which the importance of even slight evasion of predation is absolutely central to the argument, a refutation of the presumed selective forces is doubly damaging. The mimicry example had always been used to bolster the Darwinians' reliance on the power and evolutionary importance of selection, but against the arguments of McAtee and Punnett the underlying assumptions were in serious trouble.

There were two developments in ecology that I wish to discuss here, the continuation of the naturalists' tradition and the beginning of a new "self-conscious ecology" (Allee *et al.*, 1949). The classical naturalist tradition was continued, but made more rigorous, by Poulton and his students and colleagues, in a direct assault on the predator problem. Feeding trials using various natural predators and probable insect prey were a popular new experimental technique, with more sophistication in design especially evident in the work of G. D. Hale Carpenter and F. Morton Jones (Carpenter, 1921; Jones, 1932).

Throughout this time, the reality of predation on butterflies was a concern of most discussions of mimicry, and new experiments failed to settle it quickly (Poulton, 1934). As late as 1934 Francis Sumner could comment with amazement that his predation studies were the first experiments on fish coloration in "the interminable controversy regarding the protective value of animal coloration" (Sumner, 1934, p. 559). It seems that there was not widespread acceptance of the protectivenss of mimicry again until the experimental work of the 1930s was capped off by Hugh Cott's masterful compendium, *Adaptive Coloration in Animals*, in 1940. Cott was another Oxford museum worker and a student of Poulton's. Adaptation was reaffirmed only by demonstrating McAtee to be wrong in general, thus reclaiming predation as the necessary selective agent. But during the presynthesis period, ecologists had not been able to agree on the reality of constant small selection pressures because of doubts that had been raised about their traditional good selectionist example.

Perhaps one could argue that, nonetheless, it was ecologists who convinced other biologists of the reality of selection pressures. After all, R. A. Fisher was instrumental in using that selectionist viewpoint in the synthesis, and both he and his co-worker E. B. Ford were greatly influenced by Poulton and Julian Huxley (Ford, 1980, p. 336) at Oxford. Poulton and Huxley kept the adaptational selectionist position alive with ecological arguments. But it seems that they were maintaining a Darwinian heritage in theory, albeit with a modern twist, rather than demonstrating selection pressure in the field. Huxley was instrumental in pushing for experimental approaches in ecology and evolution, a wonderful result of this being apparent in the work of his student Charles Elton, the extremely influential ecologist. Elton's work, beginning in the 1920s, was primarily on population cycles and food-webs, Darwinian in a sense but certainly not demonstrating natural selection at the individual level. Poulton had, as noted, his experimentally inclined students of mimicry as well, but these efforts came to fruition only in the late 1930s. Even well into the 1950s the general opinion of mimicry students was that H. B. D. Kettlewell's (Kettlewell, 1956) and Jane Van Zandt Brower's (Van Zandt Brower, 1958) new experimental demonstrations were both necessary and finally convincing. All one can say of the 1920s is that there were more ecological data consistent with a selectionist view than there had been for Darwin, but there were also more genetical reasons against it. So, was the synthesis formed from demonstrations by ecologists? No, the synthesis meant a *genetical* theory of natural selection, to use Fisher's title, with the emphasis on the genetical aspect. By pointing to small, available variation, genetics demonstrated that Darwinism was likely; and the mathematical theories, in particular, showed that such selection would be powerful for evolutionary change. Ecology contributed in the sense of being there with ready support for selectionist theories. At least this is true for the attitude of the classical naturalists. Their ecology was, of course, consistent with neo-Darwinism because it was derived in the first place from a Darwinian point of view.

There is also the second category I mentioned, the self-conscious new ecology. Where could it have come into the synthesis as a contributing factor? That could, of course, be the adaptationist–selectionist paradigm always present in ecology. The presence of this paradigm within ecology does not explain the timing of the synthesis, that is, why other biologists began to accept the paradigm. Moreover, it was also controversial within ecology at this same time. As I have already mentioned, Richards and Robson (Richards and Robson, 1926; Robson, 1928) had convinced many biologists, not the least of these being Elton, Huxley, and Sewall Wright, that the differences between species were not adaptive. This led Wright to postulate nonselective mechanisms of evolution, as Provine demonstrates in this volume. It led Elton away from individual to group and populational selection. But that is hardly the focus or the impact of the neo-Darwinian synthesis. On the other hand, if we are referring to *specific* contributions from ecology, the candidates are evidence for the generality of adaptiveness (i.e., by implication the ubiquity of selective forces) in nature, direct evidence of the power of selection in nature, and evidence or theory about natural population structure. The first two kinds of evidence, as reflected in taxonomy and in the mimicry case, are problematic, at best as circumstantial as ever. Populations, the inherent subject of ecology, hold the most promise, especially in view of Mayr's emphasis on the importance of "population thinking" (Mayr, 1980b, p. 127) for the creation of the synthesis.

Population structure and natural dynamics were productive fields for research in ecology in the 1920s, but the laboratory work of Alfred Lotka or G. F. Gause was concerned with the mathematical forms of population growth and control (Lotka, 1924). Raymond Pearl's rediscovery of the logistic growth curve sparked his own and others' interest in and mathematical output on population questions (Allee *et al.*, 1949; Lotka, 1924). The results came to look rather like gas laws or other equations of kinetics in physics, as the ecologists involved proudly noted. There is a fundamental parallel to Fisher's rather statistical thermodynamical view of fitness in populations, and a similar nonreality of the mathematics for natural populations. Fisher was attempting with his Fundamental Theorem of Natural Selection to "combine certain ideas derivable from a consideration of the rates of death and reproduction of a population of organisms, with the concepts of the factorial scheme of inheritance, so as to state the principle of Natural Selection in the form of a rigorous mathematical principle" (Fisher, 1930, p. 22). He started with Pearl's actuarial "life table" mathematics and ended with a general statement on rates of change of fitness, defined mathematically and tied to small Mendelian variants. So the path from the ecology of populations to the synthesis is convoluted, and the signposts are mathematical. What we see here is not what we might have expected – the contribution of naturalist–systematists in terms of information on natural populations – nor is it exactly in line with Mayr's definition of populational thinking. Mayr was sympathetic to the contribution by systematists of realization of the abundance of variation within species as opposed to uniform species types. This is often pictured as the realization of character distribution rather than single species-values. Although the mathematical work in ecology is also statistical in nature, it is not totally divorced from typological attitudes. The equations can be read as an attempt to reduce the variability down to mathematically simple form and so to produce laws, or theorems, resembling those of physics. True, the variation expected of "populational thinking" is realized, but most ecology proceeded to ignore this and so to treat the species as a functional unit or to treat questions of dynamics at a species-wide level. I do not see a direct transition from the interests of ecologists in population processes to neo-Darwinism, except in Fisher's mathematical appropriation of a framework for his genetical theory.

Turning again to the mimicry case, we can dig a little deeper into Fisher. He, too, recognized the taxonomists' objections to the adaptiveness of characters, citing Robson's *The Species Problem* in his 1930 book. He further admits the lack of the ecological knowledge needed for determining actual gain in fitness of particular genetic changes. In fact, quantitative theory and its few cases are valuable because, "Our knowledge in this respect [ecology], while sufficient to enable us to appreciate the adaptive significance of the differences in organization which distinguish whole orders or families, is almost always inadequate to put a similar interpretation on specific differences, and still more on intraspecific variation" (Fisher, 1930, p. 57). But, of course, mimicry is an exception. Not only can the biologist particularize, in this case, the environmental factor and the benefits of the character, but it is also a useful case to disprove Robson's broad claims. Here is a case where the adaptive significance can be shown for differences right down to the level of local varieties. Fisher's arguments for mimicry, however, do not deal with ecological field evidence; the arguments in his 1930 book are a slight expansion of his 1927 paper on "statistical and genetic" objections to mimicry theory. The statistical objection was the one from Marshall that Punnett had used. Marshall's argument limited the conditions for a directional approach to an intermediate coloration by two unpalatable species, because of the apparent difference in selective advantage for a numerous species as opposed to a rare one. Fisher's "statistical" refutation is exactly that: He emphasizes that what we find is genetic variation in both directions about the mean for a trait, rather than a type trait with a few variants toward the mimetic pattern. In the latter case a mutation would obviously be disadvantageous when it "leaps clear outside the protective influence of its type" (Fisher, 1927, p. 271). Fisher's analysis, however, compares the fate of any mutational deviation, not with the average type, but with other deviations. This use of the immediate relative fitness among all the variants then allows a net resultant directional modification. It is a far less absolutist argument for the fitness of a variation, and in mimicry that seems likely; both the mistakes of predators and the great color variation in mimics are realistic conditions to assume. Fisher's argument reveals his use of variability in natural populations, and so, in that sense, is populational. It is an interesting question whether this stems from his Oxford connections with Darwinian ecologists or from his own Darwinian convictions and from theoretical work on variance and dominance. His genetical theory alone would emphasize the importance of small variations and guide him to such a scenario for mimicry.

The importance of genetical theory is shown in his attack on the supposed genetical limitations to mimicry theory, that is, on Punnett's saltational interpretation of the *Papilio polytes* case. With the Marshall brief against a gradual, directional mimetic tendency dismissed, what Punnett had left was the behavior of the alleles for coloration in the polymorphic butterfly. Since intermediate phenotypes were absent, Punnett had proposed absolute large mutations. Fisher challenged the assumption here that the resultant phenotype of the Mendelian factors has always been the same and that the new genes produced by the single initial mutation "have been entirely unmodified since their first appearances" (Fisher, 1927, p. 275). Drawing on W. E. Castle's work on modifier genes in piebald rats, he suggested that a similar accumulation of modifiers could have changed the expression of factors in this butterfly polymorphism as well. In a stable polymorphic population, if "selection favours different modifications of the two genotypes, it may become adaptively dimorphic by the cumulative selection of modifying factors, without alteration of the single-factor mechanism by which the dimorphism is maintained"

(Fisher, 1927, p. 277). The modifying factors, "which always seem to be available in abundance" (Fisher, 1927, p. 275), allow selection by the predators to lead to a gradual evolution of mimetic resemblance even in such an apparently discontinuous case.

Fisher does see the mimicry problem as being one of ecological adaptation, and he approaches it not as a field naturalist but through considerations of the genetics of populations. The actual population ecology is not developed. He is pursuing the theoretical point of establishing the conditions for the genetical acquisition of adaptations, and the stress is on relative advantage (fitness) and genic modification. Both demonstrate how small selection pressures can have great effect. For E. B. Ford, Fisher's 1927 paper was the "true start of the modern evolutionary synthesis" (Ford, 1980, p. 338). Revealingly, Ford says the paper "opened up the possibility that I had had in mind for some years of taking genetics into the field" (Ford, 1980, p. 338). The theory did not draw on data from natural populations; it pushed people into the field. As far as ecology goes, Fisher was, as he himself admitted, stranded with "full evidence" for natural selection in only a few cases of mimicry, although "the concurrence of independent classes of observations puts the well-investigated cases beyond possibility of doubt" (Fisher, 1930, p. 188). This is still the Darwinian argument from consilience, and mimicry is his only example to counter Robson and Richards's sweeping generalization that adaptiveness cannot be found in the majority of traits.

If we look at a better ecologist than Fisher we get a different picture. Charles Elton, who came into prominence during the 1920s and 1930s with a series of books on animal ecology, could be called one of the fathers of modern ecology. He had been one of Julian Huxley's best students at Oxford, and, in fact, it was Huxley who urged him to produce in 1927 his *Animal Ecology*, a text suitable for the modern, postmorphology curriculum. Elton's approach was thoroughly appropriate for such a role. He cared about evolutionary questions, he used quantitative methods, he was an experienced naturalist, and he set out to find generalizing principles for ecological phenomena. The topic that brought all this together was the population dynamics of natural populations, and in his treatment we see both the classical naturalist tradition and the new field of ecology. On the natural history side, Elton had field experience and an acute sense of ecological relations in the wild and, simply, the obvious love for the workings of the economy of nature that distinguishes the naturalist. Added to this, his research interests lay at the center of the newly dominant topics of ecology – community structure and population dynamics.

The rise of the field of ecology from the late 1890s on created new questions of central interest. Adaptation retained its perennial attraction, but the definitive research turned aside from evolutionary questions to a more strictly ecological time frame. Such an emphasis on present dynamics made ecology a full-fledged field of biology with its own special subject matter. Plant ecology led the way with work on population stability and community composition. The result was the theories of succession and climax for communities, and these ideas dominated all of field ecology for the next half-century. The other influential new research area was more strictly laboratory-experimental and mathematically theoretical. The mechanics of population growth and regulation were expressed in the mathematics of Lotka and Volterra (logistic equation) or Pearl and Reed (life table equations) and translated from the laboratory to field expectations. The latter development started only in the late 1920s, and one of the sparks was Elton, with his lucid demonstrations of "the numbers of animals and the ways in which they are regulated, and . . . how a great many of the phenomena connected with numbers owe their origin to the way in which animal communities are arranged and organized" (Elton, 1927, p. 102). Although Elton's own researches were not mathematical, but inclined rather to emphasize the arrangement and organization of communities, it is their application to the study of animal numbers that gives them importance. Elton's influence was early and deep in the consuming interest in population regulation characteristic of twentieth-century ecology.

The point I really wish to illustrate here is that in 1927 Elton established a view of adaptation and speciation from an ecological perspective, and this position was new and not universally accepted. Elton wedded the taxonomists' conclusion that there are nonadaptive species differences with his populational theories and produced an ecological evolutionary theory that moved distinctly counter to a coming neo-Darwinian synthesis. Of the 1926 paper of Richards and Robson on the "species problem" he wrote: "The demonstration that closely allied species have probably not been produced by natural selection, and that the differences have usually no relation to the mode of life of the animals, forms a most important step forward in the study of evolution" (Elton, 1930, p. 75). How interesting it is that this conclusion should have come from workers at the Oxford University Museum. But the taxonomic input was crucial. Elton's brief acknowledgments in his 1927 text devote a paragraph to the valuable interaction with O. W. Richards (Elton, 1927, p. viii). The message to evolutionary theory was taken to be a separation of adaptation and the mechanisms of speciation, with speciation being distinctly nonadaptive. Further, to the ecologist Elton, this means:

that adaptation is produced by selection of whole populations
rather than the selection of individuals, and while it raises
one huge difficulty by reopening the species problem, it does
away with another huge difficulty, of seeing how in practice
natural selection could ever be effective in picking out a single
individual and succeed in leading it, as it were, through the
perils of life in a fluctuating animal population. [Elton, 1930,
p. 75]

The question gradualist adaptationists had always had to answer,
on the usefulness of incipient structures, simply vanished along with
natural selection at the individual level. Everyone could breathe a
sigh of relief. Elton kept selection for some obvious cases, but denied
its generality. New traits spread in a nonselective process, isolation
of "different sections of the population from one another" (Elton,
1927, p. 185) fixes them, and only then does selection work, at a
population level. But how does a trait spread if it is not of selective
advantage? Rejecting Lamarckism and mass mutation, Elton pulled
his answer out of population dynamics, because the dramatic fluctuations of animal populations would mean periods of rapid increase
in numbers from a reduced population. That would be a period of
"practically no checks at all" (Elton, 1927, p. 187), that is, no natural
selection. The new variations, even if nonadaptive, then would have
an opportunity to spread with the expanding population.

What of the obvious adaptive case of mimicry? Elton ignores Batesian mimicry but does cast doubt on the generality of color adaptations by denying the protective (camouflage) status of most of the
claimed cases in mammals and birds. Mimicry to Poulton had always
been a special case in this broad class, presumably under the same
gradualist mechanism. Elton is swayed by the apparent greater generality of Richards and Robson's conclusion. He is proud of his theory,
since it seems to provide a solution to the Robson and Richards species
problem, and, not less important, because "it illustrates the fact that
ecological studies upon animal numbers from a dynamic standpoint
are a necessary basis for evolution theories" (Elton, 1927, p. 187).
Here is the ecological contribution to evolutionary theory, through
population considerations to nonadaptive speciation and even group,
population, and species selection.

A final example will serve to show Elton not to be alone among
ecologists with new contributions for evolutionary theory and will
take us back to Batesian mimicry itself. It also draws upon the final
thread in the weaving of a modern ecology, the input from the biology
of pest control. It must be at least noted here how the interest in
population dynamics grew also from the importance of insect control

and the successes in field and theory of economic biologists. The second father of modern population ecology was just such a biologist; A. J. Nicholson was seminal in theories of regulation of abundance in nature from the late 1920s, continued in collaboration with the mathematician V. A. Bailey in the 1930s, and figured largely in the ecological debates of the 1950s. An Australian trained at Cambridge, he returned home to a position as an important economic entomologist. Nicholson, too, had classical natural history interests. He did the usual collecting of an entomology buff and became fascinated by the mimetic series obtainable in Australia, especially of wasps and beetles.

These considerations led to the presentation, in 1927, as the Presidential Address to the Royal Zoological Society of New South Wales, of a "new theory" of mimicry (Nicholson, 1927). It is a ninety-page, rambling address, which defines the mimicry terminology yet again, examines evidence for the actuality of protective resemblance, and ventures into considerations of its evolution. I am not arguing that this was an important paper for mimicry theory or for ecology; it is merely illustrative of the evolutionary thinking of an important ecologist. In fact, although most mimicry specialists seem to know of it today, the paper was scarcely cited until John Turner's recent references to its genetical theories (see Turner, this volume). There, in the genetics, lies the interest for us. The principal criticisms had been against "the theory that mimetic resemblance has arisen by the natural selection of small favourable variations" (Nicholson, 1927, p. 16), a reference to Punnett's 1915 book. Nicholson thought arguments about mimicry confused the mechanisms of its evolution with the "undoubted fact of resemblance" (Nicholson, 1927, p. 16), and, thus, the interest in evolutionary theory meant a look at genetics.

Doubting it possible to ever produce a direct proof of mimicry in nature, Nicholson looked for simple analogies from which to draw connecting arguments. The industrial melanism of the peppered moth, usually cited as showing selection for protective crypticity, is questioned for the sensible ecological reason that the moth's melanic form is darker than the actual vegetation, erroneously pictured as sooty black. More importantly, Nicholson considered the experimental work of J. W. Heslop Harrison purporting to show a heritable environmental induction of melanism (Heslop Harrison, 1927). Although Heslop Harrison is usually dismissed as a late neo-Lamarckian, guilty of faulty experimentation and mistaken interpretation, arguments for Lamarckism were still widespread in the 1920s. Significantly, Nicholson read this work as showing mutants to follow normal Mendelian rules. Mutation can be environmentally influenced

– not such a naive interpretation, considering that H. J. Muller was busily proving just that with X-ray induction of mutations. For mimicry, the point of this lay in the apparent absence of natural selection in the spread of the mutants; all of the population must have mutated under environmental influence in similar fashion. Nicholson admits that this is a rare case, yet he is still ignoring an already growing view that the peppered moth case is proof of natural selection.

This kind of mass mutation is not likely to produce adaptive characters, and, thus, to the extent that mimicry is an adaptive character it must be preserved by selection. Nicholson shows the great influence of Punnett's book both here and in his acceptance of the general view that the key to mimicry lies in genetics. He also accepts uncritically Punnett's genetical conclusions, demonstrated in a "masterly manner" (Nicholson, 1927, p. 43), that in some complex butterfly mimicries the patterns arise suddenly and completely. The *Papilio polytes* case proves this satisfyingly for Nicholson also. He regards it as likely that the genetic constitutions of all butterflies are similar and thus yield "potencies" for large color variations of only certain types or series of types. It is also likely that simple gene changes, ramifying during development, can carry complex phenotypical character changes. Nevertheless, the operation of selection is constantly referred to in the discussion. Convinced of the reality of mimicry, Nicholson does not want to discard an adaptive viewpoint and natural selection altogether. Like Punnett, he talks a lot about its preserving the resemblance once created; its operation is not limited only to small individual variations.

Unlike Punnett, he is willing to consider the resemblances as adaptive, and the argument is in the traditional style of Poulton. Superficial similarity only, correlation of habits and appearance (especially in wasp mimics), complexity of mimetic features, coincident geographic distribution of model and mimic, and the relative rarity of mimicry all imply "response to the appearance of a model" (Nicholson, 1927, p. 69). The occurrence of suitable variation is required, and selective preservation of such variation is possible only in the model's presence. In this apparent synthesis of genetical theory with traditional ecological arguments, however, Nicholson's solution actually falls into the genetical camp. Before selection can preserve, an initial variant form must arise that has "special survival value." Thereafter, selection is possible, in Nicholson's view, because he accepts the theoretical requirements of predator discriminatory ability and model distastefulness, although he admits the experimental evidence to be "somewhat inconclusive" (Nicholson, 1927, p. 70) and in need of bolstering. Basically, he has rejected Punnett's ecological objections but has taken the genetical message to heart.

Mutations had been demonstrated recently to produce a wider range of effect, from very small to quite large, but all Mendelian. Aware of this work, Nicholson proposes a kind of two-step model, the mimetic pattern appearing as a single large mutation. This is likely, as in Punnett's account, when the genetic constitution of model and mimic are similar. Even for the case of mimicry across different orders – Poulton's most telling criticism – Nicholson considers it likely that some common coloration patterns in insects might be caused by "some very simple metabolic factor which might well be expected to occur independently in very different types" (Nicholson, 1927, p. 74). But for more complicated cross-order cases and even some highly involved butterfly cases, a second step must come in. Complex resemblance is built by selection of a number of individual mutations, mostly small but sometimes large as well. The traditional concern with incipient utility is removed by mutation, large if necessary, to the first "rough" resemblance. But Nicholson also claims that too much had been made of the amount of difference necessary for relative advantage to accrue. After the initial establishment of an approximate mimicry relationship, other mutations can be selected to complete the complex details. It is difficult to tell, given Nicholson's rather lax genetical terminology, whether or not he has in mind the same process as Fisher's "modifier" genes. He claims it is "the mimetic resemblance which is modified, not the original mutation, and modification takes place by means of the preservation of new mutations, large or small" (Nicholson, 1927, p. 79). This seems to be a more simplistic, additive process than Fisher's interactive one, but both, and apparently independently in the same year, present a kind of model using modification of the phenotype after an initial step introducing a relationship of mimicry. Although Nicholson's reason for this is his acceptance of Punnett rather than an attack on him, he does go further than Punnett had done. I think his independent invention of a genetical model demonstrates just how important questions of mechanism were to biologists of the time, including ecologists.

Nicholson even goes so far as to redefine adaptation. The input from his researches on population regulation is a belief that older views of static balance were inadequate. Numbers are fairly constant, though as a mean value about which fluctuation occurs. Nicholson never considered these fluctuations to be as drastic as did Elton; indeed, Elton criticized him for being still too tied to an outdated balance view. But they agreed at least on the existence of dynamic population control with predation as the regulator. Because Nicholson emphasizes the relative fine-tuning of the equilibrium number, the predator has to closely track the prey population with an appropriate intensity. Thus, the resultant constant population is the condition under which evolution has to operate; that is, new adaptive avoidance of predation will only bring increased checks during another part of the life cycle or from another enemy. In a rather confused argument, Nicholson reasoned that mimicry, thus, does not "increase the success of the species" (Nicholson, 1927, p. 88), and so is not an adaptation. Population constancy, which appeared from his numerical observations to be an end in itself, implied a definition of adaptation as only that which affects the level about which the population size oscillates.

Nonetheless, the preservation of mimicry was controlled by natural selection. For relative advantage under any single selective pressure would still ensure spread, even if the species gained no greater success. To retain the equilibrium, adult butterfly freedom from attack would simply be compensated for by increased caterpillar mortality. So, Nicholson declares, mimicry is not protection, despite selective replacement of nonmimetic by mimetic morphs within a species. Despite seeing relative advantage in genetic change as the important point, he can only see fitness as a species trait. By thinking at this species level, Nicholson denies the standard "teleological" significance of mimicry. Consider, for example, his statement: "Mimetic resemblance, therefore, simply serves to fit the possessors more perfectly to their natural environment, without conferring upon them any material advantage" (Nicholson, 1927, p. 90). Clearly, this can make sense only in terms of the kind of thinking, derived from concern with populations, that ascribes advantage only at the species level.

The lesson from both Elton and Nicholson is that in the presynthesis period an ecological view of evolution could take on a markedly nonadaptive cast. Population dynamics led to such conclusions even for mimicry, in which selection was occasionally important but where room had to be made for nonadaptiveness. Both moved to a kind of species-level thinking, perhaps because of the new ecology's definition of its unique questions at just that level of organization. Today such reasoning certainly seems "typological." If Elton's adaptational and evolutionary views seem the more sensible of the two, it is to the degree that they recognize the importance of the fluctuations in a populational dynamic equilibrium. With that episodic selection theory he moved, independently, close to Sewall Wright's view of evolution in nature; Elton was quick to see this in the 1930s and could recognize that the Dobzhansky–Wright theories needed the same solutions to population problems as those the ecologists were attempting (Elton, 1938).

There would seem to be a lack of ecological contribution to the synthesis precisely because of a movement, ignoring such interests, toward questions peculiar to the new field. In this context, we should not emphasize too much a "split from genetics" or ignorance of genetics (Mayr, 1980*a*), as the cases of Nicholson or even Poulton show. Indeed, the very fact that speciation and adaptation became subsidiary concerns led indirectly to new approaches to those venerable questions, based on population dynamics models and new ecological views. The concept of genetic variability, as Nicholson used it and even as Fisher expressed it, seemed to entail a sufficient supply of variability in nature to suit any of their views of the evolutionary process. Control of evolution, in the one certain continuation of Darwinian ecology, lay in ecological conditions; that may seem like keeping an adaptationist–selectionist view alive, but the ecological theorists were quite willing to include nonadaptive processes as well. Thus, the Darwinian aspects of the synthesis are the product of the theoretical geneticists. The construction of the synthesis did not draw directly on the ecological work on populations, yet the synthesis itself did reaffirm to ecologists the importance of their research into population regulation and structure. The two fields of genetics and ecology at least recognized this as a mutual problem in the genetical theory of the synthesis.

One could argue that ecology has, in fact, become evolutionary in the modern sense as it has attempted to answer the questions of population structure posed by genetical theory. A classic style for twentieth-century ecology has been to assume any genetic variation necessary for an evolutionary scenario, since that is consistent with many of the neo-Darwinian genetical models. Recently there has grown a sense that ecology can provide the information needed to discriminate between models and even to limit their form. I think John Turner's presentation in this volume demonstrates this. With ecology given direction by genetical evolutionary questions and finally in providing feedback to evolutionary genetics, perhaps it is fair to say that ecology has only recently come into the modern synthesis. Consistency has always been possible, even if the trends in ecology have not all adhered to neo-Darwinian orthodoxy. The creaking of the synthesis at present around its adaptationist and speciational ideas is perhaps to be expected as the result of a fresh ecological input.

References

Allee, W. C., *et al.* 1949. *Principles of Animal Ecology*. Philadelphia: W. B. Saunders.

Bates, H. W. 1862. Contributions to the insect fauna of the Amazon Valley. *Lepidoptera*: Heliconidae. *Trans. Linn. Soc. Lond.* 23:495–566.

Bateson, W., E. R. Saunders, and R. C. Punnett. 1906. *Reports to the Evolution Committee of the Royal Society. Report III*. London: Harrison and Sons.

Carpenter, G. D. Hale. 1921. Experiments on the relative edibility of insects, with special reference to their colouration. *Trans. Ent. Soc. London, 1921*:1–105.

Carpenter, G. D. Hale, and E. B. Ford. 1933. *Mimicry*. London: Methuen.

Cott, H. B. 1940. *Adaptive Coloration in Animals*. London: Methuen.

Crampton, H. E. 1916. Studies on the variation, distribution, and evolution of the genus *Partula*. *Carnegie Inst. Washington Publ.* 228:1–311.

Darwin, C. 1859. *On the Origin of Species*. London: John Murray.

East, E. M. 1910. A Mendelian interpretation of variation that is apparently continuous. *Am. Natur.* 44:65–82.

Elton, C. 1927. *Animal Ecology*. New York: Macmillan.

   1930. *Animal Ecology and Evolution*. Oxford: Clarendon Press.

   1938. Animal numbers and adaptation. In *Evolution*, ed. G. de Beer pp. 127–37. Oxford: Clarendon Press.

Fisher, R. A. 1927. On some objections to mimicry theory; statistical and genetic. *Trans. Ent. Soc. London* 75:269–78.

   1930. *The Genetical Theory of Natural Selection*. Oxford: Oxford Univ. Press.

Ford, E. B. 1980. Some Recollections Pertaining to the Evolutionary Synthesis. In *The Evolutionary Synthesis*, ed. E. Mayr and W. B. Provine. Cambridge: Harvard Univ. Press.

Fryer, J. C. F. 1913. An investigation by pedigree breeding into the polymorphism of *Papilio polytes*. *Phil. Trans. Roy. Soc., B* 204:227–54.

Heslop Harrison, J. W. 1927. The induction of melanism in the Lepidoptera, and its evolutionary significance. *Nature* 119:127–9.

Jones, F. Morton. 1932. Insect colouration and the relative acceptability of insects to birds. *Trans. Ent. Soc. London* 80:345–86.

Kettlewell, H. B. D. 1956. Further selection experiments on industrial melanism in the Lepidoptera. *Heredity* 10:287–301.

Kimler, W. C. 1982. Mimicry: History of an Evolutionary Exemplar. Unpubl. Ph.D. thesis, Cornell Univ., Ithaca, New York.

Lotka, A. J. 1924. *Elements of Physical Biology*. New York: Macmillan.

Marshall, G. A. K. 1908. On diaposematism, with reference to some limitations of the Müllerian hypothesis of mimicry. *Trans. Ent. Soc. (1908)*: 93–142.

Mayr, E. 1980a. Prologue: Some thoughts on the history of the Evolutionary Synthesis. In *The Evolutionary Synthesis*, ed. E. Mayr and W. B. Provine, pp. 1–48. Cambridge: Harvard Univ. Press.

   1980b. The Role of Systematics in the Evolutionary Synthesis. In *The Evolutionary Synthesis*, ed. E. Mayr and W. B. Provine, pp. 123–36. Cambridge: Harvard Univ. Press.

McAtee, W. L. 1912. The experimental method of testing the efficiency of warning and cryptic coloration in protecting animals from their enemies. *Proc. Acad. Nat. Sci. Philadelphia* 64:281–364.

Nicholson, A. J. 1927. A new theory of mimicry in insects. *Austral. Zool.* 5:10–104.

Poulton, E. B. 1903. What Is a Species? [Reprinted in Poulton, 1908.]

1908. *Essays in Evolution*. Oxford: Clarendon Press.

1916. Mimics ready-made. *Nature* 97:237–238.

1934. Termination of the controversy with Dr. W. L. McAtee. *Proc. Ent. Soc. Lond.* 9:119–20.

Punnett, R. C. 1911. *Mendelism*. New York: Macmillan.

1915. *Mimicry in Butterflies*. Cambridge: Cambridge Univ. Press.

Richards, O. W., and G. C. Robson. 1926. The species problem and evolution. *Nature* 117:345, 382.

Robson, G. C. 1928. *The Species Problem*. Edinburgh: Oliver and Boyd.

Sumner, F. B. 1934. Does "protective coloration" protect? – Results of some experiments with fishes and birds. *Proc. Nat. Acad. Sci.* 20:559–64.

Van Zandt Brower, J. 1958. Experimental studies of mimicry in some North American butterflies. *Evolution* 12:32–47.

Wallace, A. R. 1867. Mimicry, and other protective resemblances among animals. *Westminster Review, n.s.* 32:1–43.

1869. *The Malay Archipelago*. London: Macmillan.

6

"The hypothesis that explains mimetic resemblance explains evolution": the gradualist–saltationist schism

JOHN R. G. TURNER

Not always was the Kangaroo as now we do behold him, but a Different
Animal with four short legs. He was grey and he was woolly, and his
pride was inordinate: he danced on an outcrop in the middle of Australia,
and he went to the Little God Nqa.
He went to Nqa at six before breakfast, saying, "Make me different from
all other animals by five this afternoon."
Up jumped Nqa from his seat on the sand-flat and shouted, "Go away!"
Kipling: *The Sing-Song of Old Man Kangaroo*

Mendel mutation–saltation/Darwin biometry–adaptation

1

The news came through, as I was completing this chapter, that the
British had recaptured Darwin. The battle for the gentleman after
whom the port was named continues.

Evolutionary biologists are all Darwinians, as all Christians follow
Christ and all Communists, Karl Marx. The schisms are over which
parts of the Master's teaching shall be seen as central. Neo-Darwinism, or the modern synthesis, sees as Darwin's central contribution
the insight that two previously recognized but unconnected phenomena, the adaptation of organisms and their diversity, had a common
cause in the process of natural selection.

An alternative theory states that there are two processes in evolution, one of which, microevolution, accounts for adaptation and the
other, macroevolution, for diversity (Goldschmidt, 1940; Stanley,
1979; Gould, 1980). The debate about the two theories is often misrepresented as a disagreement over the speed of evolution, the neo-

I am very much indebted to my colleagues in the History and Philosophy of Science
Unit at the University of Leeds, particularly Robert Olby, Jonathan Hodge, and Isabel
Phillips for innumerable enlightening discussions on topics related to this essay, for
access to their work in progress, and for some very valuable criticism of the draft.

Darwinian theory being characterized as "phyletic gradualism" and the alternative as "punctuated equilibrium," that is, on the one hand, a belief that evolution proceeds by slow, imperceptible steps at a more or less uniform speed and, on the other, that it is constituted of long static periods (equilibrium), punctuated by periods of very rapid change. This question, which is one of fact – what is actually shown by the fossil record – is not the fundamental difference between the two schools; a variable speed of evolution can be incorporated into either theory. Rather than describing the viewpoints by the terms "phyletic gradualism," which is a straw man, and "punctuated equilibrium," which is a matter of observation, I shall, therefore, call the more traditional theory "neo-Darwinism," or the "modern synthesis," and the newer one "evolution by jerks."

Having regard to the sayings that everything in history happens twice, or that there is no new thing under the sun, I want to show how the current schism between neo-Darwinists and punctuationists is a continuation of the running battle over Darwinism and within Darwinism, which has existed ever since Darwin achieved the long-deferred triumph of Epicurus: the refutation of the Argument from Design. The realization that natural selection, just as easily as a wise Creator, could account for the apparent subordination of ends to a purpose in the construction of organisms has led to a very strong belief among Darwin's followers, especially in England, in adaptation as a central concept, and in evolution and adaptation as being virtually the same thing.

That any attempt to introduce nonadaptive processes into the theory of evolution is fiercely resisted and even ridiculed by those who see themselves as heirs to Darwinism not just as a biological theory but as the triumph of materialism and causality, not to say a fine tool for bishop-baiting, is most readily seen in the opposition first to Sewall Wright's introduction of random genetic drift and, hence, freedom of choice as a significant evolutionary force, and then to Kimura's explanation of molecular evolution as the outcome of the same stochastic process. It was not that the selectionist school denied that stochastic change in gene frequency occurred (they could not, for it is an ineluctable mathematical consequence of population genetic theory), but that they held it to be overridden always by the greater strength of natural selection. Random drift was, to use a nineteenth-century word which we will encounter again, a mere "fluctuation" around the steady trend produced by selection (Fisher, 1930).

Mimicry, the protective resemblance of one organism (a butterfly, in all the cases discussed here) to another, is particularly instructive. It is such an obvious "adaptation" that it should present no problems for evolutionists of any persuasion; in this case random drift has never been seen as an issue. Yet even aside from those biologists who have doubted either the principle of mimicry itself or most of the cases cited as examples (e.g., Robson and Richards, 1936; Shull, 1936), whose arguments and motives are another and undoubtedly interesting story, it has aroused considerable controversy and, in the process, has become one of the best understood of all adaptations. A study of the debates over the mechanism by which this adaptation arose shows that the suspicion in which Darwinists hold nonadaptive processes is much deeper, wider, and older than the question of random drift. It extends back to the discovery of Mendelian inheritance around 1900, continues rather surprisingly in the study of heterozygosity and polymorphism, and still lives on in the opposition to the theory of evolution by jerks, which, like the original mutationist theory, emphasizes both the sudden and the nonadaptive nature of evolutionary change.

The conditions for schism rather than fruitful discussion have arisen both in the old debate and the new because the incomprehension has been mutual: Darwinists have not understood that their critics were saying anything useful partly because they, the critics, overestimated the extent to which Darwinian theory required alteration and their own power to alter it. Proponents and opponents alike have believed that a radical alternative to Darwinian theory was on the table, when all that was really happening was that one of the nonadaptive elements of evolution was being uncovered. In short, both parties to the schism were, and are, guilty of overreacting.

With mutation, it took two or three decades for the schism to be healed, in what came to be called the modern synthesis. I shall suggest that a similar synthesis of the present schism is also possible.

2

Two features of the present debate that have irritated some evolutionary biologists have been the overt introduction of politics and the way in which the mass communication media have linked the theory of evolution by jerks with creationism. Thus, on the one hand, Gould (1979) has discussed evolution by jerks as a Marxist, Hegelian, or dialectical materialist theory of hierarchical organization and has pulled the legs of the neo-Darwinists by accusing them of "gradualism" (the heretical doctrine of evolutionary socialism espoused by the revisionists) and of eschewing "pluralism" (the alliance of liberals, anarchists, Mensheviks, and Bolsheviks). On the other, the creationist lobby has cited his work extensively as casting doubt both on neo-Darwinism and Darwinism, and television productions have featured his theory under such titles as "Did Darwin Get It Wrong?" Although I shall not document this so fully, I shall suggest that the journalists have perceived a connection between saltational theories and Creationism to which biologists are largely blind, that this kind of reaction also accompanied the mutationist controversy at the beginning of the century, and that, although it is now couched in the terms of dialectical materialism, the philosophical overtones of the present and of the older debate lie much deeper than that.

This, then, is a history of a continuing schism, focusing on mimicry, which, many of the disputants were agreed, constituted a test case. Ironically enough, it is through the study of a real, incontrovertible adaptation that we can see most easily how indispensable are nonadaptive processes in the theory of evolution.

3

The background to the schism requires only a brief summary, drawn mainly from Provine (1971). Darwin, Huxley, and Galton had debated the continuity of evolution and the role of what they saw as two separate kinds of variation: individual variation of a continuous kind; and sports, or saltations, like the Ancon sheep. It is not an excessive caricature to say that at the turn of the century an extensive schism developed between the school of Pearson who saw evolution as involving individual variation, driven entirely by selection, wholly adaptive, and thoroughly smooth and gradual – these adherents can be called variously biometricians and Darwinists – and the school of Bateson, who perceived selection as a settled fact and wanted to put a significantly new element into the theory in the form of the origin of the variation on which selection worked. Bateson's followers can be called variously saltationists, mutationists, or after their adoption of the rediscovered laws of inheritance, Mendelists. They regarded saltation, or mutation, as a major evolutionary force and, consequently, viewed evolution as only partly adaptive. As sports, or saltations, were large phenotypic changes, evolution must have proceeded in jerks.[1] Individual variation was considered unimportant, being written off as mere "fluctuation," that, not being inherited, could play no part in the process of natural selection.

Bateson's dismissal of selection was that of the proud man apart. Selection was Darwin's triumph: We could accept it. The new scientist must "push toward the truth through the jungle of phenomena," not be "content supinely to rest in the great clearing Darwin made long

---

[1] The word "jerks" may seem to some English readers to be a vulgar transatlanticism. Its use in evolutionary theory, in fact, goes right back to Galton (see quotes in Provine, 1971).

since" (Bateson, 1902). What profoundly irritated the Darwinists and inspired some bitter polemics was that writers, both popular and scientific, had interpreted Bateson's stance as showing that Darwinism, if not dead, was at least smelling pretty suspicious. Darwin's followers anticipated the loss not simply of a scientific theory but of a world view and a view of human nature.

E. B. Poulton, for one, was thoroughly annoyed with Bateson. A deep student of adaptation in the form of the protective colors of insects, he found a major stumbling block for him: The mimicry of one butterfly by another had been the first successful application of Darwin's new theory of natural selection when Henry Walter Bates had explained it as the outcome of natural selection, rendering the pattern of a palatable butterfly into that of a nasty one, to the confusion of the predators.[2] And it could still be used as a test case for rival theories of evolution, Bateson's, Bergson's or anyone else's. The title of this chapter is from a paper of Poulton's (1912).

Poulton (1908) issued his challenge: How could mutation create a painting? Explicitly, how could it draw on the wings of one butterfly the whole complicated and exquisite pattern of a leaf or another butterfly? It was to be taken up by Bateson's colleague Punnett, who achieved the ambition that forever eluded Poulton, a book about butterfly mimicry.

Antithesis (R. C. Punnett and R. B. Goldschmidt)

1

The mutual incomprehension between Darwinists and mutationists involves perspectives that, after the modern synthesis, we find difficult to imagine. It will help if we develop Poulton's analogy of evolution as an artist: Darwinians thought of variation as some kind of "shapeless earth" out of which selection and selection alone molded adaptations, as a potter or sculptor might create art from a lump of clay. Hence, for a thoroughgoing Darwinian an adaptively neutral character was an impossibility. The material with which selection worked was the "individual variation" that was exhibited as Wallace (1889) had said, as a "constant and necessary property of all organisms"; this view of variation as a *property* largely precluded the question, by what *process* did this variation come to be? Their view was

[2] Most readers will know that two classes of mimicry were later recognized: *Batesian mimicry*, the resemblance of a palatable *mimic* to an unpalatable *model*, and *Muellerian mimicry*, the mutually beneficial resemblance of two or more unpalatable species to each other. It is, of course, vital not to confuse the *Bates* of *Batesian* mimicry with *Bateson*.

thoroughly organic, with a fundamental blind spot for mechanics, and by whatever dimly perceived process variation arose, it was certainly not by mutation. By adopting Darwin's definition of natural selection as everything strictly analogous to artificial selection (Hodge, 1983), and thus embracing the survival of the fittest *plus* heredity with variation *plus* the resulting adaptation *plus* evolutionary diversity, in a fuzzy kind of way, the Darwinians conceived of natural selection as creating the very varieties on which it worked. As Punnett (1915) noted, selection was barely distinguishable from Paley's watchmaker.

We can think of the Mendelians' artistic analogy of evolution as much more of the pointillist or the creator of a mosaic. It was all very fine and acceptable that selection should decide which little stones or blobs of paint to use, and where they should be placed, but these elements had to come from somewhere, and the Mendelians saw much more clearly than the Darwinists that the creation of the paint and the stones was a concept to be separated from natural selection, which they came to define in its modern, restricted sense of the differential survival of the more fit. Punnett's epitome of this view in his textbook of 1912, that selection was a conservative, not a creative force, constitutes an early statement of the current fundamental paradigm of population genetics: "We cannot regard mutation as a cause likely by itself to cause large changes in a species. But I am not suggesting for a moment that selection alone can have any effect at all. The material on which selection acts must be supplied by mutation. Neither of these processes alone can furnish a basis for prolonged evolution" (Haldane, 1932). But if new varieties arose by mutation, and selection merely had the job of preserving those that were advantageous and rejecting those that were harmful, then a mutation of neutral survival value was also possible. The mutationists trod firmly on the toes of the Darwinians by rejecting their thorough, global adaptationism.

This subtle but profound difference of approach between the Mendelians and the Darwinists goes a long way toward explaining why the modern synthesis was so slow in arriving, even when it was realized so early in the century that Mendelian genetics and continuous variation were not incompatible. The mutual dismissal of "sports" by the Darwinists, and of individual "fluctuation" by the Mendelians, as being of no significance in evolution, because on the one hand they were maladaptive monstrosities, or on the other nonheritable, or as we would now say environmental variation, was seen to be a false antithesis by F. A. Dixey: "If . . . a well-adapted type must have arisen, not by one or more large mutations, but by a series of mutations both numerous and minute, we would wish to know how such mutations are to be distinguished from continuous variations" (Dixey, 1907; quoted by Poulton, 1908). With the experiments by Castle and Wright on the modification of the hooded gene in rats and East's experiments on continuous variation (see Provine, 1971) it became rapidly apparent that individual variation was, in part, due to Mendelian genes, as even Punnett admitted in the 1919 edition of *Mendelism*, where he incorporated a discussion of size in poultry, which he attributed to four genetic factors. What died hard was not the question of inheritance, but of evolution. Right up to the final edition of 1927, *Mendelism* continued to proclaim the unreal distinction between inherited "mutations" and noninherited "fluctuations" without any adequate separation of the latter category into its heritable (genetic) and nonheritable (environmental) parts; Punnett continued to give the impression that only large changes were important in evolution. One's views on genetics, one might say, are subject to evidence; evolution is more profoundly a matter of belief.

2

Punnett's treatment of mimicry, first in *Mendelism* (1912) and then, considerably expanded, in *Mimicry in Butterflies* (1915) constitutes a milestone in evolutionary theory. Bateson had asserted that "the conception of Evolution as proceeding through the gradual transformation of masses of individuals by the accumulation of impalpable changes is one which the study of genetics shows immediately to be false . . . Genetic variation is a phenomenon of *individuals*. Each new character is formed in some germ cell of one particular individual at some point in time" (Bateson, 1909; italics added). The tension here between the idea of evolution as a transformation of populations, and of mutation as arising in individuals, was resolved when Punnett argued that natural selection would cause a favored mutation to increase its *frequency* in the *population*, a concept now so familiar to us that its originality is hard to perceive.

This then was Punnett's explanation of mimicry: Accepting the view that a close mimetic resemblance to the model would confer protection, he proposed that mutations producing such mimicry would spread through the population. "A rare sport is not swamped by intercrossing with the normal form, but . . . on the contrary if it possesses even a slight advantage, it must rapidly displace the form from which it sprang" (1915). Just how rapidly was shown by the classic table, which Punnett asked H. T. J. Norton to prepare for him, showing the time required for an allelic substitution in a population under natural selection.

In laying, thus, the foundation for the contributions which Haldane was to make from 1924 onward, Punnett had established a fundamental cornerstone of the modern synthesis. Ironically, he has been treated dismissively by most later evolutionists, partly because of the honesty with which he followed the implications of his own theory. Norton's table showed that an advantageous form would replace its allele quite rapidly in evolutionary time. Yet the mimetic butterfly on which Punnett had genetic data, *Papilio polytes,* was polymorphic, having one nonmimetic and two mimetic forms, and it was known that this polymorphism had persisted for several centuries. As one form had not replaced the others, Punnett was forced to conclude that all were now of neutral selective value where they coexisted in Ceylon, a conclusion that he found embarrassing as his whole thesis depended on mimicry having some selective value. The solution, as Fisher (1927) was to show, was that the forms at the equilibrium point, but not at other frequencies, must be of *equal* selective value, and that this was an inherent property of mimetic polymorphisms resulting from a decline in the fitness of a Batesian mimetic form as it becomes commoner in the population. But the first theoretical demonstration of this kind of balanced polymorphism was not to be given until Haldane (1926) showed that it could be produced by heterozygous advantage; Punnett's attempts to grapple with the problem have led later readers to believe that he totally dismissed natural selection as a force producing mimicry and attributed its existence solely to mutation. Nothing could be further from the truth.

3

Nonetheless, for all the diplomacy of his style, intended to sleek over the rugged looks of Bateson and Poulton, Punnett plainly saw himself as supporting the new, rigorous mutationist view of evolution against extreme adaptationist Darwinism. To this end he had to show not only that his views were plausible but also that the mimetic patterns had not arisen gradually. Here Punnett's grasp of theoretical points paid off against Poulton's superior knowledge of butterflies. The view of mimicry as an oil painting built up slowly on a wing by natural selection had already encountered a difficulty, one that had been faced and temporarily solved by Darwin. It was a gift to Punnett, a problem not confined to mimicry, but potentially present in the discussion of all adaptation, made famous by Mivart and much beloved to this day by creationists: the "unbridgeable gap."

This difficulty proved to Punnett's satisfaction that mimicry arose by a single mutation, and that the gradual molding of the phenotype perceived by the biometric school was ruled out completely. His argument uses all the proper criteria: the theoretical impasse for the opposing school, experimental evidence in favor of his own view, and his own theoretical model to account for the observations.

1. *Theoretical difficulties* with the gradualist approach to the evolution of mimicry arose from reasonable assumptions about the behavior of birds. If they had learned that a butterfly with orange and yellow stripes was to be avoided and that a white one was to be eaten, then a very small increase in the amount of orange and yellow on the white butterfly could not be expected to deceive them: a telling example of an unbridgeable gap. For mimicry, Darwin had solved the problem by postulating that the mimetic resemblance arose when the model had developed its defensive flavor but not its full warning color. At this stage both model and potential mimic had had similar dull patterns, and gradual convergence was possible. Later the model developed its bright colors, also gradually, pulling the mimic along with it. Punnett showed that this theory became barely tenable for models that were mimicked by several unrelated mimics. If we accepted that the ancestral pattern of the mimics was still exhibited by their males (he chose cases where the mimicry was sex-limited), then the model might once have been like one of them, but hardly like all of them. This argued strongly that mimicry had arisen in a single large mutational step.

2. *Experimental evidence* from the mimetic butterflies *Papilio polytes* and *Hypolimnas dubius* confirmed the existence of this single mutational step. If the pattern of a mimic had been built up gradually as the biometric school would maintain, then all the intermediates should appear when a mimic was crossed with a nonmimetic form or one mimetic form with another. Yet in *Papilio polytes* both the mimetic forms and the nonmimetic form segregated clearly as single alleles.

3. *A theoretical model* was needed to explain how a single mutation could produce the same pattern in two distantly related species. Punnett found the solution in the "mimicry" exhibited by certain mammalian coat-color mutants: the mouse, rabbit, and guinea pig, for instance, all exhibited agouti, black, chocolate, blue-agouti, blue, and fawn varieties, as a result of having the same hereditary factors. Although the situation in butterflies was no doubt more complicated, Punnett did not see that it was difficult to imagine that model and mimic shared enough hereditary factors for mutation to produce a straightforward resemblance.

4

In thinking about the Darwinist–Mendelist debate it is important to distinguish "a variety" from "variation." Both sides agreed that "variation" was *not* produced by mutation. The difference was over what became of it: According to the Mendelists, nothing; according to the Darwinists, natural selection built it up into well-marked forms, or "varieties." Punnett took the mutationist view, that, on the contrary, "varieties" were the result simply of mutation. But his masterly treatment of the way natural selection would preserve advantageous varieties had still not finally solved the question: Was the complete, complex mimetic pattern really achieved by a single mutation? Although the theory of genetic homology was ingenious, it became less and less likely the less closely the model and mimic were related. Certainly it could not explain insects that mimicked leaves, although perhaps in that case there was no unbridgeable gap to be crossed.

But evolution, writ large, is not the generation of varieties, but of species; the realization that these were two different categories was a slow and complex business, leading from the early Mendelist belief that the whole problem had been solved when de Vries showed, as he supposed, that new species of *Oenothera* arose by saltation, to, on the one hand, the familiar modern synthesis view of gradual speciation through the accumulated effects of microevolutionary processes; and, on the other, to the views of Richard Goldschmidt, now becoming familiar once more because of their importance for punctuational theory. This theory seeks to reconcile the modern synthesis with de Vriesian mutationism by proposing that evolution should be seen as two distinct processes, the one adaptation or microevolution that was gradual and involved individual variation; and the other, speciation or macroevolution that occurred by saltation (Goldschmidt, 1940).

In seeking evidence for macroevolution, Goldschmidt (1945) turned to the unlikely source of butterfly mimicry, unlikely because all the existing genetic work had been done on mimetic polymorphisms, which on Goldschmidt's own argument, were adaptations within populations, and, therefore, should have occurred by microevolutionary processes. "I wonder whether [he writes in *The Material Basis of Evolution*] . . . the simple Mendelian difference between mimetic forms is in fact a difference of a whole chromosome, the architecture of which has been changed by systemic mutation, thus demonstrating *within* a species in a case of special adaptation the happenings otherwise found *between* species" (Goldschmidt, 1940). Mimetic polymorphism thus became an honorary kind of speciation, produced by systemic mutation, the process normally reserved for the generation of new species. As evidence for the occurrence of such systemic mutations, or "hopeful monsters," Goldschmidt cited the very considerable changes that could be produced, in *Drosophila* as well as in the moths he worked with, by single laboratory mutations, and made the telling point that only a few mutational changes were needed to produce what to us is a profound alteration in pattern: a series of stripes on a snail's shell would be turned through a full right angle simply by combining two mutations, the first breaking the stripes into dots, the second joining the stripes together.

From this, Goldschmidt, using some sophistry about what one meant by "the same," proposed again Punnett's theory that model and mimic achieved their similar patterns by means of "the same," or in Goldschmidt's term, "parallel" mutations. He proposed not necessarily that the genes involved were homologous, but that the mimic's development was being switched into pathways homologous with those in the model. His definition of homology, operational rather than explicit, is roughly the conventional one used by morphologists. The question about the uncovering of old, lost developmental pathways is clearly a most important and interesting one; for reasons that may have little to do with Goldschmidt's particular style of advocacy, it has yet to be taken up. But given the hit-and-miss nature of mutation, many evolutionists simply found it improbable that the whole of a mimetic resemblance, which frequently involved elaborate and extraordinary detail, could be the result of a single mutational change.

Synthesis (E. B. Poulton and A. J. Nicholson)

1

Gradualistic Darwinism and saltational mutationism were both unsatisfactory as explanations of mimicry, the one by failing to bridge the unbridgeable gap, the other by making excessive demands on the ability of mutation to hit a predetermined phenotypic target bang on. The simple solution was to combine the two approaches, by postulating that the evolution of mimicry was not *either* mutational *or* gradual, but both. If a single mutation could achieve an approximate resemblance, close enough to deceive some of the predators some of the time, then once this pattern was established in the population it could be further improved, right up to the limits of perception of the predators, by the gradual selection of continuous, heritable variation. This, the current theory (known as the "two-phase" model) requires two ideas for its acceptance: first, that Mendelian genes produce continuous variation (what, as it predicts the conversion of quality into quantity, we could call the "backward Hegelian principle"); and second (and harder), that the effects of one gene can be altered or modified by other genes, and (particularly hard) that in a polymorphic species the modification of the different forms can take place simultaneously and independently.

And yet for all its difficulty, this theory made an early, but obscure, appearance. In the first issue of *Bedrock*, Poulton (1912) put forward his synthesis, almost as an aside in the middle of a refutation of Bergson:

> If it be unreasonable to suppose that all these mimetic features arose spontaneously and together, what is the probable explanation of their origin? It is probable that by spontaneous variation a white band [on the hindwing of the butterfly *Acraea alciope*] appeared in the ancestral form . . . and that this was from the very first sufficient to confer some advantage by suggesting the appearance of a dominant Model . . . From this point Natural Selection acting on further variations produced the detailed likeness which we see in the white band itself and in the other mimetic features.

Poulton, therefore, rejected the saltatory origin of species, which as he said looked ridiculous to anyone acquainted with systematics and biogeography (Poulton, 1908), but was willing early to accept mutation as the origin of new varieties. He had, furthermore, seen clearly that the evolution of mimicry at least was best explained by a mixture of saltational and gradual change. His fate, like that of all moderates, was to be forgotten.

2

Poulton's theory next surfaced in the likewise neglected[3] writings of A. J. Nicholson (1927). Accepting the modern synthesis view that natural selection could operate successfully on *any* heritable variation whether the mutation was large or small, Nicholson considered the twin questions: How large are the mutations that produce mimetic resemblances, and how accurate are the resemblances that they produce?

Nicholson suggested that the magnitude of the initial change would depend on how similar model and mimic were already in their superficial appearance, only a very small change being needed if

> The ancestral form of the mimic had, quite accidentally, some resemblance to the insect which later served as the model . . . This first "rough" resemblance must . . . be purely fortuitous . . . The mutations on which natural selection commences to operate may be very small, if the normal appearance of the species is close to that of the new model, or must be large [a very great change from the normal, probably a large muta-

---

[3] I first encountered this paper when I purchased it from a dealer in a lot of reprints, and was thoroughly surprised to discover what Nicholson had written.

tion] if the insect about to become a mimic is very unlike the model. [Nicholson, 1927]
The direction of mutation was, as we would now say, independent of selection, and if an "unbridgeable gap" confronted the mimic it would, if it was so fortunate, cross it by means of a single, large mutation. As regards the accuracy of the mimicry first produced "it is . . . evident that an incipient mimetic pattern must resemble some model sufficiently closely to cause the insect bearing it to be confused sometimes with the model before natural selection can commence to operate in its favour." However, the resemblance did not have to be very close, on account of what we now call generalization by the predator: "a more perfect resemblance would have a superior survival value and would be selected at the expense of the earlier and vague resemblance. In this manner a very perfect resemblance would be built up . . ." (Nicholson, 1927).

Thus Nicholson proposed again Poulton's compromise model, that mimicry could, but did not necessarily, commence with a large gap-bridging mutation, which was then subject to continuous improvement. The likelihood that the first mutation achieved high accuracy was dependent on the taxonomic relationship of the organisms, close relatives being more likely to produce the same pattern with the same gene than unrelated species. At one extreme, mimicry of leaves by insects could in no way be regarded as genetic homology; at the other, he accepted Punnett's contention about homologous genes in the mimicry of one butterfly by another, and was convinced by Punnett's evidence from *Papilio polytes* that, in this case, only one mutation was involved in the mimicry. For although Nicholson gives an account of the modification of the effects of the first mutation that would delight any disciple of Fisher, he had not conceived, as Fisher had, that two patterns both in the same population might be modified independently for their resemblance to two different models; that required one to conceive of modifiers whose action was specifically confined to the effects of one gene.

3

Both Poulton's and Nicholson's thoughts have surely been more influential than their citation rating suggests. Punnett (1915) was well aware of Poulton's theory, and without attributing it to anyone in particular ("there are writers . . . who adopt a view more or less intermediate between [gradualism and saltationism]"), gives a very fair account of it. He rejected it for two reasons.

First, there was his argument against all gradualism in the evolution of mimicry, drawn from considering the behavior of predators. For predators simultaneously to maintain the sport in the population *and* to improve its resemblance to the model seemed to require that the predators simultaneously do two incompatible things: confuse the mimic with the model, so that they avoid it, and distinguish the mimic from the model, so that they attack it. Paradoxically, Punnett solves this problem himself:

> If birds help to bring about the resemblance we must suppose
> that it is done by different species – that there are some
> which do the rough work, others which do the smoothing
> and others again which put on the final polish and keep it up
> to the mark. This is, of course, a possibility, but before it can
> be accepted as a probability some evidence must be forthcom-
> ing in its favour. [Punnett, 1915]

If we substitute for "different species" the words "different individ-uals or the same individuals at different times" then we have a good summary of what recent experimental work has found – that the generalization of experience by predators is statistical, or that if a rather poor resemblance gives some protection to a mimic, then more protection is given by a better resemblance. But Punnett's deeper reason for rejecting the mutation-improvement theory seems to have been his fundamental antipathy to the ideas of the biometricians, revealed in one of those giveaway asides that are so rare in his book: "Those who take this view [the mutation-improvement theory] hold also that the continued action of natural selection is necessary in order to keep the likeness up to the mark. They suppose that if selection ceases the likeness gradually deteriorates owing to the coming into operation of a mysterious process called regression" (Punnett, 1915). "Regression" was the law discovered by Galton that the offspring of extreme parents "regressed" toward the mean. Punnett does not elab-orate this criticism further. He simply seems to have had a complete block against thinking of continuous variation as having anything to do with evolution. Such were the results of two decades of schism.

4

By Goldschmidt's time the concept of modification was much clearer: Goldschmidt's rejection of it arises from his need to regard large mu-tations as *sufficient* for some forms of evolution and is based mainly on a simple logical fallacy. Because we know that mutations can pro-duce large changes of phenotype, he says, there is no difficulty in imagining that they produce a *particular* large change, namely, a fully developed mimetic pattern. But as the pattern required is predeter-mined by the pattern of the model, this is like arguing that because our artillery has no problem in firing at a town ten miles away (in the course of which bombardment they will hit *some* houses), there is no difficulty in hitting the house at 55 High Street. All macromutations must produce *some* phenotype, but what probability is there that they will produce a *particular* phenotype? Goldschmidt has confounded a priori and a posteriori probabilities.

He then tries to argue the case both ways. If parallel mutations are exploiting homologous developmental pathways, then it might be argued that they are quite likely to produce patterns that are so similar between model and mimic that the mimicry is perfect within the predators' power to know the difference. This is not unreasonable: We might expect such an effect in Muellerian mimicry between closely related species. But as the species become less related, even homologous pathways are likely to produce divergent effects, and non-homologous mutations, admitted by Goldschmidt in the case of leaf-mimicry, will play an increasing part. But these nonhomologous mutations, he argues by means of the "large equals accurate" fallacy, will also produce accurate mimicry. In other words, some mutations require no modification because they produce accurate mimicry by being homologous; others require no modification because they are not homologous. We are here brought somewhat near to that great proof that the Achaeans are fools: "Agamemnon is a fool to offer to command Achilles; Achilles is a fool to be commanded of Agamemnon; Thersites is a fool to serve such a fool; and Patroclus is a fool positive."

With this odd compromise between saltationism and Darwinism, Goldschmidt drew attention away from the valuable points he had made about the way the timing of action of genes during development could cause them to have large and diverse phenotypic effects. This was a pity, for Fisher had been led, no less than Goldschmidt, into an eccentric interpretation of mimicry through his enthusiasm for his own grand scheme of evolution. Yet it is Fisher's reputation that stands.

Thesis again (R. A. Fisher)

1

Of the triumvirs of the modern synthesis, only Fisher discussed mimicry, reckoning in the *Genetical Theory* (1930) that it merited two color plates and a whole chapter, itself a lightly edited version of an earlier paper (Fisher, 1927). A fashionable explanation for the importance attached to the subject would be that it belonged to the slightly dotty tradition of English bug-hunting, alien to Wright and too effete for

Haldane's robust radicalism. The truth is that mimicry presented a serious challenge only to Fisher's particular evolutionary theory, which insofar as simple labels are helpful, we can categorize as a synthesis of Mendelism and Darwinism by means of *gradualism, adaptationism,* and *reductionism*; a picturing of evolution as a series of minimal changes guided by selection, with a minimal interference from nonadaptive processes, and a minimum of interaction between different genes.

Fisher's model of evolution is bold and general: It imagines a population perpetually under natural selection, because it is perpetually confronted with the "deterioration" of the environment produced by undirected physical changes and the evolutionary advances, also under natural selection, of other species. The population, or species, is, therefore, on what Wright called a treadmill, or in what is now called the "Red Queen dilemma," running as hard as it can to stay, roughly, in the same place. Mutation supplied, and selection consumed, the genetic variation that allowed the population to move forward in this evolutionary caucus-race, which Fisher pictured as involving the population as trying to approach an "ever-removed" *ignis fatuus*, a point of maximum fitness. As selection removes the less "fit" individuals from the population, the average fitness of the population will be perpetually increasing (although just as perpetually frustrated by environmental deterioration), and the rate of increase, Fisher showed by an all-too-brief mathematical proof, would be equal to whatever variance in fitness mutation was able to produce – this is the "Fundamental Theorem of Natural Selection."

In deriving the Fundamental Theorem, Fisher made the reductionist assumption that the total variance could be obtained by adding the variances generated by individual mutations; this is not merely a mathematical convenience, for it assumes that there is no significant interaction between gene loci, and from this consideration flow Fisher's extreme gradualism and adaptationism. For if there is no significant degree of interaction between genes, then there is only one adaptive optimum for the organism, whether individual or species. It follows immediately that random genetic drift can only have the role of spoiling or frustrating the improving work of selection: Stochastic change in gene frequency is, at the best, random noise and can have no creative function, unless there are multiple optima within the field of genotypes or gene frequencies, as Wright proposed. Reductionist assumptions thus led to the elimination of this nonadaptive process from any serious role in evolution.

The same reductionist assumption also leads immediately to Fisher's gradualism, the statistical elimination of all but the smallest mutations from an important positive role. For if we imagine the organism removed from its optimum point in the hyperspace that represents all dimensions of the phenotype or the environment (Fisher does not make it clear which, neither is it clear whether we are still discussing a population or an individual organism),[4] then if fitness declines uniformly in all directions from that optimum, and mutation is truly random in direction, then a very small mutation has an equal chance of moving the organism in the direction of improved or of reduced adaptation. A larger mutation, on the other hand, because of the curvature of the adaptive function round the optimum, has an increasing chance of reducing the organism's adaptation and a decreasing chance of enhancing it, much as a small adjustment in the alignment of a piece of machinery may equally improve or deteriorate its performance, whereas kicking it is overwhelmingly likely to disarrange it. Thus, with statistically rare exceptions, only very small mutations will play an important part in evolution. Again, the argument depends crucially on there being one optimum: If there are two, then a large mutation may do what a small mutation cannot achieve, the transfer of the organism to the region of a new optimum. With this theorem, Fisher solved the outstanding worry about mutations that had been revealed by experimental genetics – that most of them were deleterious. Indeed, they mostly were, and the larger and the more easily observed they were, the more deleterious they would be. Only the very small ones, with their limiting probability of 50 percent for improving an adaptation, were the real stuff of evolution.

Within this theory, only one particular kind of interaction was permitted – *specific modification*. By this was meant, that if we had three genotypes, *AA*, *Aa*, and *aa*, another gene might improve the adaptation of *AA* and *Aa*, but have no effect at all on the phenotype of *aa* (or, of course, improve *aa* and leave the others unaffected). Nonspecific modification, which meant improving *AA* and *Aa* while reducing the fitness of *aa*, was the excluded interaction that produced multiple peaks! Alternatively, the modifier might affect the expression of *Aa* only, in which case it was a modifier of dominance. It is widely known how Fisher used this theory to explain dominance; but specific mod-

---

[4] The same ambiguity between individual organisms and whole populations occurs in other parts of the mathematical theory: A continuous field of gene frequency and a discontinuous grid representing all possible genotypes were, at this time, not distinguished clearly either in Wright's or Fisher's models. This led Fisher to equate, incorrectly, the average effect of a gene substitution on the phenotype with the derivative of mean fitness on gene frequency, producing some infelicities in the derivation of the Fundamental Theorem, the least, but most obvious of which was the omission of the factor 2.

ifiers were also of crucial importance in his consideration of mimicry, which presented a very considerable challenge to the whole gradualist, adaptationist thesis.

2

Poulton, the arch-Darwinist, had originally thrown down mimicry as a challenge to the mutationists; Punnett had answered it with such success that mimicry was now, in turn, an embarrassment to Fisher. The problems were threefold: polymorphism in mimetic butterflies, each form being clearly adapted to the local environment, suggested that a population could simultaneously be in the region of two or more adaptive optima; the bridging of the unbridgeable gap suggested evolution between one optimum and another; and, most serious because empirical, the genetics of mimetic butterflies indicated evolution by single large mutational steps. This was indeed a challenge worthy of Fisher's formidable mind. His solution, presented with some convolution, is brilliant and, to most later workers, unconvincing. Its outstanding and major contribution is Fisher's extension of the Backward Hegelian Principle into the notion of specific modification. Those who like reductionism in the history of science should note that Fisher's solution of the problem is surprisingly dialectical, and has the mimetic species, to some extent, generating two adaptive optima for itself!

Citing Castle's experiments with hooded rats, Fisher argued that the phenotypic effects of a gene could be considerably altered by the selection of other, independent genes, while leaving the major gene itself unaltered. The fact that complete mimetic patterns segregated as single Mendelian genes did not prove, as Punnett had maintained, that mimicry had arisen, fully complete, by saltation. Contrariwise, if "selection favours different modifications of the two polymorphic genotypes, the species may become adaptively dimorphic by the cumulative selection of modifying factors, without the alteration of the single-factor mechanism by which the dimorphism is maintained" (Fisher, 1927).

This concept is very important: The present effects of a gene were not necessarily the effects that it had when it first appeared. Fisher, therefore, proposed that the mimetic dimorphism of *P. polytes* had started as a simple nonmimetic dimorphism, with the two forms controlled by a single gene. One form had gradually been modified toward mimicry of the model *Pachlioptera hector* and the other to mimicry of a different distasteful butterfly (*Pachlioptera aristolochiae*).

Fisher's way of synthesizing Mendelism and Darwinism led him then not to the compromise solution of Poulton, in which the effects of a major mimetic saltation were later improved, but to a thorough-going evolutionary gradualism, in which the mimetic pattern was built up slowly from a nonmimetic one just as Darwin had supposed. The major mutation that one detected in polymorphic mimics had itself nothing whatever to do with the mimicry: It was simply the residue of a polymorphism which was originally present for totally different reasons.

All this left the problem of the bridgeless gap untouched, and Fisher had to expend some ingenuity in demolishing Punnett's argument from mimicry rings. The crucial point in Punnett's argument had been that where two or more unrelated mimics copied a single model, gradual evolution was excluded because the ancestral model would not have looked like *both* of the ancestral mimics (their patterns being revealed by their still nonmimetic males). But, Fisher triumphantly observed, the Sinalese mimicry ring used by Punnett to illustrate this point, actually contained *two* model species, which were mutual Muellerian mimics. Therefore, once upon a time model A had looked like mimic a, and model B had looked like mimic b, all the butterflies being dull in color. As Darwin had originally supposed, the models then each developed their warning colors, pulling their mimics along with them, and formed into a mimicry ring because the two models gradually converged to become Muellerian mimics as well.

Fisher was able to show that two distasteful and warningly colored species, already sufficiently similar so that generalization by predators caused them to be mistaken one for another, would, given that their variability was inherited, gradually converge in just this way to become good Muellerian mimics. In thus completing his thesis, Fisher set a trap for later investigators, for this demonstration is very clear and prominent, whereas his equally firm and important belief that the evolution of Batesian mimicry was likewise entirely gradual is hard to disentangle. It has generally been believed that Fisher showed Muellerian mimicry to evolve gradually, and Batesian mimicry to involve an initial major mutation. The growth of scientific myths is a subject worth more study; Fisher's views on Batesian mimicry have been consistently misdescribed.

Synthesis by stealth

1

Wright's multipeak model, which assumes extensive genetic inter-action that puts a species within striking distance of several simul-taneous adaptive optima, is clearly not challenged by mimicry. Even convinced selectionists were later to use just this model of Wright's, minus random genetic drift, to explain the evolution of mimicry in *Papilio dardanus* (Clarke and Sheppard, 1963). His conception of the adaptive optima being maxima of a surface of mean fitness was also to be used, particularly in Dobzhansky's notion of "adaptive poly-morphism" to cope with the difficult notion that a polymorphic pop-ulation was somehow adapted to the same environment in two ways at once. With the simpler forms of selection, the equilibrium gene ratio occurred at the maximum value of mean fitness: Hence, it was held that the polymorphic population, occupying *one* peak of the sur-face, was the fittest, and that not merely were the polymorphic forms adapted to the environment, but the polymorphism *in itself* consti-tuted an adaptation. Thus, even the more or less large mutations that Fisher had had to admit in the case of polymorphism, and which were an anomaly in his overall gradualism, became assimilated into the doctrine of global adaptiveness: Their presence adapted the pop-ulation (see criticism of this concept by Cain and Sheppard, 1954, passim).

Less obviously, Haldane's approach in *The Causes of Evolution* (1932) was fundamentally one of multiple optima. As would be expected both from the way he had expanded Punnett's ideas on gene sub-stitution and his conviction that gene mutation was translated into the phenotype via biochemical changes, he was very much more in-clined than Fisher to attribute evolutionary change to mutations of quite large effect, rather than to the summation of many minor genes. Like Fisher, he believed that such large mutations would usually have rather disastrous effects on fitness even when one of the effects they produced was adaptive. Unlike Fisher he did not argue that this ruled out large mutations in evolution, but proposed, instead, that where, let us say, a mutation provided a beneficial camouflage effect while causing physiological dislocations, the deleterious effect of the gene would be removed by further modification. What would be required for the initial major mutation to become established would be that the camouflage benefitted the organism more than the physiological effect dislocated it: If we imagine that this means a gain in fitness along one dimension of the hyperspace offsetting a loss in other di-mensions, it becomes clear that the population must be in the region of *two* optima of the adaptive function, and without emphasizing this point, Haldane is using an approach radically different from Fisher's.

2

If the modern synthesis were Fisher's alone, then the criticism that it is thoroughly gradualist, adaptationist, and reductionist would be fairly made. The mutationists had asked that mutation have a maximal effect in evolution; Fisher, that its role should be reduced to the opposing minimum. That Haldane and Wright should take a different view is not surprising. What is significant is that Fisher's closest and staunchest associates could not accept his minimal mutationism and immediately subjected it to specific modification in the direction of the Poulton–Nicholson compromise, while attributing the theory entirely to Fisher through laying great stress on modification as the key idea. What was to Fisher originally a nonmimetic polymorphism, one or both of whose alleles became the basis of a mimetic form, became transformed into the initial major mutation of Nicholson's two-phase system, which then rose to a polymorphic equilibrium and was improved, as was Fisher's allele, by gradual modification.

One of the major factors undermining Fisher's extreme gradualism was the discovery of selection coefficients of up to 50 percent in melanic moths (Ford, 1964). These would have presented no difficulty for Fisher's views if the mutant form had been disadvantageous, but here it was the new advantageous black form that was being selected with such great strength. This cast an empirical shadow over Fisher's theory that the vast majority of positively selected mutants were of very small effect, and made the acceptance of large mutations much easier in evolutionary theory. But the introduction of large mutations into Fisher's mimicry theory occurred two decades earlier, when the print was barely dry on the *Genetical Theory*.

In *Mendelism and Evolution* (1931), a highly influential, because lucidly written account of the Fisherian version of the modern synthesis, a kind of reply to the now outdated *Mendelism* of Punnett, Ford recounts the Poulton–Nicholson compromise:

> We have no reason to assume that when genes such as these first appeared their effects were similar to those which they produce today. On the contrary, we may suppose that in a given palatable species a gene arose which chanced to give some slight resemblance to a protected form. This would gradually be improved by selection of the gene-complex and the consequent alteration of the effects produced by all the genes acting together, until an accurate mimicry had been attained. Such a process would be one of slow continuous change, but at the end the profound difference so produced would still be under the control of a single factor; yet this would not mean that the mimic had arisen from the nonmimetic form suddenly by a single act of mutation.

Ford's later writings (e.g., 1964) are sometimes a little ambiguous, but they can all be taken fairly as describing the Poulton–Nicholson system rather than Fisher's.

The full transformation of Fisher's theory had come from what most scientists would think of as the best source: the interpretation of experimental results. The unraveling of the genetics of *Papilio dardanus* by C. A. Clarke and P. M. Sheppard showed that all its multiple mimetic forms were controlled from a single site on a single chromosome. To attribute this to the evolution of mimicry by each of a series of nonmimetic morphs would be stretching credulity a long way, especially as there are no widespread nonmimetic morphs that one could suggest as the ancestors of mimetic forms with restricted distributions, and the two nonmimetic races (in Madagascar and the Comores) are not polymorphic. The Poulton–Nicholson–Ford version of the modification story thus came to be favored – each mimetic form was a separate, approximately mimetic mutation that had been improved; Fisher's nonmimetic polymorphism dropped from prominence.

3

The experimental evidence, obtained over two decades from genetic work, chiefly on *Papilio* and *Heliconius*, has largely supported the modern synthesis view of mimicry that was developed from such diverse sources as Punnett, Poulton, Nicholson, Fisher, and Ford. For the full genetic details, which are extensive, I must refer the reader to other reviews (Sheppard, 1961; Turner, 1977) but, in brief, there is good evidence that in both Muellerian and Batesian mimicry both large (saltatory in some sense) and small (biometrical in some sense) mutations are involved, and that the latter can very fairly be described by Fisher's term "modifiers." The most impressive experiment, repeated many times, has been to cross a mimetic form into a race where it does not occur; although the major gene is present in the hybrid, the mimicry becomes considerably deteriorated, showing that, as expected, the modifiers are absent in the nonmimetic race. But even the modifier theory does not explain everything.

Goldschmidt's systemically altered chromosome, that he held controlled the polymorphism in some mimics has been shown – particularly by work on *Papilio memnon* – to be a tightly linked cluster of some half dozen genes each controlling an aspect of wing pattern, body color, or wing shape. The usual explanation for the existence of this "supergene" has been by an extension of Fisher's modification idea, which he himself originated, to the linkage of interacting genes: It has been proposed that because recombinants are at a disadvantage (which they undoubtedly are, as a result of combining the body color of one model with the tails or hindwing color of another), then selection will have tightened the linkage between the loci, or even (using ideas drawn from C. D. Darlington) have caused unlinked loci to be moved onto the same chromosome by the preferment of structural interchanges. Theoretical work (Charlesworth and Charlesworth, 1976), however, makes it likely that a stable polymorphism for all the loci can be maintained only if they are already fairly tightly linked and strongly suggests that the supergene is built up, not by modification of the genetic architecture but by a "sieve" allowing polymorphism to develop only at loci that have, in advance, the appropriate linkage relations.

This will immediately suggest such very un-Darwinian ideas as preformation, directed evolution, or even Goldschmidt's transmogrified chromosome. But the solution is probably simpler. It is only in those butterflies that happen, for other fortuitous reasons, to have the appropriate linkage between the loci that spectacular mimetic polymorphisms develop; and it is, naturally, only in these species that we carry out genetic experiments.

Similarly, some of the outcrossing experiments demonstrating the presence of the specific modifiers that improve the mimicry, indicate that some of the "modifiers" were present in advance of the major mutation. In this instance, a "sieve" has allowed only those major mutations to enter the population that gave a reasonably accurate mimetic pattern in the local ambient genetic background (Turner, in press). We are giving a habitation and a name to Goldschmidt's otherwise tautological assertion that only that which is capable of mimicry will become mimetic.

4

Mimicry is quite a complex phenomenon and requires a complex synthesis of ideas for its explanation. Fisher's modifiers, Haldane's or Punnett's major mutations, and Wright's multiple peaks are all present. One might want to go along with Gould and Lewontin (1979) and praise this aspect of the modern synthesis for its "pluralism," or side with Gould (1980) and criticize it for using so many hypotheses as to be incapable of refutation! It is difficult to draw a clear moral from the story, for although it is now clear that a satisfactory theory had to be forged from diverse elements, there was no way of telling in advance which parts of the conflicting theories would need to be kept and which discarded, nor indeed that certain of the theories should not be discarded altogether (this necessarily partial history

having omitted those that were).[5] But if pluralism as a scientific program is simply naive, synthesis in a better theory is not, and it is
plain that schism, encouraging as it does a too rigid adherence to the
views of one's own school, can severely limit one's own insight (as
happened to Punnett and to Fisher), or even delay the general acceptance of the synthesis (Provine, 1971). The perception that more
was at stake than a scientific theory can have done little to make views
more flexible.

As mimicry seemed less and less able to provide simple answers
to naive questions, it dropped from sight as a central issue and was
mentioned in very few textbooks. Not only was the problem of adaptation seen as settled, but the complexity of the phenomena made
it difficult to explain in a few pages, and geneticists became increasingly prone to consider evidence from only two of the higher eukaryotes: one of them was *Drosophila*, and the other wasn't.[6]

Speciation as mutation – new theories and old

1

In various ways, and to varying degrees, the ideas of the mutationists
became fused into Darwinian theory to form the neo-Darwinian synthesis. But when confronted with what several of the actors were
agreed on as a crucial test case, the experimental evidence on the
evolution of mimicry, the necessarily naive mutationism of Punnett
and the perversely naive saltationism of Goldschmidt both failed. So,
too, did such extreme forms of gradualism as the one so ingeniously
constructed by Fisher. The modern synthesis, far from being the extreme gradualistic theory of Fisher, admits, in varying degrees, the
possibility both of large mutations and of rapid evolutionary change.

Now that evolutionary theory has "circled the dolorous path," its
schismatic wounds healed only to be reopened on the cutting edge
of the new saltational, punctuational theory, it is time to question
that theory yet again with the old riddle. How will it cope with a real
case of adaptation that we do, to a considerable degree, understand?

2

The theory of evolution by jerks (Gould, 1980; Stanley, 1979) aims to
provide an alternative explanation to the allegedly gradualistic theory

---

[5] Who these days has heard of Reinhard's theory that there is an innate tendency within organisms to become brightly colored, which was deflected by
selection toward dull cryptic coloring if the organism was subject to predation
(quoted in Shull, 1936)?

[6] Most geneticists believe, for instance, that sex-linkage was first described in
Drosophila; it was, in fact, first discovered in moths.

now espoused by a majority of evolutionary biologists. In that sense it can be seen as in the tradition of earlier saltational theories. Like Goldschmidt's theory, it invokes developmental models to explain the saltations, and like Bateson's theory it has caused writers on science in the popular media to announce the demise of modern evolutionary theory (creationists have announced the demise of the whole of Darwinism, but as one of Britain's more famous daughters put it "They would, wouldn't they?"). The theory originated from the suggestion that the fossil record showed an extremely uneven rate of evolutionary change, which was explained, at first, in largely neo-Darwinian terms (Eldredge and Gould, 1972), and then progressed to the presentation of a serious alternative to neo-Darwinism (Gould, 1980), seen by some commentators as a paradigm case of the Kuhnian paradigm-shift (Lewin, 1980), which with Mallarméan self-reflexiveness, the theory itself resembled.

The "gradualism" of the new synthesis certainly does not exclude from evolution mutations of rather large phenotypic effect, nor does it make any strong prediction about evolution occurring at a constant speed. Hence the saltationist element in the new theory – which it must have in order not to be seen simply as a variant of neo-Darwinism that emphasizes the importance of rather short periods of "rapid" change, lasting no more than, say, fifty millennia – is to be found in the belief that evolution consists of two processes, one of which is the adjustment of gene frequencies in populations familiar in neo-Darwinism, and the other a separate process of speciation which accounts for most of the divergence and diversity produced in the course of evolution (Figure 1a). These events that generate the branching of the tree in both senses – isolating the branches genetically and generating most of their phenotypic distinction from other branches – are held to be analogous to mutation, in being nonadaptive, and are followed by a higher order form of selection on whole species, selection that determines the long-term trends of evolution. This nonadaptive element is an important feature of the theory, separating as it does the two processes that Darwin brought together – individual adaptation and species diversity – and harks back to the coolness to adaptationism shown by Punnett and other Mendelians, just as the erection of a two-tiered system of evolution harks back to Goldschmidt. All three theories, saltationism, Goldschmidtism, and evolution by jerks, agree in regarding speciation as a mutational event.

3

We might expect, in this attempt to reassert saltational speciation, a confusion like the one that dogged the early Mendelians and their

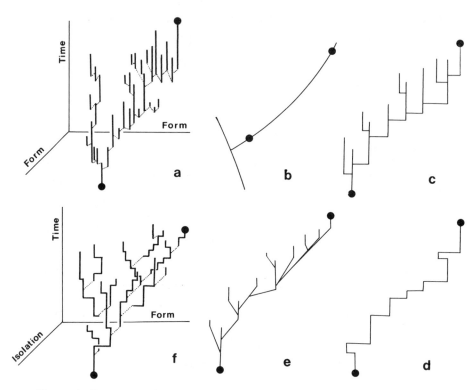

Figure 1. Various evolutionary trees, representing punctuational (a, c) and gradualistic change (b, e) or patterns that fit neither simple model (d, f). See text for explanation. [a, b, c after Gould and Eldredge (1977).]

Darwinian opponents, between varieties and species. Thus, Gould and Eldredge (1977) argue that although asexual species, viewed on a coarse scale, appear to show smooth, gradual evolutionary trends (Figure 1b), when examined closely they reveal a series of punctuational steps, produced every time a successful new clone replaces the old. As clone replacement takes place by selective substitution, as understood by Punnett, Haldane, and all later neo-Darwinists, the modern synthesis and evolution by jerks do not differ at this level, except in the terms used in the description. Clone substitution simply *is* a punctuational event. A new clone is *both* a new variety *and* a new species. But then evolution in a sexual species can also be described as the substitution of one clone by another, provided we forget about the diploid individuals and think only of the DNA. Again this is precisely the approach of Haldanian population genetics. This suggests that the theory of evolution by jerks is simply neo-Darwinism disguised in a new terminology. Now a new variety is *not* a new species, but if species are defined morphologically, as they must be in the fossil record, then any phase of rapid evolution, whether it be produced by a single gene substitution or by alterations in the frequencies of a number of genes, generates what is, by definition, a new species, and the theory of evolution by jerks is true by strict tautology.

In order to avoid this tautology, proponents of the theory have to insist that when a rapid change occurs, the evolutionary tree *must* branch at the same time, and to deny quite explicitly that rapid evolution without branching (punctuated phyletic evolution) can occur (Gould, 1979; Stanley, 1979). The confusion is thus complete: In the diagram of punctuated equilibrium in an asexual species (Figure 1c) the upright lines represent only the old clone that is being removed by selection, not a true branching of the tree. Yet without them (Figure 1d), the diagram would be a clear representation of the allegedly nonexistent punctuated phyletic evolution! What, then, are we to make of the diagram (Figure 1a) of punctuated equilibrium in the sexual species? Are the upright, unaltered branches *really* branches of the tree, that is, the old species persisting in some geographical location, or are they merely an old genotype that is being replaced by the new one? And in that case, in both diagrams, there is something odd about the new branches starting off horizontally. The new form does not become the established phenotype of the species until it has *replaced* the old one, in which case a period of transition or polymorphism must be represented. This could be done approximately by sloping the new branch up to the level at which the old one disappears (Figure 1e). But what we have now is a tree that is uncomfortably close to depicting phyletic gradualism. The theory of evolution by jerks is being largely created by juggling with definitions, and by using a diagram whose vertical axis, time, is clear enough, but whose horizontal axes conflate, in a most confusing way, phenotypic change and geographical separation. That they are further incapable of representing the process of substitution of one genotype (or phenotype) for another in a population seems innocent enough (how could one represent that kind of complexity?), but further compounds the ambiguity about whether the new branch and the old are in the same geographical locality or not. These diagrams, and the theory they represent, undo eighty years of progress in population genetics and take us back to the position that Bateson was in in 1902, unable to distinguish clearly not merely species from varieties, but the origin of a new type by mutation from its substitution in a population, or to the equivalent confusion of his Darwinist opponents, who had not separated clearly in their minds, mutation from selection, nor selection from isolation.[7]

No two taxonomists have ever agreed on what a "species" is, and it does not seem likely that we can find any single process by which species are formed. What we call "speciation" is likely to be the combined result of a large number of processes operating at different speeds, relative to one another, in different cases, allowing us, for example, to find "species" that differ morphologically but are not reproductively isolated, and "species" that have no detectable morphological distinction but that will not interbreed under any circumstances. To look for an event called "speciation" is like looking for an event called "evolution of vision." If evolution occurs according to neo-Darwinian principles within phyletic lines, but at a variable speed, and has superimposed on it a process of branching generated as a complex outcome of those same processes of change (Figure 1f) then the pattern observed in morphological diversity and in the fossil record would be one of punctuated equilibria, differing in no way from the one predicted by the simplistic, two-tiered systems of Goldschmidt, Gould, and Stanley.

4

The neo-saltational theory might seem to predict nothing in particular about mimicry. But it does make predictions about diversity, and it is to the diversity of patterns in mimetic butterflies that experimental geneticists have now turned their attention. This allows us to test the theory against real data (Turner, 1981).

The South American genus *Heliconius* consists of approximately fifty species (including some in related genera), exhibiting relatively few color patterns. The rife Muellerian mimicry within the group made them a collector's paradise and a taxonomist's purgatory.[8] A table showing color patterns on one axis and taxonomic groupings on the other (Brown, 1981) reveals that much of the mimicry cannot derive, as one is at first inclined to assume, simply from common

---

[7] The extent of the regression toward the infancy of our science is seen when Gould (1980) discovers the following new principle: "Adaptation may determine whether or not a hopeful monster survives, but primary constraint upon its genesis and direction resides with inherited ontogeny, not with selective modelling." This is no more than the mutationist view that mutation is neither adaptive nor a molder of adaptations, and that evolution is restrained within the bounds of gene action, long ago (we thought) absorbed into the modern synthesis (cf. quotations by Punnett and Haldane, above).

[8] The evidence that the butterflies are distasteful, and successfully protected from birds by their mutual mimicry, is reviewed elsewhere (e.g., Brown, 1981; Turner, 1981).

ancestry, but must involve a considerable amount of convergence as well, for most cells of the table are filled. Within taxonomic groups, the patterns are diversifying, at the same time as they converge between the taxonomic groups, producing a paradigm case of the familiar cycle of adaptive radiation and convergence.

What the theory of evolution by jerks predicts in a case like this is that the diversity, or most of it, has been produced in punctuational events during which morphological change and speciation went hand in hand, and not, as neo-Darwinian theory would have it, by adaptation within phyletic lines. Now it may indeed be true that some species have arisen with their own new patterns, but as a general prediction this is easily refuted. The diversification of pattern commences *within* the species, producing, in most of the more widespread of them, races that can differ as much in pattern as do different species. It might be argued that these races are, in fact, species. That they are indeed races within one species, as that is usually understood, is shown by the normal criteria: allopatric replacement with natural hybridization at the zones of contact, and a high level of interfertility in the laboratory. It might, on the other hand, be thought that the pattern differences between species are of a different kind from those between races. Again, this is not so: Appropriate combinations of the genes found to produce pattern differences between *races* are found to produce patterns very close to those of related, although superficially very dissimilar, species.

The prediction that follows from the main distinguishing feature of the new theory, phenotypic diversity should arise mainly at speciation, is refuted. But things are worse than that. Much of the interracial diversity arises, as one might expect from what has been said about the evolution of mimicry, from genes of rather large effect on the phenotype. The most plausible reconstruction of the time scale involved suggests that the substitution of one pattern for another would have occupied rather a limited time span in the whole history of the race, or that if we could find fossil butterfly patterns we would discover long periods of stasis separated by comparatively rapid punctuational events. The butterfly palaeontologist would, willy-nilly, have to call the different phenotypes "species," and would claim a spectacular confirmation of the theory of evolution by jerks.

It is not necessarily the case that all punctuational events involve single gene substitutions; many probably involve a number of genes. Elsewhere I have suggested that what generates the alternation of stasis and change is the alternate saturation of the ecological niche space, which largely inhibits evolution (a return to Charles Lyell's old antievolutionary idea of "occupancy"), and the emptying of individual niches by extinction that induces a rapid response in the species which are left (a return to Darwin's emphasis, in Chapter 4 of the *Origin*, on new places in the "polity" of nature) (Turner, 1982). It is probable that the mimetic patterns of *Heliconius* are responding to such extinction cycles at one remove, by tending to mimic whatever pattern has been rendered most abundant and, hence, best protected, by extinctions of competitors, parasites, parasites of competitors, etc., both at the same and at other trophic levels in the ecosystem. The stasis of the patterns of *Heliconius* present no theoretical problem – as is to be expected with warning color, minority deviants are removed by predators.

5

In summary, there is little here for schism. The modern synthesis incorporated many of the ideas of the mutationists and, as we have seen explicitly in the case of mimicry, admitted the role of mutations of major effect. To set up neo-Darwinism as a simple continuation of nineteenth-century phyletic gradualism, sustained by the intellectual milieu of liberal capitalism allied to gradualistic socialism, is naive. On the point at which it can claim to be truly different from neo-Darwinism, the existence of a separate nonadaptive process that generates both species (branches) and diversity, the theory of evolution by jerks suffers from the old confusion between species and variety, and when put to the mimicry test, is refuted.

Some neo-Darwinists, it is true, have been overly impressed by Fisher's argument that large mutations would be maladaptive and have neglected the contemporaneous and rather more mutationist and multipeak approach of Haldane and Wright. Some of the major events in evolution could well have involved rather large phenotypic shifts produced by one, or a few mutations, acting as Goldschmidt proposed at an early stage in ontogeny, in much the same way that mimicry is initiated by a major mutation. For them to survive when still something of a mechanical, energetic, or physiological disaster would require only that they produced an adaptation to a relatively empty niche in which there was, initially, no competition. (Mimicry operates in a kind of inversion of this principle – the mimetic pattern "niche" must be fully occupied by distasteful butterflies.) The biggest mistake that any punctuationist could make would be to assume, as Goldschmidt did, that this "hopeful monster," if that is what one wants to call it, would be perfect to the point of undergoing no further modification. Goldschmidt did his better ideas a great disservice by attaching them so firmly to this perverse view.

The new saltationism fails in much the same way that the old saltationism failed, and however inevitable was the schism between that theory and Darwinism in the first decades of the century, reopening the schism now is a waste of energy. The theory of evolution by jerks may not quite fit the description of "bringing no new insights, and citing on its behalf not a single unambiguous fact" (a bon mot applied originally to another controversial modern theory), but it comes preciously close to being simply a restatement of the modern synthesis in different terms.

Why then has a theory in which what is right is not new, and what is more or less new is not right (Charlesworth, Larder, and Slatkin, 1982) attracted, as mutationism did at the turn of the century, the claim (e.g., Lewin, 1980) that it is overturning or radically altering current evolutionary theory? "The rise of creationism" would be a popular answer, "Marxism" an unpopular one (see Gould, 1979; Halstead, 1980). But it is something at once nearer to a real scientific question than a belief in God, and nearer to real human concerns than questions of collectivist economics, that is really at stake. There is also one very important scientific point concealed within the theory.

How the gorilla lost his hair – a Just So Story

1

Ism-schisms may be unprofitable, but they are seldom stale or flat. Scientists, let us be frank, find straw men useful for structuring their own arguments and developing their own careers. But the various needs of scientists do not explain why the public has taken such an interest in evolution by jerks. That comes from a permanent need the nonbiologist has for an alternative to Darwinism, with its uncomfortable implications about our ancestry and the process that brought us to our present pass. That Lyell's discovery of the old philosophical problem of pain in the form of the struggle for existence generated much angst long before the *Origin* will be understood by anyone who has read *In Memoriam*. "Seeing that of fifty seeds, she [Nature] often brings but one to bear," and seeing further that this was not merely a woeful imperfection in the order of things, but the central creative force of evolution, may have contributed even to the malaise of Darwin's middle age (Fleming, 1961). Shaw rejected natural selection as damnable for its reduction of order and beauty to nothing but the artistry of a railway accident, and Wallace was concerned enough to devote many pages of *Darwinism* to a demonstration, backed by the first-person testimony of no less than Dr. Livingstone, that to be eaten by a lion, or a similar fate, was not as painful as one might think. The eugenics movement might seek to strive upward, working out the beast by improving on the works of God and the negligence of man, but to most people the rise of mutationism offered the exciting prospect of laying the ghost of nature red in tooth and claw, and particularly of exorcising the thought that the claws might be ours.

For the other horror on which Darwin riveted our gaze so firmly, "not of woods only and the shade of trees," has, like pain, a long pedigree (Midgley, 1979). From his incarnations as the born devil on whose gross nature nurture would never stick, and as Swift's Yahoo, he stalked through the pages of *Punch,* in the years following the *Origin,* as a gorilla, now hideous ("*Am* I a man and a brother?"), now suave in white tie and tails, scaring the wits out of a footman ("Oh, Mr. G-g-g-o-o-o-rilla"). Was that *really* us in the mirror? Our options for believing that it wasn't became sharply limited by the development of sociobiology. Originally intended to explain the more human aspects of brute behavior, it was inevitable that it should be used to explain, even excuse or justify, the beastlier parts of our own. To some that idea, as Lady Bracknell might have said, reminded them of the worst excesses of the Third Reich. Sociobiology was located in the line of intellectual descent leading from the eugenists to Hitler (Allen *et al.,* 1975; papers in Rose, 1982).

The concern of radical scientists and creationists alike was summed up by a British creationist: "If you tell people they are animals they will behave like animals." The need if not to bury Darwin, then at least to re-inter him in a more suitable place, became urgent once again. Robbed of its technical language, the view of scientific radicals and social anthropologists is that no "important" human behavior is influenced by genes – that is to say that no behavioral differences between individuals arise from genetic differences – and that however helpful sociobiological theory is in explaining animal behavior, it says nothing whatsoever of interest about human societies, whose structure is *sui generis* and created only by social and economic forces.[9] Both the human psyche and the human genome are, in some sense, blank – the one a *tabula rasa*, the other in possession of an array of genes that have decoupled brain function from the effects of other genes.

Thus it is not a coincidence, and not entirely due to Wright's unassuming manner, to find of the three theories offering serious challenges to the ruling Darwinian adaptationism of their time – Bateson's, Wright's, and Gould's – that it is the first and last, offering us a clean break from a cruel process and an animal past, where Wright

---

[9] "Some few behaviours, such as suckling by infants, are highly canalized, but those that *really matter* to us are not" (Futuyma, 1979) (Italics added).

offered us difficult mathematics and free will, that have evoked the popular conclusion "Darwinism is dead."

2

There are six ways out of the dilemma that evolutionary and genetic considerations might apply to the sensitive areas of our own being.

1. *Creationism:* Simply to deny evolutionary theory is a solution that has always been used, and which is gaining in popularity. Its adherents are generally seen to be, and probably are, toward the opposite end of the political spectrum from adherents of the other solutions.

2. *Lamarckism/vitalism:* Accept evolution but deny Darwinism. Optimistic philosophers (Samuel Butler, Henri Bergson), Fabian socialists (George Bernard Shaw), and supporters of the green revolution in biological thinking, who want to place something nonmaterial and nondeterminist between genotype and phenotype (contributors to Rose, 1982), have espoused this approach. As Gould (1980) puts it, we can get the organism back into biology; and the organism, with some of its own autonomy, is no longer a slave of its genetic history, nor entirely prey to the cruel, external force of selection (although lions eat missionaries, nonetheless).

3. *Lysenkoism:* Accept Darwinism but deny the validity of Mendelian genetics. This philosophy has, obviously, a lot in common with the previous one, and in its extreme Soviet form is barely acceptable to any scientist not under ideological pressure.

4. *Anti-Jensenism:* The softer alternative to (3) is to deny that Mendelian genetics is applicable to humans, or at least to their behavior. The first of these denials is, in effect, the scientific policy of the Soviet Union, where even medical genetics barely exists as a subject; the second, particularly in respect to IQ, has been a very popular approach in the West, where fear of "eugenic" laws has led to the espousal of a doctrinal belief in rigid environmental determinism as a way of getting us as far as possible from any suggestion of "taints of blood."[10]

5. *Spiritualism (Wallace's line)/theistic creationism:* Acceptance of Darwinism and of Mendelian genetics, coupled with a denial of their applicability to humans, can be uncomfortable intellectually; why

[10] This view can become in its extreme form the conviction, very firmly held, not that the heritability of IQ is zero, but that IQ cannot have a heritability any more than a sulphur atom can have a color or a minor third a taste. Contrast "The project of innateness will not allow us to go . . . towards denying that intelligence or any subcomponent is the sort of thing for which it makes any sense to seek direct, quantifiable genetic transmission" (Barker, 1982).

should we be different from other animals? A popular approach with "rational" faiths such as spiritualism, and with many adherents of liberal theology, has been the belief that somewhere in our evolution something uniquely human, perhaps nonmaterial, was given us. Alfred Russel Wallace believed this. "There is no reason to believe that the process of 'spiritual' creation would alter *Homo sapiens* physically or physiologically, although obviously Christians believe that it makes humans distinct from the rest of the animals and gives [us] a special relationship to God" (Berry, 1982).

6. *Prometheism:* As a materialist alternative to divine breath, accept Darwinism in its general outline and Mendelian genetics as scientifically sound, but deny their applicability to humans because of a unique evolutionary event in our history. This has been called the "hypothesis of the Promethean genes," the genes that freed us from our genes (Lumsden and Wilson, 1981). The more orthodox approach is to argue for the existence of Promethean genes from existing neo-Darwinian theory (Futuyma, 1979).[11] The more exciting approach is to attempt to ditch neo-Darwinism altogether, and to set up an alternative, genetically based Darwinism: the theory,of evolution by jerks, or perhaps "Gouldschmidtism."

3

The problem for scientists is that sociobiology, call it what one likes (and who doesn't?), is a strong discipline. If it successfully explains much of animal behavior in terms of neo-Darwinism, it is hard work indeed to convince oneself, not that humans have extra behaviors which do not fit well with the model, but that they have no behavior whatever which *is* explained by it. But that is what is demanded by the Promethean program. Ho and Saunders (in Rose, 1982) point out that the major problem for many critics of sociobiology has been their implicit acceptance of neo-Darwinism.

The solution is straightforward. Sociobiology depends heavily on the view that the phenotype is adaptive, including its behavioral aspects, and that we have some kind of continuity with our evolutionary past. Enter the theory of evolution by jerks, to pull the neo-Darwinian rug out from under sociobiology: Evolution is not continuous, and much of it is not adaptive. Hence, we find Gould (1978) writing that "In Darwin's day, an assertion of our similarity [to animals] broke through centuries of harmful superstition. Now we may need to em-

---

[11] The scenario is, in brief, that the need to respond to stimuli differently according to a very complex context, selected for the removal of any inbuilt, instinctive responses, left the human brain, apart from behaviors like infant suckling that do not really matter, a *tabula rasa* entirely open to learning.

phasize our difference as flexible animals with a vast range of potential behavior," and Edmund Leach (1981), a convinced believer in environmental determinism in human behavior, saying that had Darwin lived today he would certainly have preferred the neocatastrophic theories of Gould to the neo-Darwinism of E. O. Wilson.

That Gouldschmidtism gives us no detailed account of the neocatastrophic Promethean event is a strength rather than a weakness. Attempts to create a Promethean scenario (Futuyma, 1979) out of neo-Darwinism raise awkward questions: If we were selected for highly flexible behavior that ultimately decoupled our brains from any genetic influence, and if that means selecting for intelligence, then there must have been genes governing intelligence; so how did the heritability plummet to zero once we came unto man's estate? (By homozygosity of all the genes, obviously, but then what about back-mutation in a polygenic system?) With Gouldschmidtian Prometheism the vagueness of the evolutionary mechanism leaves us free to imagine, or not to imagine, what we will. Like the catastrophic theory of Cuvier (we await a modern Byron to incorporate it into literature) it reconciles for us two uncomfortably jarring ideas – in Cuvier's theory, biblical authority and the evidence of the rocks; in the present case, a man with both genes and a *tabula rasa* brain – without our having to consider too deeply how they might fit together.

4

The English school of population genetics has tended to follow the early Darwinians in viewing all aspects of organisms as fundamentally adaptive. But while the view that selection is omnipotent and can mold organisms in any way, without genetic restraint and with equal attention to all parts at all times, is clearly naive, it was not a good tactic for the proponents of the theory of evolution by jerks to lampoon adaptationism as a series of *Just So Stories* (Gould and Lewontin, 1979). As the epigraph shows, Kipling's originals all concern not gradual evolution, but a series of punctuational events in which kangaroos gain long legs; camels, humps; and wives a proper submissiveness to their lords, in a period of less than twenty-four hours. The theory of evolution by jerks is, in its own right, a Just So Story, one of a long line designed to answer Disraeli's question "Is man an ape or an angel?"; this time in a way that will keep the scientific evolutionist "on the side of the angels." Yet the odd thing about Disraeli's question is its exclusiveness. The previous century knew very well that the answer was "yes"; he was "placed on this isthmus of a middle state, a being darkly wise and rudely great."

And evolution is not gradual *or* saltational, either.

A modest proposal

1

If there really are close analogies between the present schism and the Mendelian–Darwinist controversy then the lesson of history is that the punctuationists are saying something of value which neo-Darwinists find hard to perceive, just as punctuationists do not see that their rejection of many central neo-Darwinist principles throws out the baby with the bathwater. It was not, in the old controversy, that a choice had to be made between new mutation plus discontinuity, on the one hand, and preexisting plus continuous individual variation, on the other; but that a way had to be found of including all *four* categories (for mutation was not always large, neither was continuous variation a substance independent of mutation) into a coherent theory. It is even now a matter for discussion among population geneticists just how much of evolution depends on variation that is being fed in by mutation, or on variation that is held in the population by selection, and what balance is struck between Haldane's rather large individual mutations and Fisher's numerous small ones. What then are the mutual blind spots in the current debate?

The aspect of neo-Darwinism mistakenly rejected by the punctuationists is amazingly close to the one mistakenly rejected by the original mutationists – it is *fluctuation*. For the mutationists, this meant individual variation, which they thought was not inherited and which played no role in evolution. For the punctuationists, the fluctuations, which also play no role in evolution, are the adjustments that natural selection makes to an evolving lineage, adapting it to its environment (Gould 1979, 1980 – the actual term "fluctuation" is used) and they are held to play no part in evolution for precisely the same reason: They are not heritable – that is, in some way they are not handed on to the daughter species when the lineage splits. Why they are not handed on is unclear, but the idea seems to be connected with their smallness, just as the small individual variations of the Darwinists were once thought to be of no significance. The folly of this new rejection, as of the old one, is obvious. What is to stop phyletic changes from being large? (A point raised by all critics, e.g., Levinton and Simon, 1980; Charlesworth, Lande, and Slatkin, 1982.) Certainly those that occur will not be reversed as a matter of course at speciation, and, hence, they must play a role in evolution.

What constitutes the other half of the *folie à deux* is a perception held by some, but not all, neo-Darwinists that species are some kind of superorganism, which are themselves adapted. The writings of Dobzhansky, for example, give a strong sense of an integrated gene pool, protected from foreign genes by isolating mechanisms, as an organism is protected from viral DNA by its immune defenses. The punctuationists are right in insisting that speciation is nonadaptive. This is easily understood by considering two of the extreme cases: the establishment of a major chromosomal rearrangement by random drift, and isolation by a geographical barrier, which, at least temporarily, makes a population genetically independent of the rest of the species just as effectively as any other hindrance to cross-fertilizing. Neither the possession of the translocation, nor the colonization of a remote island, can be seen as "an adaptation" in any sensible meaning of that term, nor are they the result of natural selection.

The reason that this point is not seen by either side in its correct perspective is that there is a deep confusion over the meaning of "speciation" or "cladogenesis" or "splitting," analogous to the old and forgotten confusion over "natural selection." The three terms, particularly when used in discussions of phylogeny, conflate the ideas of the separating evolutionary lines developing reproductive isolation *and* of their becoming phenotypically different. The confusion of these two effects, which have no *necessary* connection and which, as every biology student knows, are often separate, allows one side to feel that "speciation is adaptive" and the other, equally incorrectly, that "cladogenesis is nonadaptive."

2

The beliefs that adaptation does, or does not, account for organic diversity are mistaken in exactly the same way that the beliefs that natural selection did, or did not, account for evolution were mistaken in the period around 1900. The diverse mimetic patterns of *Heliconius* butterflies are manifestly the outcome of selection; or, if one likes to use the result as a metaphor for the cause, they are "adaptations." But the diversity of pattern cannot be explained solely in these terms, for there had to be at least two nonadaptive processes as well: mutation, which produced the altered wing patterns, and speciation, or cladogenesis (and I here include simple geographical separation), which split the group into genetically independent lines. In polymorphic Batesian mimics we can discern a further set of mechanisms, not adaptive in themselves, that allow some of the diversity to exist within a single population (it was another neo-Darwinian error to believe that polymorph*ism*, rather than just the *polymorphic forms* themselves, was an adaptation). And remembering the principle adumbrated by Nicholson (1927) that an individual adaptation is not necessarily "good for the species," the fate of the evolving, diverse *Heliconius* species may not, in any simple way, be determined by their mimetic patterns. As lineages go extinct *as lineages,* taking all their adaptations, superb as well as merely so-so, with them, any long-term trends in the group, such as the one for the black, red, and yellow, long-winged species to outnumber the orange, short-winged species, cannot necessarily be said to be adaptive trends. A mimetic pattern enables an individual, not a whole species, to survive.

To sum up, to think that diversity is *purely* an outcome of selection is as foolish as to say that it is not *in any way* a product of adaptation. Evolutionary change and diversity are indeed generated by natural selection, together with a series of nonadaptive processes. These we can now recognize are (1) mutation; (2) random stochastic drift; (3) speciation or cladogenesis; and (4) a set of mechanisms, discovered by Fisher and Haldane, involving mutation, selection, and the mating system, for which we have no collective name, that maintain hetero-zygosity or stable gene ratios in populations. We should probably add here the extinction of phyletic lines, which, although punctua-tionists call it "species selection," cannot be called "selection" any more than can genetic drift (both extinction and drift involve differ-ential survival: that is *not* the criterion for selection!). Darwinians have an inborn reluctance to recognize these processes, either by mini-mizing them – in the case of mutations by maintaining that only small ones were important, in the case of drift by maintaining that it was always overridden by selection – or by trying to convince everyone, in the case of speciation and the maintenance of heterozygosity, that the processes were, *in themselves,* adaptive. Extinction is attributed, by extreme adaptationists and punctuationists alike, to "group se-lection." The drawing of these nonadaptive processes into the Dar-winist philosophy was what the modern synthesis was all about – the recognition of cladogenesis as one of these is the latest phase in generating what may one day be seen as modern synthesis II. For those who like more or less simple diagrams, I have tried to portray this theory as an evolutionary tree (Figure 1f); it differs from con-ventional, punctuational trees by separating cladogenesis and phe-notypic change on separate axes, a small but absolutely crucial dis-tinction.

3

As regards the program which runs parallel with punctuationism, that of "getting the organism back into biology," which embraces structuralist theories from embryology to cladistics,[12] I do not think

---

[12] The suggestion by Halstead (1980) that cladistics was Marxist is a bit of a "red hennig," but did not deserve the ingenuous (and disingenuous) ridicule it re-ceived. Cladistics is undoubtedly structuralist.

that anyone who thinks about mimicry for five minutes would disagree. Who but God, knowing only the laws of quantum physics, or even the properties of DNA and mitochondria, would *predict* the creation of what can only be described as works of art like the mimetic beetles figured in the frontispiece of the *Genetical Theory of Natural Selection*? Between the DNA and the mimicry there is a vast network of feedback loops that involve not just the chemistry of pigment synthesis in model and mimic and in the retina of the predator, but the structure of the nervous system of the one and of the populations of the others. But to hope that that consideration will in itself get back for us something which, taken from its political packaging, is nothing less than free will, is as vain as Shaw's intuitive feeling that if only he could banish Darwin the lion would lie down with the missionary.

References

Allen, E., B. Beckwith, J. Beckwith et al. 1975. Against "Sociobiology." *New York Review of Books* Nov. 13; reprinted 1978 in *The Sociobiology Debate*, ed. A. L. Caplan. New York: Harper & Row.

Barker, M. 1982. Biology and Ideology: the Uses of Reductionism. In *Against Biological Determinism*, ed. S. Rose. London and New York: Allison & Busby.

Bateson, W. 1902. *Mendel's Principles of Heredity: a Defence*. Cambridge; Cambridge Univ. Press.

  1909. *Mendel's Principles of Heredity*. Cambridge, Cambridge Univ. Press.

Berry, R. J. 1982. Darwin cleared – official. *Biologist* 29:100–4.

Brown, K. S., Jr. 1981. The biology of *Heliconius* and related genera. *Ann. Rev. Entom.* 26:427–56.

Cain, A. J., and P. M. Sheppard. 1954. The theory of adaptive polymorphism. *Amer. Natur.* 88:321–6.

Charlesworth, B., R. Lande, and M. Slatkin. 1982. A neo-Darwinian commentary on macroevolution. *Evolution* 36:474–98.

Charlesworth, D., and B. Charlesworth. 1976. Theoretical genetics of Batesian mimicry. II. Evolution of supergenes. *J. Theor. Biol.* 55:305–24.

Clarke, C. A., and P. M. Sheppard. 1963. Interactions between major genes and polygenes in the determination of the mimetic patterns of *Papilio dardanus*. *Evolution* 17:404–13.

Eldredge, N., and S. J. Gould. 1972. Punctuated Equilibria: an Alternative to Phyletic Gradualism. In *Models in Paleobiology*, ed. J. M. Schopf, pp. 82–115. San Francisco: Freeman, Cooper.

Fisher, R. A. 1927. On some objections to mimicry theory; statistical and genetic. *Trans. Roy. Ent. Soc.* 75:269–78.

  1930. *The Genetical Theory of Natural Selection*. Oxford: Clarendon Press.

Fleming, D. 1961. Charles Darwin, the anaesthetic man. *Victorian Studies* 4:219–36.

Ford, E. B. 1931. *Mendelism and Evolution*. London: Methuen.

1964. *Ecological Genetics*. London: Methuen; New York: Wiley.

Futuyma, D. 1979. *Evolutionary Biology*. Sunderland, Mass.: Sinauer.

Goldschmidt, R. B. 1940. *The Material Basis of Evolution*. New Haven: Yale Univ. Press.

  1945. Mimetic polymorphism, a controversial chapter of Darwinism. *Q. Rev. Biol.* 20:147–64, 205–30.

Gould, S. J. 1978. *Ever since Darwin*. New York: Norton.

  1979. The episodic nature of change versus the dogma of gradualism. *Sci. and Nat.* 2:5–12.

  1980. Is a new and general theory of evolution emerging? *Paleobiology* 6:119–30.

Gould, S. J., and N. Eldredge. 1977. Punctuated equilibria: the tempo and mode of evolution reconsidered. *Paleobiology* 3:115–51.

Gould, S. J., and R. C. Lewontin. 1979. The spandrels of San Marco and the Panglossian paradigm: a critique of the adaptationist programme. *Proc. Roy. Soc. Lond. B* 205:581–98.

Haldane, J. B. S. 1926. A mathematical theory of natural and artificial selection. Part III. *Proc. Cambridge Phil. Soc.* 23:363–72.

  1932. *The Causes of Evolution*. London, New York, Toronto: Longmans, Green.

Halstead, B. 1980. Museum of errors. *Nature* 288:208.

Hodge, M. J. S. 1983. In *Evolution from Molecules to Men*, ed. D. S. Bendall, pp. 43–62. Cambridge: Cambridge Univ. Press.

Leach, E. R. 1981. Men, bishops and apes. *Nature* 293:19–21.

Levinton, J. S., and C. M. Simon. 1980. A critique of the punctuated equilibria model and implications for the detection of speciation in the fossil record. *Syst. Zool.* 29:130–42.

Lewin, R. 1980. Evolutionary theory under fire. *Science* 210:883–7.

Lumsden, C. J., and E. O. Wilson. 1981. *Genes, Mind, and Culture. The Coevolutionary Process*. Cambridge, Mass.: Harvard Univ. Press.

Midgley, M. 1979. *Beast and Man. The Roots of Human Nature*. Hassocks, Sussex: Harvester Press.

Nicholson, A. J. 1927. A new theory of mimicry in insects. *Austral. Zool.* 5:10–104.

Poulton, E. B. 1908. *Essays on evolution 1889–1907*. Oxford: Clarendon Press.

  1912. Darwin and Bergson on the interpretation of evolution. *Bedrock* 1(1):48–65.

Provine, W. B. 1971. *The Origins of Theoretical Population Genetics*. Chicago and London: Univ. of Chicago Press.

Punnett, R. C. 1912, 1919, 1927. *Mendelism*, 4th, 5th, and 7th eds. London: Macmillan.

  1915. *Mimicry in Butterflies*. Cambridge: Cambridge Univ. Press.

Robson, G. C., and O. W. Richards. 1936. *The Variation of Animals in Nature*. London, New York, Toronto: Longmans, Green.

Rose, S. (ed.) 1982. *Towards a Liberatory Biology*. London and New York: Allison & Busby.

Sheppard, P. M. 1961. Recent genetical work on polymorphic mimetic Papilios. In *Insect Polymorphism*, ed. J. S. Kennedy, pp. 20–9. London: Royal Entomological Society.

Shull, A. F. 1936. *Evolution*. London and New York: McGraw-Hill.

Stanley, S. M. 1979. *Macroevolution. Pattern and Process*. San Francisco: Freeman.

Turner, J. R. G. 1977. Butterfly Mimicry: the Genetical Evolution of an Adaptation. In *Evolutionary Biology* 10:163–206, ed. M. K. Hecht, W. C. Steere, and B. Wallace. New York: Plenum.

  1981. Adaptation and evolution in *Heliconius*: a defense of neoDarwinism. *Ann. Rev. Ecol. System.* 12:99–121.

  1982. How Do Refuges Produce Biological Diversity? Allopatry and Parapatry, Extinction and Gene Flow in Mimetic Butterflies. In *Biological Diversification in the Tropics*, ed. G. T. France, pp. 309–35. New York: Columbia Univ. Press.

  In press. Mimicry: The Palatability Spectrum and Its Consequences. In *Butterfly Biology*, ed. R. I. Vane-Wright and P. R. Ackery. New York: Academic Press.

Wallace, A. R. 1889. *Darwinism. An Exposition of the Theory of Natural Selection with Some of Its Applications*, 2nd ed. London: Macmillan.

PART III. THE GERMAN PALEONTOLOGICAL AND MORPHOLOGICAL TRADITION

7

Evolutionary theory in German paleontology

WOLF-ERNST REIF

Introduction

Only after eighty to ninety years of discussion between Darwinians and anti-Darwinians was it shown by the founding fathers of the synthetic theory (see Mayr and Provine, 1980) that Darwin's view of evolution was the uniting basis for a broad range of fields of biology. However, it took several more decades before the neo-Darwinian, synthetic view found wider acceptance in German-speaking countries.[1] In fact, as will be shown later in this essay, anti-Darwinian theories still have a significant number of adherents. The different historical events in Germany and in the English-speaking countries require explanations which go much beyond the usual "paradigm shift." Rather, there were influences of tradition and culture which have had this long-lasting effect.

Furthermore, there are not only language barriers separating specialists of different nations, but there are also professional barriers that largely separate specialists of even adjacent fields of science. Thus, German paleontology developed, to a certain degree, independently of German zoology as well as of the paleontology of other countries. My goal here is to explain the triumph of the anti-Darwinian, antiuniformitarian, orthogenetic, and saltationist theories that dominated a large part of this century. However, in order to find the philosophical roots of this phenomenon, one has to go back to the nineteenth century, as I have done elsewhere,[2] grouping important authors on evolution into six schools of thought (traditionalists,

---

[1] Because language barriers are often a much higher barrier for communication in science than political boundaries, I will include all three German-speaking countries Austria, Switzerland, and Germany.

[2] I have discussed this in "The search for a macroevolutionary theory in German paleontology," manuscript submitted to *J. Hist. Biol.*

early Darwinians, pluralists, neo-Lamarckians, orthogeneticists, and typostrophists).[3]

In the present context, I shall begin with the last of these.

Typostrophist school: general remarks

The typostrophist epoch in German paleontology began with Beurlen (1932); the term "typostrophism," however, was introduced only in 1945 by Schindewolf. Assimilating orthogenesis (evolution regarded as an orderly, autonomous process of unfolding), typostrophism added several major aspects to the orthogenetic theory. Most important are saltationism (macroevolutionary novelties do not evolve gradually, but are produced in jumps) and cyclism (evolutionary history has proceeded in cycles that are analogous to the life cycle of individuals). Most typostrophists were enthusiastic "idealistic morphologists."

It was very important for "idealistic morphologists," in general, and for typostrophists, in particular, that they thought they could trace their concepts and approaches back to the authorities of the late eighteenth and early nineteenth centuries (Cuvier, Lamarck, Goethe, Schelling, Oken, and other German Romantics and *Naturphilosophen*). Most important among these concepts are cyclism and the morphological "type." According to idealistic morphologists, morphological "types" can be understood only intuitively. They cannot be defined. These concepts necessarily brought a mystical, speculative (and all too often a strongly antiuniformitarian) touch into evolutionary theory, and it was an obvious strategy to defend this mysticism by referring back to the Romantics and their contemporaries. It is impossible to go into details, but one can say bluntly that modern accounts by philosophers and some biologists (e.g., Weber, 1958) have shown that none of the above authors (with the partial exception of Oken) can be used as a key witness to mysticism and vague metaphysical speculations.

Two philosophical extremes are often referred to in biology. One extreme is identified with experimental approach, analytic approach, objectivity, materialism, mechanism, causality, and rationalism; the other extreme is identified with descriptive approach, intuition, subjectivity, holism, vitalism, mysticism, art, and poetry. There is no

---

[3] The names per se of the schools should not be taken too seriously. I will try to show that there were indeed strong interconnections within each school. This does not necessarily mean that the members of a school discussed problems together and agreed with each other; on the contrary, it could even mean that an author found his strongest opponents among the members of his own school.

doubt that these alternatives are extreme simplifications. Each of the terms would require a special definition and none should be used as a catchword. This is, however, exactly what was done over and over again: Catchwords were used with a positive connotation to criticize the opponent's view. For the sake of brevity the first extreme will be called the "objectivity paradigm" and the second the "subjectivity paradigm." The polarity between these paradigms was strongly felt in the time of declining *Naturphilosophie*, in the 1840s. This can be shown in the critical review by the physiologist Carl Ludwig of Rudolf Leuckart's *Über die Morphologie und Verwandtschaftsverhältnisse der wirbellosen Tiere* (1848; summarized by Kühn, 1948). Ludwig criticized Leuckart precisely for mysticism, that is, for adhering to the subjectivity paradigm.

Twenty-four years earlier, in 1824, the physiologist Johannes Müller gave his inaugural lecture in Bonn: *Von dem Bedürfnis der Physiologie nach einer philosophischen Naturbetrachtung* (reprinted with commentary by von Uexküll, 1947). Müller develops a concept of biology that is not split into the two paradigms but integrates the experimental–analytic and the holistic approach. Nevertheless, authors in the second half of the nineteenth century tried to trace the polarity between the objectivity and the subjectivity paradigm back to the Romantic period and beyond. Very influential was Haeckel (1866), who described four periods in the history of biology: (1) first empirical epoch: Linnaeus; (2) first philosophical epoch (*Naturphilosophie*): "fantastic-philosophical morphology," Lamarck, Goethe; (3) second empirical epoch: Cuvier; and (4) second philosophical period: Darwin.

The phylogenic morphology of Haeckel and his followers dominated zoology in the 1870s and 1880s. By the end of the century there was the well-known shift in biology against Darwinism and in favor of neo-Lamarckism.[4] Together with this decline, Haeckel's phylogenetic morphology was regarded as being too straightforward, too naive, too speculative. Since the rise of physiology, morphology was regarded as a field which could not boast of logical, methodological, and experimental rigor. However, it took another two or three decades before the attempt was made to formulate a new methodology for morphology.

The situation after the turn of the century is clearly reflected in Emanuel Radl's book: *Geschichte der biologischen Theorien der Neuzeit* (1905, 1909; 2nd ed., 1913, 1918). Radl was a biologist, not a historian, and he presents a rather biased view. He is strongly against Darwinism and against Haeckel and has a strong inclination toward the subjectivity paradigm. He tacitly and sometimes explicitly regards the

---

[4] Neo-Lamarckism in paleontology began later than in zoology.

subjectivity paradigm as a typical German tradition. Radl adopts Braun's (1862) term "idealistic morphology" for the period of the late eighteenth/early nineteenth century. Radl used the term in a very positive way and made it well known. He sees a strong "contradiction between theory and practical results" (Radl, 1909, p. 45) in Goethe's methodology. Goethe's results are regarded by Radl as being rather simple speculations on the relations between forms (the vertebrate theory of the skull; the discovery of the human intermaxillary; the theory of the archtypical plant, *Urpflanze*; the concept of "metamorphosis"). Goethe's theoretical ideal, on the other hand, was, according to Radl, to learn about the forces that form organisms. Radl explains this contradiction by the antagonism between the subjectivity and objectivity paradigm. In other words, Goethe should have been much more subjective! (It should be repeated, that such an interpretation is not justified by modern philosophical studies!) Radl lays the foundation for a new understanding of morphological methodology. He claims that "Darwinian morphology" (i.e., evolutionary morphology) can never be a historical science (Radl, 1909, p. 328), because, in contrast to human history, evolutionary morphology never has historical data at its disposal. (This, according to Radl, was Haeckel's basic error.) Rather, morphology is restricted to the search for homologies and analogies by the method of comparison. Radl (p. 28) regrets that the vitalistic part of Goethe's theory of metamorphosis was later forgotten and that idealistic morphology was supplanted by evolutionary morphology. In Radl's mind (p. 50) it was unjustified that the terms of idealistic morphology (homology, analogy, metamorphosis, differentiation, progress, architecture, natural classification, work specialization) were reinterpreted in an evolutionary sense.

A new foundation for the methodology of idealistic morphology was finally laid by the Swiss invertebrate zoologist Adolf Naef in 1919. His book, *Idealistische Morphologie und Phylogenetik* (*Zur Methode der systematischen Morphologie*), was very influential for several decades. Though Naef does not quote Radl in his book, it is obvious that he must have known him. Naef's first argument reminds one directly of Radl: There is no acceptable methodology for a phylogenetic systematics (or evolutionary morphology, for that matter). This methodology could only lead to the "naive phylogenetics" of Haeckel, which is plagued by too many uncontrollable assumptions. Instead, one has to start with pure morphology, which is not encumbered by presuppositions. The only pure morphology that is available is idealistic morphology. Its success is shown by the fact that it had erected a natural classification of organisms long before an evolutionary theory was proposed, let alone accepted. The new methodology of ideal-istic morphology can be based directly on Goethe's theoretical writ-ings. The basic methodological tool of Goethe was the term "type." The basis of Naef's new idealistic morphology is the concept of "typ-ical similarities." One finds, according to Naef, "typical similarities" by a strictly geometric analysis of organic shape. The term "typical similarity" is unclear, even mystical, in Naef's writings because it is never clearly defined (actually it is undefinable), but is only circum-scribed in sentences like: "Typical similarity exists between complex natural objects, if we can imagine that they arose by a stepwise mod-ification from a common archtypical shape [the type], in other words, if they can be derived from the type." "Typical similarity appears at first as the similarity between complex totalities, which consist of similar parts in a corresponding arrangement." "Typical similarity is consequently an ideal relationship between shapes, which expresses itself in the homology of the parts." The *tertium comparationis* in the comparison of individual shapes is the type; it has the property of an objective reality. As Goethe has already formulated the principle, individual shapes behave similarly to the type, as do individual cases in relation to a law. Organisms are grouped in Naef's methodology according to their typical similarities in a "natural classification." If one wants to learn about phyletic relations between groups of or-ganisms, one must, in a second step, *interpret* this "natural classifi-cation."

Naef was an excellent comparative morphologist and probably had no vitalistic, mystical inclinations. It has to be acknowledged very positively that he tried to give morphology a new foundation. But Naef unfortunately had no understanding of functional morphology, of the distinction between primitive and advanced characters, and of the now undoubted fact that overall morphological similarity is often only a poor indicator of phylogenetic relationships. The zoologists Hennig (1950, 1966) and Weber (1958) refuted Naef's methodology from a biological point of view.

A historical–philosophical analysis shows that Goethe was by no means the idealistic mystic or speculative metaphysician he was often considered to be later by biologists. The question whether he was, as a biologist, an adherent of the subjectivity paradigm or the objec-tivity paradigm simply makes no sense. Goethe was very instrumen-tal in overcoming the static analysis of nature and made a significant contribution to the dynamic world view of the Romantic period. With his concept of morphology (he coined the term in 1795) Goethe showed that a rational methodology existed by which one could com-pare organisms and understand their shapes in a dynamic framework.

At the time when Goethe introduced the concept of morphology (1784–95) it would have been regarded as purely speculative if he had claimed that the similarities between organisms could be explained by common descent in a process of evolution that occurred over hundreds of millions of years. At that time, one had just begun to discover that the earth was older than a few thousand years (Schmidt, 1918). Hence, it was a rational and very fruitful undertaking to search for *Baupläne* (types) and to show how given forms could be transformed into one another by a process which Goethe called "metamorphosis." Additionally, Goethe cannot be used, as was done by Naef and others, as a key witness for pure, geometry-oriented morphology, which detests functional analysis. One has to face the fact that Goethe rejected mystical teleology in biology (Zimmermann, 1953) but at the same time had a good understanding of functional morphology (Goethe, 1830).

One of Goethe's statements on morphology is reminiscent of the concept of "theoretical morphology," which was introduced and fruitfully applied by Raup and Michelson (1965). Goethe wrote in a letter: "The archtypical plant will become the most wonderful creature in the world, for which even Nature must envy me. With this model and the key to it [i.e., the rules of metamorphosis] one can then invent additional plants *ad infinitum*; of necessity they will be consistent with the type, that is even if they do not exist, they could exist and they are not painterly or poetic shades and illusions, but they have an inner reality and necessity. It will be possible to apply the same law to all other organisms" (Goethe, 1787, from Naef, 1919, p. 9). From this discussion it follows clearly that the type concept of the twentieth-century idealistic morphologists cannot be defended by referring it back to authorities of the Romantic period. It would require too much space to document the impact that idealistic morphology had on German zoology and botany. It suffices to refer to the works of the botanist Troll and the zoologist Remane (e.g., 1955), who strictly followed Naef's ideas. They and their schools influenced their whole fields for several decades before and after World War II. It is characteristic that Troll defined morphology as *Urbildforschung* (research on archetypes).

The second important concept of typostrophism, namely saltationism, has no roots in either Goethe or the Romantic period. Just the opposite is true. Goethe argued against Linnaeus's species as absolutely discrete entities and against the concept of Cuvier that there were four discrete *Baupläne* in the animal kingdom that could not be transformed into one another. Saltationism was not forced upon the typostrophists either by mutation theory or by the genetics of the 1920s and 1930s. It will be shown later that typostrophists themselves decided for macromutations (which create evolutionary jumps) *after* they had accepted saltations. Rather, saltations were the necessary outcome of a dogmatic application of the type concept, in the sense of Naef. If types were objective realities, as was believed by the ty-postrophists, then the borders between types could be crossed only by jumps. In order for the species b to evolve from species a an evolutionary jump was necessary, because all members of species a belong to type a and all members of species b belong to type b; *tertium non datur*. The same is true for the evolution of genus B from genus A, of family 2 from family 1, and all the way up to the evolution of phylum II from phylum I.

We now come to the third issue of typostrophism, namely, cyclism. The model that all historical processes are controlled by laws which are the goal of any historical investigation cannot be traced here. Popper (1960) calls this concept "historicism." Historicism has played an important role in evolutionary paleontology, especially in typostrophism. Numerous texts can be found with predictions of future evolutionary success (needless to say, these predictions are untestable) or with prediction-like arguments that attempt to explain past evolutionary processes as a necessary outcome of certain laws and certain boundary conditions. One historicist theory is cyclism, that is, the notion that the life cycle of individuals is a model for patterns and order in nature and for events in the history of societies, cultures, and races. Popper (1960) traced this idea back to Plato. Haeckel's (1866) "epacme, acme, paracme" concept was probably the first cyclistic model in evolutionary theory. Cyclism in human history was made famous by Oswald Spengler (1880–1936) in his book *Der Untergang des Abendlandes. Umrisse einer Morphologie der Weltgeschichte* (1919–1921). The first sentence of the book clearly expresses Spengler's historicist approach: "In this book, the attempt will be made for the first time to determine history in advance" (Spengler, 1919–1921, p. 3). The basic tenet is that the number of historical phenomena is limited and that they are grouped into "types." Spengler purposely chose Goethe's term "morphology" for the subtitle; he talks of a *Formensprache der Geschichte* (form language of history, p. 6) and says: "I owe the philosophy of this book to the philosophy of Goethe" (Spengler, 1919–1921, p. 69). It need not concern us here whether or not he does justice to Goethe with this statement. There is no doubt, however, that Spengler owes much to other cyclistic historians and philosophers of history, including Goethe's contemporary, Herder. The merits of cyclistic historical philosophy cannot be discussed here; suffice it to say that at one time or another it was very instrumental in pushing back the dominating theories of progressivism and teleology in history. Spengler's book went through numerous editions and was widely read and discussed among biologists in the 1920s (Beurlen, oral comm.) and it clearly influenced the idealistic morphologists and the future typostrophists.

Beurlen and Schindewolf

For reasons of space, such predecessors of typostrophism as Stille, Salfeld, Dacqué, and Kühn must be passed over here.[5]

The founder of typostrophism (i.e., the combination of orthogenesis, cyclism, and saltationism) is Karl Beurlen.[6] Between 1926 and 1940 he was one of the most prolific authors on topics of theoretical paleontology, including evolutionary theory. Only a few of his papers and his book of 1937 will be quoted here. Many of Beurlen's conclusions were based on his own data from the evolution of Crustacea and Ammonoidea. Beurlen is primarily a systematist, but also a stratigrapher and a paleobiologist. He is interested in adaptations, but these pose no major problem because they can be accounted for by a regular, lawlike interaction between environmental factors and inner causes (orthogenesis). In his paper of 1926, he continues with the statement that orthogenesis can explain lawlike patterns in evolution, like (1) the explosive radiation at the evolutionary beginning of a taxon; (2) the fact that stem groups have a low, and derived groups have a high evolutionary rate; and (3) the fact that groups evolve from the adaptive realm to the inadaptive realm by size increase and by developing hypertrophied organs.

The type concept again necessitates (as in Dacqué's and Kühn's models) the acceptance of saltations. The first taxa in a new type are unspecialized; they can react to environmental changes by an inner flexibility. This leads to explosive radiation. In later phases of evolution, flexibility is reduced; consequently the type (i.e., the higher taxon) has to retain its original direction. Inner causes predetermine the future evolutionary fate, which eventually leads to overspecialization. The type finds itself in a cul-de-sac and will eventually be doomed to extinction. A truly remarkable sentence is found in Beurlen (1932). He states "that it is a very general rule, that the pathway of evolution within a taxon – irrespective of whether it is a unit of higher or lower rank – is cyclic; evolution starts with a first phase with rich,

---

[5] A number of these figures deserve fuller treatment than I have been able to give them here. See reference list, this chapter.

[6] Born 1894; invertebrate paleontologist; student and assistant professor under Hennig in Tübingen; professor of geology and paleontology in Kiel, Königsberg, and Munich; after 1950 in Brazil.

saltationistic and explosive creation of forms, it goes on over a phase of orthogenetic continuation (in which evolution is directional and purposive and does not produce new form types) to a final phase of degeneration and disintegration of forms and thus of extinction." This sentence, published in the very important journal *Die Naturwissenschaften*, is the beginning of typostrophism. Beurlen acknowledges in this context that the vertebrate paleontologists Hummel (1929) and von Huene (1931) had found the same kind of cyclism independently of him. The major issue for discussion that remains for Beurlen – after the laws of evolution have been explained by referring them back to the laws of the individual organism (birth-differentiation-maturity-old age-death) – is the origin of new types. New types evolve by introducing drastic sudden changes into the developmental program of early ontogenetic stages, which are more plastic than later ontogenetic stages. The stimulus for the change comes from the environment. The answer of the organism is a new developmental program; the answer is directionless and is independent of the specific cause. It is not a direct adaptation to environmental stimuli, but is determined only by the norm of reaction of the organism. This norm of reaction is not explained rationally by Beurlen, but is only approached by vitalistic expressions (*schöpferische Wirklichkeit, Leben*, creative reality, life, 1936, p. 456). Adaptation is not the driving force and goal of evolution but rather a means of progressive evolution. Progressive evolution allows the organism to exercise increasing control over its environment (1936, p. 457).

In his 1937 book Beurlen summarized his numerous papers on evolutionary theory. He calls the origin of new types "neomorphosis," but does not offer new mechanisms to account for this phenomenon. Ultimately, the driving force of all evolution is a vitalistic force for which Beurlen uses the terms *Wille zum Dasein* (will to existence) and *Wille zur Macht* (will to power; but this term, well-known from the writings of Nietzsche and Spengler is used here in a completely different context). In a more recent publication (1978) Beurlen repeats his evolutionary theory, though in a weaker form, but fully accepts Schindewolf's concepts of typostrophism, typogenesis, typostasis, and typolysis.[7]

Beurlen is very much interested in philosophy and in many of his papers of the 1920s and 1930s he comments on his views of cultural and intellectual traditions. Already in 1926 he regards the tools of the

---

[7] It is important to note, especially with respect to Beurlen, that typostrophism is *not* a fascist theory. I hope I have made clear that typostrophism, as such, is only the consequence of idealistic morphology as reintroduced by Naef (1919).

objectivity paradigm, which provide an "exact, mechanistic explanation of the world" (p. 247), as being intellectually insufficient. His
philosophical ideas are summarized in his book of 1937. He simplifies
history of philosophy to such an extent that he sees only two world
views: the mechanistic (objectivity paradigm, in my terminology) and
the nonmechanistic. He finds the former especially among philosophers of the Renaissance and the French Enlightenment, but also in
the works of Linnaeus, Lamarck, Darwin, and other French and English authors (Darwin is characterized as mechanistic, empiricist, and
liberal). He regards "objective thinking" as scientifically unproduc-
tive, because this approach is not interested in "*dem Wesen* (the essence) of reality" (1937, p. 255). The nonmechanistic world view is
largely a German tradition which can be traced back to medieval
times; important authors are Paracelsus, Kepler, and finally (of
course) Goethe, the founder of idealistic morphology. An important
foreign author is Buffon, an opponent of Linnaeus. Later biologists
of that tradition are von Baer, Driesch, Spemann, and Troll. Beurlen
calls the antithesis to "objective thinking" *sachliches Denken* (subject-
matter-related thinking) and claims that *sachliches Denken* is deeply
rooted in the "world view of the Germanic man" (1937, p. 254). His
ultimate simplification at the end of the book is that the opposition
between "objective thinking" and *sachliches Denken* is the expression
of the difference of two "strange and opposed types of races, the
Judaic Oriental and the Germanic Nordic" (p. 256). These conclusions
are undoubtedly the expression of political opportunism; they play,
however, no role in Beurlen's evolutionary theory. Yet the conclusions show how far the antagonism between the objectivity paradigm
and the subjectivity paradigm (which had originated in the second
half of the 1800s and which produced vitalism, antimaterialism, mysticism, and also, to a certain degree the new idealistic morphology)
was carried in biological *Weltanschauung* at that time.

O. H. Schindewolf (1896–1971) not only gave typostrophism its
name, but was its most important proponent.[8] Schindewolf's main
field was stratigraphy and the evolution of ammonoids and corals
(with important contributions on their morphogeny), but he also
worked on regional geology, paleogeography, theoretical principles
of stratigraphy and taxonomy, and on evolutionary theory. Of his
many publications on theoretical biology only the most important will
be discussed here. In his first general paper (on the principles of
systematics, 1928) Schindewolf defends Naef's principles of classifi-

---

[8] Schindewolf, a student of Stille in Göttingen, received his Ph.D. under Wedekind in Marburg, was associate professor in Marburg, director at the Geological Survey in Berlin, professor in Berlin and in Tübingen.

cation, based on "typical similarities," which are nowhere defined. Like Naef, he regards a phylogenetic systematics as an impossible task for logical reasons. Systematics is a pure ordering science (p. 126). "Another methodology than that which is built up from the principles of idealistic systematics is impossible" (Schindewolf, 1928, p. 145). Schindewolf emphasizes the common occurrence of parallel evolution and, at the same time, he is ready to neglect it by defining taxa that arose by parallel evolution rather than by splitting up of lineages (polyphyletic taxa in contrast to monophyletic taxa).

This polyphyly is not only a methodological issue, but also plays an important role in Schindewolf's later model of evolution. In his paper "Ontogeny and Phylogeny" (1929) Schindewolf redefines Haeckel's palingenesis, that is, the mechanism that leads to a recapitulation of phylogeny in ontogeny. He adds new aspects derived from paleontological data and regards Haeckel's caenogenesis as a special case of palingenesis. The main purpose of the paper is to give a theoretical account of "proterogenesis" (a term which he had already introduced in 1925). Proterogenesis is the opposite of palingenesis in that new characters are introduced at the beginning of ontogeny. These characters anticipate the mature characters of descendants, because they are gradually shifted from early ontogeny to late ontogeny (or they spread from early ontogeny to ontogeny as a whole). For Schindewolf these new characters are clearly *prophetic characters*. The purpose of the proterogenesis concept is obviously to reduce the importance of "Haeckel's law" as a tool in phylogenetic reconstruction. Additionally (this is clearly expressed by Schindewolf), some apparent saltations in the fossil record can be reinterpreted as gradual series in the light of this concept. Schindewolf's first book (1936) had the ambitious aim of arriving at a synthetic theory of evolution by harmonizing the data of paleontology and genetics.[9] The idea itself, however, was not quite original, as is shown by the book of the geneticist and developmental biologist Dürken and the paleontologist Salfeld (1921) and by the 1929 Tübingen meeting on genetics and paleontology (see Federley, 1929; Weidenreich, 1929). Yet genetics plays only a minor role in Schindewolf's book. He quotes only four German and four American geneticists; his main authority is the macromutationist R. Goldschmidt. It has often been claimed that Schindewolf devel-

---

[9] "Schindewolf, in a small book published in 1936, was the first person to attempt a genuine general synthesis of evolutionary genetics and paleontology. For its time, that was an original and brilliant achievement . . . His initial work in this field was one of the stimuli for the modern synthesis, to which he is himself so bitterly opposed" (G. G. Simpson, *Q. Rev. Biol.* 27:388–9, 1952; quoted from Schindewolf, 1969).

oped a saltationistic theory *because* he happened to know and to read
Goldschmidt (a famous author at that time) and was strongly influenced by him. This, however, is not true. Schindewolf did *not* discover Goldschmidt for paleontology. Goldschmidt was already
quoted by Schindewolf's professor, Rudolf Wedekind (1916) and by
Beurlen (1932). Additionally, as was shown earlier, saltationism is the
almost inevitable result of an application of Naef's concept of types
and typical similarity to evolutionary questions. Hence, it was only
logical that Beurlen (1932) and Schindewolf should use Goldschmidt's
macromutations to defend saltationism. The small role that genetics
plays in Schindewolf's book is not surprising, because Schindewolf
states clearly in the introduction (p. iv) that only paleontology can,
with certain corrections, "read the course of the real evolutionary
process and derive the ruling laws." Schindewolf takes a clear stance
against all mysticism and vitalism (that is why he rejects certain
models of Dacqué and Beurlen). However, this position forces him,
at the same time, into a very schematical description of the phylogenetic process and of evolution as a whole – in which the question
of the origin of adaptations is never seriously asked. Schindewolf's
concepts of idealistic morphology are, in fact, fully based on Naef.
One should stress that Schindewolf had no understanding of gene
pools and populations. Each species, each genus, and each other
higher taxon represents a type of its own. Ontogeny is programmed
in such a way that the most general characters (i.e., the characters of
the phylum and class to which the specimen belongs) develop first.
Then the characters that characterize the order develop, then those
of the family, the genus, and the species. A transition from species
a to the descendant species b is brought about by a parallel deviation
in numerous specimens. They develop class-to-genus characters in a
regular way and then develop the new characters of species b instead
of the characters of species a. The same is true for all transitions. If
a family boundary is crossed, phylum-to-order characters develop
normally and then new family, genus, and species characters develop. It is remarkable that the boundary of higher taxa is crossed
not only by individuals of a single species, but in parallel. In other
words, several species convergently cross the boundary to a new
genus, several genera to a new family, several families to a new order,
and so forth. The reason for this phenomenon is that it is the *type* of
the ancestral taxon that produces the descendant taxon. "At the root
of a species there is hence not a concrete individual with fully developed organization, but a general archetype which is deprived of
all individual characters, namely the type of the ancestral species"

(Schindewolf, 1936, p. 16). This model of the jumps between taxa fully documents Schindewolf's saltationism, because it is obvious that there can be no gradual transition ("The first bird hatched from a reptile's egg" p. 59). (On p. 22 he claims, however, that sometimes some characters of the old taxon are retained in the new one. This slightly weakens the model and leaves space for *Übergangsformen*, "transitional forms.") Schindewolf's model of deviations in ontogeny, which produce the evolutionary jumps, is a generalization of his proterogenesis concept. He calls this generalization: *Gesetz der frühontogenetischen Typenentstehung* ("law of origination of new types early in ontogeny"). This law states that early ontogenetic changes have a strong influence on future ontogenies, that is, on future phylogeny; changes that take place late in ontogeny, on the other hand, have only a weak effect. This leads in consequence (p. 63) to a dichotomy in phylogeny: A short phase of an "explosive, sudden origin of a type" is followed by a long phase of a "quiet, orthogenetically progressing organization" of the new type. Before the extinction of the type one finds anomalies, degenerations, and hypertrophies. In other cases, a new jump can lead to a new type (p. 64). It is obvious that Schindewolf is talking of cyclism here, but he does not formulate it as clearly or as strongly as Beurlen (1932) had done. In a subsequent discussion, Schindewolf rejects both Lamarckism and Darwinism, because they cannot explain the phase of explosive origin of the new type. Macromutations dominate the first, explosive phase; micromutations, the second, "orthogenetic" phase. Selection plays a role only in the orthogenetic phase, as one directing cause. Orthogenesis is not explained. The new type that is produced in the explosive phase migrates into that habitat for which it is preadapted. Evolution, as a whole, is not completely autonomous and independent of environmental factors. These induce mutability and, hence, cause periods in which numerous new types originate.[10]

In 1942, Schindewolf emphasizes cyclism and introduces three phases: explosive origin, orthogenetic phase, and decline of a taxon, which strictly parallel the life cycle of an individual. (He does not refer to Beurlen here.) The orthogenetic phase is characterized by an increasing "specialization." It is predetermined and independent of the environment. It cannot be explained by selection. Phylogenetic size increase is a typical phenomenon of the orthogenetic phase. To account for orthogenetic trends Schindewolf finally (p. 382) reintroduces Abel's law of inertia, in a different wording but without referring

---

[10] The war prevented the English version of Schindewolf's book from being printed (Schindewolf, 1940).

to Abel.[11] Schindewolf now began to take a strongly cyclistic, anti-adaptationist, antiselectionist stance.[12]

The origin of Schindewolf's terminology has an interesting history. Heberer (1943), a gradualist who will be discussed later, introduced the term "additive typogenesis" (gradual origin of types) when he proposed a gradualistic, nonidealistic model for the origin of higher taxa. Schindewolf (in papers given in November 1943 at the Hungarian Geological Society, the Hungarian National Museum, and the Hungarian Biological Research Institute, which were published in 1945), used Heberer's term "typogenesis" for his explosive phase and introduced "typostasis" for the orthogenetic phase and "typolysis" for the degenerative phase. One cycle is called a "typostrophe."[13]

After the 1942 paper, Schindewolf published three more papers on evolutionary theory before he completed (in 1944) the manuscript of his 500 page book *Grundfragen der Paläontologie*. This was to appear only in 1950 and it includes, in 340 pages, the final version of the typostrophic theory. Schindewolf refers to numerous examples from the fossil record to demonstrate "explosive" typogenesis, linear orthogenesis during typostasis, and hypertrophies and degenerations during typolysis. The neck of the giraffe and the head of the musk are used as Recent examples of hypertrophies that are doomed to extinction in the near future. The model of typostrophes is refined by introducing typostrophes of different orders of magnitude. For example, the typostrophic cycle of a class can include typostrophic cycles of families belonging to this class. These smaller cycles take place during the typostasis of the class. Typostasis does not produce anything new. The orthogenetic pathway is determined by whatever evolutionary direction prevailed at the beginning of typostasis. Evolutionary progress is made only during typogenesis and the beginning of typostasis. "These processes occur independently of the mode of life of their carriers and without direct cooperation of selection (which is responsible for the origin of races). Only adaptational characters within *Baupläne* are selected and never the *Baupläne* themselves" (Schindewolf, 1950, p. 397). Selection, thus, operates only at a very low level of taxonomic diversification ("microevolution"); it is irrelevant with respect to the origin of higher taxa, genera, and so forth, and it has nothing to do with anagenetic processes. Indepen-

---

[11] It is quite obvious that between 1936 and 1942 Schindewolf's theory had hardened.

[12] Schindewolf published at least eleven papers of theoretical interest or on theory between 1936 and 1942.

[13] The terms "macroevolution" and "microevolution" were traced back by Schindewolf (1945) to Philiptschenko (1927).

dently of this low-level selection, typostasis leads to increasing adaptation and specialization. Schindewolf states clearly that at the beginning of typostasis organisms are not completely unspecialized and unadapted. However, it shows his lack of interest in adaptation when he uses the two terms "adaptation" and "specialization" as synonyms. His biological information was insufficient. Otherwise he would not have claimed that certain organs, like the placenta, could have evolved only in a jump. "A placenta cannot at the same time be missing and be present" (p. 201). There is no doubt that the placenta of mammals evolved gradually; the different kinds of placentas in pelagic sharks give a model of such a gradual evolution.

Between 1950 and 1970 Schindewolf wrote twenty-two papers dealing with evolutionary theory. Several of these publications were printed papers that he had given at meetings, others were written for a nonspecialist readership. Four of the papers (1956, 1963, 1964, 1969) are of great historical interest. In 1956 Schindewolf showed that evolution as a whole is a largely autonomous process. He rejected the assumption of numerous geologists that tectonic processes in the earth's crust (orogeneses, transgressions–regressions) controlled major events of evolution like the origin of new higher taxa, extinctions, and sudden changes in diversity. Seven years later (1963) Schindewolf extended his hypothesis (which has its roots in his 1950 book) that cosmic radiation (ionizing radiation) caused the phases of rapid extinction and species turnover at the end of the Paleozoic and of the Mesozoic. In another context he attributed the sudden increase in intensity of the cosmic radiation to the explosion of a supernova. However, the basic idea of this hypothesis was anticipated by Wilser (1931).

In one of his last papers (1969) Schindewolf defended the typological approach ("types" as basic units of investigation) as the most appropriate and fruitful approach to problems of morphology, systematics, and phylogenetics. He pays lip service to uniformitariansim in paleontology and to the basic importance of population genetics for evolutionary biology. The purpose of the paper is to answer numerous critics, and Schindewolf defends typostrophism in a slightly weakened form. Although typolysis has lost in importance, nonadaptive typogenesis and orthogenetic typostasis are still important, and selection has gained a little bit. However, Schindewolf expresses great doubts whether the synthetic theory in its current form is able to explain all the phenomena observed by paleontologists.

Schindewolf was always a very clear, pragmatic, and sober thinker, who attacked any kind of vitalism and who never could understand why he was considered a Platonist by other authors. He had extensive experience in the comparative morphology of invertebrates and in stratigraphy and, additionally, a broad knowledge of the systematics of vertebrates and plants and of geology. Schindewolf's theories can be fully explained from this experience and from his upbringing in the tradition of idealistic morphology (*sensu* Naef). His papers are free of any vitalism, mysticism, Romanticism, and *Weltanschauung*. He never had any sympathies with the Nazis.[14] Only in one paper (1964) given at a meeting of the Academy of Sciences and Literature, Mainz, did he summarize his philosophical ideas. He shows that there are close parallelisms between the history of the earth, the history of life, and human history. One of the major indications of this parallelism are cycles in earth history (Stille 1913, 1924, 1941, 1949), in the history of life (Schindewolf refers to his 1950 book), and in human history (Spengler, Toynbee). Schindewolf does not give an explanation of this parallelism, but he indicates that cycles may be expressions of all historical processes in general. Human history proceeds in cycles possibly because the development of the human mind is, like the bodily-organic development of mankind, a part of nature.

Schindewolf's theory was much more successful than Beurlen's. In fact, typostrophism is usually attributed to Schindewolf, especially to his 1950 book. In most of his publications Schindewolf refers extensively to the literature, including Beurlen's papers. However, it is obvious only from a historical perspective, and not directly from Schindewolf's writings, that Schindewolf's theory is a corrected version of Beurlen's theory. The major correction that Schindewolf made was to expel vitalism. This caused him to relegate adaptationism to a minor role and to stress the importance of internal (orthogenetic) clocklike mechanisms in his cycles. Schindewolf's evolutionary theory was *the* important evolutionary theory in German paleontology after the war. The reasons for this success (in comparison with Beurlen's theory) are manifold. Schindewolf's 1936 book is well presented, with numerous illustrations, whereas Beurlen's 1937 book has no illustrations and is at certain places overburdened with philosophy. Another reason may be Beurlen's vitalism, which was not accepted by numerous readers. Schindewolf wrote many papers after 1936 and during the war, whereas I found only two theoretical papers by Beurlen published after his book and before 1945. Beurlen failed to introduce a new terminology, whereas Schindewolf's terminology ("typostrophe," etc.) was rapidly picked up by his colleagues. Schindewolf's reputation as a paleontologist was already great before 1945. He was

---

[14] For political reasons, he did not get a full professorship until 1945. He reported to J. Kullmann and to J. Wiedmann (Tübingen) that he was politically attacked by K. Beurlen (J. Kullmann, oral. comm., January 25, 1982).

founder and editor of a new handbook of paleontology and of *Fortschritte der Paläontologie* and editor of *Paläontologisches Zentralblatt*. After 1945 he became the leading German paleontologist for two decades, whereas Beurlen emigrated to Brazil. The great success among German paleontologists of Schindewolf's ideas of evolution and of stratigraphy must, at least partly, be attributed to his superb personality, his warm-heartedness, his capabilities as an organizer, and his diplomatic talent for convincing other people.

The synthesis

We move now from paleontology to attempts to arrive at an evolutionary synthesis by harmonizing the data from genetics, systematics, and paleontology.

The study of the historical processes that led to the evolutionary synthesis has only just started in the book of Mayr and Provine (1980). As this is not an integrated text but a collection of articles, different readers will understand it differently. The most common interpretation of the book will be that it shows that the synthesis was brought about by the books of Dobzhansky (1937), Mayr (1942), Simpson (1944), and Stebbins (1950), in America and, to a certain degree independently by Huxley (1942) in England and also by Rensch (1947, written earlier) in complete isolation during the war in Germany (see, for example, Ruse, 1981). After the publication of Mayr and Provine's book, there are undoubtedly many historical questions still open. The three most important questions for this essay are:

1. Exactly when was a synthesis accomplished in Germany?
2. How independent were the processes in Germany during the 1930s and during the war from those in other countries?
3. What influence did the synthesis have on German paleontology after the war?

Four German efforts at synthesis are worth mentioning; apart from Rensch, they are those of Ehrenberg, Heberer, and Zimmermann. Ehrenberg (1943) edited a collection of three articles that had been given as papers at the Zoological-Botanical Society in Vienna in 1938. His attempt at a synthesis clearly failed. All three authors (including Ehrenberg) spend most of their time demonstrating that there was no scientific reason whatsoever to doubt that evolution had actually taken place. The botanist and geneticist Fritz von Wettstein introduced only the most basic principles of genetics and population genetics. Wilhelm von Marinelli, a zoologist and comparative anatomist, presented a moderately orthogenetic view of evolution. Ehrenberg himself, also a moderate orthogenesist, claimed that paleontologists rightly expect that genetics will eventually find "directed mutations." None of the three authors discussed the really pertinent problems of a synthesis.

One book that is usually overlooked in the context of the history of the synthesis is that of G. Heberer (1943): *Die Evolution der Organismen.* Heberer was a zoologist and later a paleoanthropologist. In the 1920s he had cooperated with Rensch on problems of biogeography. Heberer succeeded in bringing together a team of nineteen authors, none of whom were Lamarckians and all of whom were selectionists. The topics in the book range from a philosophical analysis of the theory of descent, through "biological proofs" of the theory of descent, methods of phylogenetics, ethology, phylogeny of plants, animals and man to population genetics, selection theory, and macroevolution. Although there is no space here to go into detail, it may be asserted that these authors really did succeed in achieving a synthesis, exactly the goal of Heberer. Heberer's own article "Das Typenproblem in der Stammesgeschichte" (the type problem in macroevolution) is of central interest for the present discussion. First of all, Heberer redefines the ambiguous term "type" in a nonidealistic way and equates it simply with "taxon." Boundaries between types are artificial (p. 548). One has to state very clearly that Heberer purposely used a terminology that was intelligible to his paleontological colleagues. He calls the concept of the typostrophists a "two-phase paleontology." The "explosive" phase is called by him "typogenesis," and the "orthogenetic" phase "adaptogenesis." ("Adaptogenesis" is the equivalent of Schindewolf's later "typostasis." The degenerative phase, "typolysis," is not taken seriously by Heberer and, hence, not discussed). Heberer shows clearly that there is no difference whatever between the origin of taxa (typogenesis) and their adaptation (adaptogenesis). Types (= taxa) do not evolve in jumps but rather gradually (Heberer calls this "the hypothesis of the additive origin of types"). Evolutionary phases of rapid diversification, like all other events in micro- and macroevolution, can be explained by the mechanisms of selection theory. They are not an artifact, but they do not require special evolutionary mechanisms as advocated by the typostrophists.

If one reads Heberer's 1943 book and refers to question 1, one must state that Heberer and his nineteen co-authors clearly succeeded in accomplishing an evolutionary synthesis that was supported by a wide spectrum of different biological disciplines. The second question can at least partly be solved by reading the reference lists of the individual papers. There is no doubt that among Heberer's authors themselves were the leading German Darwinians. All important foreign authors from genetics and paleontology are quoted. Dobzhansky's book of 1937 appeared in 1939 in a German edition and was widely read and quoted. Surprisingly, the literature lists do not end with the year 1939, but include American, English, Finnish, Italian, and Russian publications of the years 1940 and 1941. This means that, in contrast to my earlier expectations (Reif, 1980), German scientists received the foreign literature not only during the late 1930s but at least some of them even managed to get some foreign literature in the first years of the war.

Of Rensch's contribution to the synthesis, only his book of 1947 is usually quoted (especially after it was translated into English in 1960). It is forgotten that this book (written during the war) was preceded by three papers (Rensch, 1939; Rensch in Heberer, 1943; Rensch, 1943). The four publications form one coherent argument and have to be seen in context. The seed for the paradigm shift toward a Darwinian interpretation of macroevolution was laid by the 1939 paper, published in *Biological Review* and, hence, readily available to an international audience. It was quoted by Huxley (1942), many of the authors of Heberer (1943), but not by Simpson (1944), and not by any of the authors of Jepsen, Mayr, and Simpson (1949).

Starting with biogeography at low taxonomic levels, Rensch had studied geographic speciation, polytypic species, and also "climatic rules." These rules describe trends of body shape, body size, and coloration within one taxon from lower to higher geographical latitudes. The success of these rules made Rensch ready to accept the "evolutionary rules" ("rule" as a mild version of "law") of paleontologists. Moreover, Rensch is largely willing to accept the rules at face value. He does not ask the questions: How much of the rules is an artifact; How much of the rules is the result of the application of certain biased taxonomic procedures, caused by misinterpretation of the fossil record, or due to the fact that paleontologists are largely inexperienced in ecology and developmental biology of Recent organisms. Rensch's research project is best summarized by the last sentence of his 1943 paper: *"All the evolutionary rules discussed can be explained by directionless mutation and natural selection without the help of inner evolutionary forces of unfolding or trends of formation.* Only the change in environment is important for evolutionary novelties. Hence, evolution presents itself altogether, not as an autogenesis, but rather as an ectogenesis. Nevertheless, the possibility of autogenetic processes on principle is not to be denied. For a proof of their efficiency we would require a different set of facts than those which are provided by paleontology so far" (Rensch, 1943, p. 53). Like Heberer, Rensch is not a scientific revolutionist; he is not aware that the paradigm shift away from antiuniformitarianism, antiselectionism to selectionism requires drastic measures. Rather, he tries to correct the theory, including the language of Beurlen, Schindewolf, and others, as well as he can. As far as possible, Rensch tries to understand and accept the opposing theory in his own framework of thought and he goes as far as to state (1943, p. 49) that in his mind Schindewolf's (1942) paper "has further reduced the differences in the viewpoints of paleontologists and zoologists."

Thus, Rensch's strategy is to find selectionist explanations of Cope-and-Deperet's rule of size increase, of orthogenesis (in Rensch's definition the more or less rectilinear evolution of certain body shapes or organs), of increasing specialization that leads to extinction, of irreversibility, of parallel evolution, and of anagenesis. Despite his selectionist credo, Rensch's language sounds internalistic ("autogenetic") at times. Even a careful analysis leaves some doubts. Rensch is correct in trying to avoid "atomism," that is, the study of individual, artificially separated characters, and he attempts a holistic approach (no metaphysical connotation!). An important concept in his holistic biology is "compensation." Organs increase in size or decrease in size at the expense of or for the benefit of other organs. In frogs, increase in leg length is compensated for by reduction of the vertebral column; in snakes, increase in number of vertebrae is compensated for by the loss of the legs; in snails, increase in size of spines is compensated for by the reduction of spine number. This sounds at first glance convincing, but it is a rather geometry-oriented (not to say "typological") argument. It fails to give functional explanations for the trend. Another case where Rensch's argumentation sounds internalistic is that of extinction. Selection leads to increasing specialization (in one case, 1943, p. 21, Rensch says even "increasing adaptation"). The trend from eurytopy to stenotopy is concomitant with the loss of evolutionary potential. It produces overspecialized or excessive organs or taxa and finally causes extinction.

> *Senile variation, or degeneration,* as the case may be, at the end of some lineages, is in most cases brought about by orthogenesis (gigantic size, excessive forms, . . .) and the concomitant disintegration of normal correlations in the body. Too narrow specialization is in most cases the reason for an inevitable *extinction* of many phyletic lineages. This specialization itself, however, is inevitable, since it almost always provides selective advantages; the dangers of such a development are in the future and are thus independent of selection. [Rensch, 1943, p. 52]

Rensch thus tries to find selective explanations for all macroevolutionary rules and he does not accept any other positive mechanism for evolution than those provided by selection theory.

Independently of Rensch, the botanist W. Zimmermann published a book (1948) on the philosophical background and principles of a united theory of micro- and macroevolution. Like his colleagues, he comes to the conclusion that selection theory suffices to explain evolution. Schindewolf quotes the three books (Heberer, 1943; Rensch, 1947; and Zimmermann, 1948) in some of his later publications; Simpson (1953) includes them in his reference list. Nevertheless, the content of these books has never been evaluated sufficiently from the point of view of a unification of evolutionary theory.

Many authors (e.g., Heberer in Simpson, 1951; the German version of Simpson, 1944; Simpson in his introduction, 1951) give the impression that the synthesis was accomplished "in parallel" (but rather independently) in America, England, and Germany. Heberer says rightly in his introduction (Simpson, 1951, p. 4) that books like those of Mayr (1942) and Huxley (1942) remained unknown to German authors during the war. This is true, but if one takes into consideration that the manuscripts of Heberer and his co-authors were probably finished in 1942, the two books mentioned could hardly, in any case, have had any influence on Heberer (1943). Yet the isolation and independence of the architects of the synthesis who wrote their books during the war should not be overemphasized. The idea of a synthesis was already there before the war, and the international literature up until 1939 (empirical and theoretical papers, review articles) was a good basis for the synthesis. This certainly does not reduce the importance of the great efforts of individual authors who are now rightly known as the architects of the synthesis.

Postwar evolutionary theory in Germany

Schindewolf's version of the typostrophic theory dominated German paleontology in the following decades. Heberer (1943), Simpson (1944, translated by Heberer 1951), Rensch (1947), and Zimmermann (1948) had no perceptible influence, although these books were surely in the stacks of departmental and university libraries. Simpson's later work (1953) was never translated into German. No original publication was written that either criticized typostrophism or contrasted typostrophism and synthetic theory. The *Paläontologische Gesellschaft* was re-founded in 1948 and the first annual meeting was held in Munich in 1950. Of the several hundred papers presented at the meetings between 1950 and 1970, very few had any bearing on theoretical discussions. In the reports on the meetings, I found only fourteen titles of theoretical interest. To judge from the titles, almost all of them were within the framework of typostrophism; in some cases, the authors tried to make minor corrections in this theory by introducing new empirical data. Evolutionary theories themselves were never explicitly discussed. Only some of the papers mentioned were later published in *Paläontologische Zeitschrift*.

"I think it is correct to say that all younger evolutionary biologists [in Germany] have now adopted the selectionist interpretation, including, for example Schindewolf's students" (Rensch, in Mayr and Provine, 1980, p. 298). This sentence is largely correct, but it needs some qualifications. The fourteen or more former students of Schindewolf who are now associate or full professors (in Germany, Canada, and Switzerland) are mainly working on stratigraphy and facies geology. Very few devote part of their time to paleobiological problems. In the writings of most of them, it is very difficult to find clear selectionist or antiselectionist statements. The only author who publishes extensively on evolutionary theory is the former Schindewolf student (now professor of paleontology in Bonn) H. K. Erben. Of his work, his two books and one article will be discussed. They are written for a wider readership. Their titles clearly demonstrate Erben's research program: *Die Entwicklung der Lebewesen. Spielregeln der Evolution*, 1975 (The Evolution of Organisms. Rules of the Game of Evolution); "Uber das Aussterben in der Evolution," 1979 (On Extinction in Evolution), and *Leben heisst Sterben. Der Tod des Einzelnen und das Aussterben der Arten*, 1981 (To Live Means to Die. The Death of the Individual and the Extinction of the Species). Erben pleads for a weak typostrophism. He continues to see origin, longevity, and extinction of taxa as analogous to birth, life, and death of an individual. He accepts Haeckel's law of recapitulation in a weakened form, namely, as a rule rather than a law. Schindewolf's typogenesis is regarded as being almost identical with Simpson's quantum evolution. Erben attributes the scarcity of links between higher taxa in the fossil record in some cases to evolutionary jumps, in other cases to the imperfection of the fossil record. Although selection is widely discussed, it is not fully accepted and an internalistic, orthogenetic view prevails. Like Schindewolf, Erben confuses adaptation and specialization. Cope's rule (new higher taxa arise from unspecialized ancestors) is as strongly defended as Rosa's law (gradual reduction of evolutionary potential within lineages). "Overspecializations" and "anatomic decorrelations," which inevitably lead to extinction, are accepted as facts. Erben uses genetic and epigenetic mechanisms (pleiotropy and Waddington's canalization) in evolution. (Waddington's canalization is an ontogenetic and *not* an evolutionary mechanism.) It is obvious that this is only a different wording for an antiselectionist, internalistic hypothesis. What Erben really means is a weak form of typolysis, but he states that typolysis in its strong form – namely, as a regular pattern or rule, has no empirical basis and, hence, must be rejected. Erben quotes numerous examples of overspecializations from the vertebrates and rejects the attempts of functional morphologists to explain the adaptive significance of overspecialized organs.

The current situation is such that modern evolutionary theory is rarely taught in a geological-paleontological department in Germany, and few professional paleontologists show much interest in evolutionary theory. This has several reasons:

1. The lasting influence of Schindewolf and the reluctance to destroy his authority or the authority of others.
2. The strong influence of geology, which offers more job opportunities and guarantees more status. This leaves very few paleontologists working in paleobiology, let alone evolutionary theory.
3. The language barrier between German and English. This leads to specialization, because most German paleontologists read only English papers in their own special area and scarcely have time to read the vast literature on different areas of evolutionary biology that has been published in English.
4. A general lack of interest in theoretical discussion. (This is probably an important reason why plate tectonics took so long to be accepted.) This lack of interest is partly due to a prevailing empiricism.
5. A hostility toward theory. One reason for this is that rigorous theoretical studies, whether they are mathematized or not, are usually confused with mysticism and personal confessions of all-encompassing world views. This is partly the fault of the beholder but partly the fault of the author, who obscures the fact that he is transgressing the boundary between science and philosophy or metaphysics. These "world-view books," which are often the sum total of the experiences and views of very successful scientists, have had a long tradition but are also written today and include those of C. Bresch, *Zwischenstufe Leben. Evolution ohne Ziel* (1979); K. Lorenz, *Die Rückseite des Spiegels* (1973); and W. Heisenberg, *Der Teil und das Ganze* (1969).

The situation of evolutionary theory in German paleontology can also be illustrated by the available textbooks. The most widely used textbook on general paleontology was published by a former student of Schindewolf, B. Ziegler (1972). The chapter on evolution is 30 pages long in comparison to the 240 pages of the whole book. The account of selection theory is very short and partly misleading ("Selection can influence mortality, in rare cases also the number of progeny," p. 85). Most of the chapter is devoted to *Gesetzmässigkeiten* (i.e., lawlike patterns of evolution), such as evolutionary trends ("from unspecialized to highly specialized types"), irreversibility, Haeckel's law, typostrophism, and rates of evolution.

The paleontologist E. Flügel (Erlangen) contributed a chapter to R. Siewing's interdisciplinary, multiauthor book *Evolution* (1978). The topics in the book range from cosmic evolution, through chemical and biological evolution, to the evolution of cultures and of technology. Flügel does not discuss mechanisms of selection theory, but suggests instead that evolution as seen in the fossil record can be explained by the lawlike patterns we observe: orthogenesis, irreversibility, and typostrophism (in the version provided by Erben, 1975). Siewing's book appeared in a series of university textbooks.

As far as I know, evolution is not widely taught in faculties of biology. Courses of evolution are not required for a M.Sc. (Diploma) in Biology in Germany. In Tübingen, for example, evolutionary theory is taught only as a side aspect in courses on population genetics, taxonomy, and comparative morphology. During the last two semesters, professors of different areas of biology have presented the contributions of their own field to evolution in a *Ringvorlesung*, a series of public lectures for students of all faculties, held in the early evenings.

The most widely used textbook in Germany on general biology is the Springer-Verlag textbook edited by Czihak, Langer, and Ziegler (1978). The chapter on evolution is only 50 pages long, out of a total of 860 pages. The evolution of new *Baupläne* is explained by "additive typogenesis." In other words, Heberer's (1943) terminology, which was originally very important but is now completely outdated, is still used, despite its misleading connotations. Selection is dealt with mainly in a short basic account of quantitative population genetics; important concepts and current controversies in selection theory, such as selectionism–neutralism, kin selection–group selection, density dependent–density independent selection, r- and K-selection, and species selection, are missing.

Epilogue

One of the conclusions of this study is that gradualist, selectionist models for macroevolution were published in Germany before the classic book of typostrophism (Schindewolf, 1950) appeared. If one tries to explain the long-lasting success of typostrophism, one important aspect is that the four morphologists who contributed to the synthesis were outsiders in paleontology. Rensch is a zoologist and he never gave a talk at a meeting of the Paläontologische Gesellschaft (which united all German-speaking paleontologists) after the war. Heberer was a paleoanthropologist. He never delivered a paper either and he was probably never integrated into the paleontological community. Weigelt (in Heberer, 1943) was a professional paleontologist. He never worked on theory, but was strongly influenced by Rensch (1939) and by Heberer personally. His own field was fossilization and the origin of fossil deposits, where he made classic contributions, internationally known. It must be borne in mind that he died in 1948. Zimmermann was a botanist; he had no connections with the earth-science faculty and he was more interested in phylogenetic models than in patterns of macroevolution.

If one surveys the whole historical process since 1859,[15] one finds that, like anywhere else, those paleontologists who became enthusiastic Darwinians were very young. The enthusiasm for Darwinism faded very soon, partly because Würtemberger, Hilgendorf, and Kowalewsky never became established university paleontologists (Neumayr died young and did not complete his magnum opus); partly because apparently Darwinism (i.e., selection theory) did not hold what it had promised. It turned out to be impossible to apply Darwinism to the interpretation of the fossil record. There were only two solutions to the problem: Lamarckian mechanisms (*sensu lato*) and internalism (orthogenesis, etc.). It was well'known that Darwin himself had, in some cases, had recourse to Lamarckism. Large-scale patterns of evolution were not taken into consideration by Darwin, but they were anticipated by Haeckel (1866). Only at the turn of the century had enough been learned about the fossil record so that one really found phases of origin and rapid diversification of taxa, phases of long-term stability of taxa, and phases of decline of taxa (Jaekel, 1902). From Jaekel, through Wedekind to Beurlen and Schindewolf, this model hardened and was formulated as an extensive theory. Internalism started to dominate and there was soon no necessity for Lamarckian mechanisms. Nobody regretted that uniformitarianism was thrown overboard, because nobody knew what bearing observations made of the Recent biosphere could possibly have for the interpretation of the fossil record. Those few authors who took adaptationism seriously (Jaekel, Abel, Dacqué, and Beurlen) make com-

[15] See footnote 2.

plex efforts to explain adaptation without giving selection a major role in the drama. Selection simply could not explain the major events of macroevolution.

Naef's idealistic morphology with its seemingly clear concepts of types and typical similarity soon became an important rationalization of paleontology's morphological methods. What else was there to group fossils but typical similarity? Biological species concept, mating behavior, and so forth? Unobservable! Sexual dimorphism? Untestable! Morphogenesis? Not preserved in many groups; very seriously studied by Schindewolf in cephalopods. Variability? Fossils are mainly too rare to study variability. Additionally, nonheritable and heritable mutations can never be distinguished in fossils. Yet Dacqué and Wedekind took variability very seriously. Wedekind even tried to find Mendelian ratios and used them for a species discrimination among ammonoids.

Typological thinking as opposed to process-oriented thinking was not only very successful in grouping fossils, but also in geology, at least up until the advent of plate tectonics (in the 1960s). There was no mistake in grouping rocks, minerals, tectonic phenomena (faults, folds, geosynclines, etc.), and ore deposits according to their typical similarity. As in Naef's methodology, the explanation of the phenomena was made *after* they had been grouped. That this method caused some artificial questions was overlooked. Paleontology developed out of geology in the nineteenth century and has never become closely attached to biology. All paleontologists in Germany have been trained as geologists and stratigraphers. Real interdisciplinary work with biologists not only provided methodological problems, but also practical problems, because of the wide separation of the fields and because morphology was more and more desiccated after the 1920s, when the classical books on comparative anatomy of vertebrates, invertebrates, and plants had been written.

Evolution was never an important subject in German biology in the first five or six decades of this century. This is the reason why there were so few German evolutionary biologists of international reputation. It is very difficult to explain this trend. There is no indication that it had anything to do with religion, in a country where the influences of Catholics and Protestants approximately balance each other. Biologists usually had no problems accepting the theory of descent, but they hesitated to go along with Darwin's selection theory, which they regarded as too mechanistic and materialistic. The tradition of a misrepresented, misunderstood Romanticism led them to search for organismic laws rather than physicalistic, mechanistic laws. An explanatory model (selection) for the theory of descent that did not account for all aspects and complexities of life could not be satisfying.

Evolution was still less important in paleontology than in biology. Paleontology had many other practical and theoretical tasks than to provide a model for macroevolution or to consider whether selection theory could be a unifying concept for macro- and microevolution. Hence, a conspicuous scientific revolution, a paradigm shift from internalism to Darwinism (synthetic theory), has never taken place. Paleontologists were right to concentrate on solid, empirical problems of earth science, because whenever a trend spilled over from biology to paleontology it was not a challenge for a rational research project, but was a vague idea like neo-Lamarckism (organisms directing their own evolution by their instincts), the materialism–vitalism dispute or the Nazi idea that evolutionary theory was *undeutsch* and *westlerisch* (Zimmermann, 1948).

The reason for the rather conservative trends in German paleontology of this century are sociological structures. The language barrier is only one explanation. Between the two World Wars English started to become the international language of science. This language barrier acts in both directions. German articles are no longer read as extensively by foreigners, and it is difficult for German scientists to keep up with the huge English-language literature. However, hierarchical structures in Germany seem to be still more important than the language barrier. Those few paleontologists who knew anything about evolutionary theory were accepted as authorities. They, like other famous scientists, were regarded as *Gelehrte* ("educated men" is a very weak translation!). *Gelehrte* had played an enormous role in the 1800s, but also in this century. *Gelehrte* are men (never women?) who are accepted as absolute authorities, because of their immeasurable knowledge and their superior world view, judgment, and understanding. *Gelehrte* are able to draw the essentials from their own subject and go much beyond their own fields to develop a *Weltanschauung* which gives answers to general and philosophical questions (see also Hamburger in Mayr and Provine, 1980, p. 303). The hierarchical structure of German universities, in which one full professor dominates and in which dissemination of the ideas of the young faculty members is slowed down, was described by Mayr (in Mayr and Provine, 1980, p. 281). There is a high premium on good conduct because of the professor's tremendous influence on the future career of his students and young faculty members.

Only very recently is more attention being paid in German paleontology to modern problems of evolutionary theory. These trends are too recent to study now. Only ten years hence can the results be evaluated.

References

Beurlen, K. 1926. Paläontologische Beiträge zur Entwicklungslehre. *Die Erde* 4(5):241–66. Braunschweig: Verlag von Friedr. Vieweg und Sohn.
　　1930. Vergleichende Stammesgeschichte. Grundlage, Methoden, Probleme unter besonderer Berücksichtigung der höheren Krebse. *Fortschr. Geol. Pal.* 8(26):317–580. Soergel, W., ed.. Berlin: Borntraeger.
　　1932. Funktion und Form in der organischen Entwicklung. *Naturwissenschaften* 20:73–80.
　　1936. Das Gestaltproblem in der organischen Natur. *Z. Ges. Naturwiss.* 11:445–57.
　　1937. *Die stammesgeschichtlichen Grundlagen der Abstammungslehre.* Jena: Fischer.
　　1978. Die Entfaltungsgeschichte des Lebens. *Ber. Naturforsch. Ges. Bamberg.* 53:38–99.
Braun, A. 1862. *Über die Bedeutung der Morphologie: Rede zur Feier des 68. Stiftungstages des Friedrich-Wilhelm Instituts.* Berlin: Unger.
Bresch, C. 1979. *Zwischenstufe Leben. Evolution ohne Ziel?* Frankfurt: Fischer Taschenbuch, No. 6802.
Czihak, G., H. Langer and H. Ziegler, eds. 1978. *Biologie.* Berlin, Heidelberg, New York: Springer-Verlag.
Dacqué, E. 1921. *Vergleichende biologische Formenkunde der fossilen niederen Tiere.* Berlin: Borntraeger.
　　1935. *Organische Morphologie und Paläontologie.* Berlin: Borntraeger.
　　1951. *Die Urgestalt. Der Schöpfungsmythos neu erzählt. Neue erweiterte Auflage.* Wiesbaden: Insel.
Dobzhansky, T. 1937. *Genetics and the Origin of Species.* New York: Columbia Univ. Press. [German translation: *Die genetischen Grundlagen der Artbildung*, Jena: Fischer, 1939.]
Dürken, B. 1923. *Allgemeine Abstammungslehre.* Berlin: Borntraeger.
Dürken, B., and H. Salfeld. 1921. *Die Phylogenese.* Berlin: Borntraeger.
Ehrenberg, K., F. V. Wettstein, and W. V. Marinelli. 1943. Der heutige Wissensstand in Fragen der Abstammungslehre. *Palaeobiologica* 7:153–211.
Erben, H. K. 1975. *Die Entwicklung der Lebewesen. Spielregeln der Evolution.* Munich-Zürich: Piper.
　　1979. Über das Aussterben in der Evolution. *Mannheimer Forum 78/79*:23–122. Mannheim: Boehringer Mannheim.
　　1981. *Leben heisst Sterben. Der Tod des einzelnen und das Aussterben der Arten.* Hamburg: Hoffmann und Campe.
Federley, H. 1929. Weshalb lehnt die Genetik die Annahme einer Vererbung erworbener Eigenschaften ab? *Paläont. Z.* 11:287–310.
Goethe, J. W. von. 1830. *Principes de Philosophie Zoologique.* In J. W. von

Goethe, *Werke*, Hamburger Ausgabe, Vol. 13:219–50. Munich: V Beck, 1975.

Gross, W. 1943. Paläontologische Hypothesen zur Faktorenfrage der Deszendenzlehre. Über die Typen – und Phasenlehre von Schindewolf und Beurlen. *Die Naturwissenschaften 31*:237–45.

Haeckel, E. 1866. *Generelle Morphologie der Organismen*. 2 vols. Berlin: Reimer.

Hallam, A. 1973. *A Revolution in the Earth Sciences. From Continental Drift to Plate Tectonics*. Oxford: Clarendon Press.

Heberer, G., ed. 1943. *Die Evolution der Organismen. Ergebnisse und Probleme der Abstammungsgeschichte*. Jena: Fischer.

1959. Johannes Weigelt, 1890–1948. In *Die Evolution der Organismen*, ed. G. Heberer, 2nd ed. vol. 1, pp. 203–5. Stuttgart: Fischer.

Heisenberg, W. 1969. *Der Teil und das Ganze*. Munich: Piper.

Hennig, W. 1950. *Grundzüge einer Theorie der phylogenetischen Systematik*. Berlin: Deutscher Zentralverlag.

1966. *Phylogenetic Systematics*. Urbana, Chicago, London: Univ. of Illinois Press.

Huene, F. von. 1931. *Abh. Senckenberg. naturf. Ges. 42*, No. 4.

1940. Die stammesgeschichtliche Gestalt der Wirbeltiere, ein Lebenslauf. *Paläont. Z. 22*:55–62.

Hummel, K. 1929. Die fossilen Weichschildkröten (*Trionychia*). *Geol. – Paläont. Abh. N.F. 16*:357–488.

Huxley, J. 1942. *Evolution. The Modern Synthesis*. London: Allen Unwin.

Jaekel, O. 1902. *Uber verschiedene Wege phylogenetischer Entwicklung*. Jena: Fischer.

Jepsen, G. L., E. Mayr, and G. G. Simpson. 1949. *Genetics, Paleontology and Evolution*. Princeton: Princeton Univ. Press.

Kühn, A. 1948. Biologie der Romantik. In *Romantik, ein Zyklus Tübinger Vorlesungen*, ed. Steinbüchel, Th. Tübingen and Stuttgart: Leins.

1942. Typologische Betrachtungsweise und Paläontologie. *Acta Biotheoret. 6*:55–96.

1943. Die Deszendenztheorie. Eine kritische Ubersicht. *Z. Kathol. Theol. 67*:45–74. Würzburg: Wegner.

1979. Entelechie und Evolution. *Ber. Naturforsch. Ges. Bamberg 53* (for 1978):145–94.

1981. *Die Evolution. Ergebnisse und Probleme*. Altötting: Gebr. Geiselberger.

Leuckart, R. 1848. *Über die Morphologie und Verwandtschaftsverhältnisse der wirbellosen Tiere*. Braunschweig.

Lorenz, K. 1973. *Die Rückseite des Spiegels*. Munich, Zurich: Piper.

Mayr, E. 1942. *Systematics and the Origin of Species*. New York: Columbia Univ. Press.

Mayr, E., and W. B. Provine. 1980. *The Evolutionary Synthesis: Perspectives on the Unification of Biology*. Cambridge, Mass.: Harvard Univ. Press.

Mertens, R. 1947. Die "Typostrophen-Lehre" im Lichte des Darwinismus. *Aufs. u. Reden senckenberg. naturf. Ges. 2*, Frankfurt.

Naef, A. 1919. *Idealistische Morphologie und Phylogenetik*. Jena: Fischer.

Philiptschenko, J. 1927. *Variabilität und Variation*. Berlin: Borntraeger.

Popper, K. R. 1960. *The Poverty of Historicism*, 2nd ed. London: Routledge & Kegan Paul.

Radl, E. 1905, 1909. *Geschlichte der biologischen Theorien*. 2 vols. Leipzig (2nd ed., 1913 and 1918).

Raup, D. M., and A. Michelson. 1965. Theoretical morphology of the coiled shell. *Science 147*:1294–5.

Reif, W. E. 1980. Paleobiology today and fifty years ago: A review of two journals. *N. J. Geol. Paläont. Mh.* 1980:361–72.

Remane, A. 1955. Morphologie als Homologienforschung. *Zool. Anz. (Supp. Bd.) 18*:159–83.

Rensch, B. 1939. Typen der Artbildung. *Biol. Rev. 14*:180–222.

1943. Die paläontologischen Evolutionsregeln in zoologischer Betrachtung. *Biol. General. 17*:1–55.

1947. *Neuere Probleme der Abstammungslehre. Die transspezifische Evolution.* Stuttgart: Ferdinand Enke Verlag [English ed.: *Evolution above the Species Level*. New York: Columbia Univ. Press, 1960.]

Ruse, M. 1981. Origins of Modern Synthesis. A review of Mayr & Provine, 1980. *Science 211*:810–11.

Salfeld, H. 1919. Über die Ausgestaltung der Lobenlinie bei Jura und Kreide-Ammoniten. *Nachr. Königl. Ges. Wiss. Göttingen, Math. -physik. Klasse* 1919.

1924. *Die Bedeutung der Konservativstamme für die Stammesentwicklung der Ammonoideen. Grundlagen für die Erforschung der Entwicklung der Ammonoideen der Jura- und Kreidezeit.* Leipzig: Max Weg.

Schindewolf, O. H. 1925. Entwurf einer Systematik der Perisphincten. *N. J. Min. etc., Suppl.* 52B:309–43.

1928. Prinzipienfragen der biologischen Systematik. *Paläont. Z.* 9:122–66.

1929. Ontogenie und Phylogenie. *Paläont. Z.* 11:54–67.

1936. *Paläontologie, Entwicklungslehre und Genetik. Kritik und Synthese.* Berlin: Borntraeger.

1940. Zur Theorie der Artbildung. *Sitzungsber. Ges. naturforsch. Freunde* 1939:368–85, Berlin.

1942. Evolution im Lichte der Paläontologie. *Jenaische Z. Med. Naturwiss.* 75:324–86.

1945. Darwinismus oder Typostrophismus? *Kulönnyomat a Magyar Biol. Kutat. Munkaibol. 16*:107–77.

1947. Fragen der Abstammungslehre. *Aufs. u. Reden senckenberg. naturforsch. Ges. 1*, Frankfurt/Main.

1950. *Grundfragen der Paläontologie.* Stuttgart: E. Schweizerbart'sche Verlagsbuchhandlung.

1956. Tektonische Triebkräfte der Lebensentwicklung. *Geolog. Rundschau* 45:1–17.

1963. Neokatastrophismus? *Z. deutsch. geol. Ges.* 114:430–45.

1964. Erdgeschichte und Weltgeschichte. *Akad. Wiss. Lit. Abh. mathem.-naturwiss. Klasse 1964* (2). Wiesbaden: Steiner.

1969. Über den "Typus" in morphologischer und phylogenetischer Biologie. *Akad. Wiss. Lit., Abh. math.-naturwiss. Klasse* 1969 (4). Wiesbaden: Steiner.

Schmidt, H. 1918. *Geschichte der Entwicklungslehre.* Leipzig: Kröner.

Siewing, R. ed. 1978. *Evolution: Bedingungen, Resultate, Konsequenzen.* Stuttgart, New York: Fischer.

Simpson, G. G. 1944. *Tempo and Mode in Evolution.* New York: Columbia Univ. Press [German translation with an introduction by G. Heberer: *Zeitmasse und Ablaufformen der Evolution.* Göttingen: Musterschmidt, 1951.]

1953. *The Major Features of Evolution.* New York: Columbia Univ. Press.

Spengler, O. 1919–1922. *Der Untergang des Abendlandes. Umrisse einer Morphologie der Weltgeschichte.* Munich: Beck.

Stebbins, G. L. 1950. *Variation and Evolution in Plants.* New York: Columbia Univ. Press.

Stille, H. 1913. *Tektonische Evolutionen und Revolutionen in der Erdrinde. Antrittsvorlesung gehalten am 22. Januar 1913, Universität Leipzig.* Leipzig: von Veit.

1924. *Grundfragen der vergleichenden Tektonik.* Berlin: Borntraeger.

1941. *Einführung in den Bau Amerikas.* Berlin: Borntraeger.

1949. Das Leitmotiv der geotektonischen Erdentwicklung. *Deutsch. Akad. Wiss. Berlin, Vortr. u. Schriften 32.* Berlin: Akademie.

Tschetverikow, S. S. 1915. Waves of Life. *Dnevn. Zool. Otd. Moskau 3.*

Uexküll, J. von. 1947. *Der Sinn des Lebens, Gedanken über die Aufgaben der Biologie mitgeteilt in einer Interpretation der zu Bonn 1824 gehaltenen Vorlesung des Johannes Müller: Von dem Bedürfnis der Physiologie nach einer philosophischen Naturbetrachtung.* Godesberg: Küppers.

Weber, H. 1958. Konstruktionsmorphologie. *Zool. Jahrb. Abt. 3, allg. Zool.* 68:1–112.

Wedekind, R. 1916. *Über die Grundlagen und Methoden der Biostratigraphie.* Berlin: Borntraeger.

1918. Beiträge zur progressiven Entwicklung der Organisation der Organismen. *Ges. Beförderung d. ges. Naturwiss. Marburg, Sitzungsberichte, No. 2, July 1918,* 25–6.

Weidenreich, F. 1929. Vererbungsexperiment und vergleichende Morphologie. *Paläontologische Z.* 11:276–86.

Wilser, J. L. 1931. *Lichtreaktionen in der fossilen Tierwelt. Versuch einer Paläophotobiologie.* Berlin: Borntraeger.

Ziegler, B. 1972. *Allgemeine Paläontologie.* Stuttgart: Schweizerbart.

Zimmermann, W. 1938. *Vererbung "erworbener Eigenschaften" und Auslese.* Jena: Fischer.

1948. *Grundfragen der Evolution.* Frankfurt/Main: Klostermann.

1953. *Evolution. Die Geschichte ihrer Probleme und Erkenntnisse.* Freiburg-Munich: Alber.

8

The role of morphology in the theory of evolution

RUPERT RIEDL

This theme was suggested to me by the editor of this volume, and I am glad to expound my position on this contrast, for the matter is as paradoxical as the epistemological position of contemporary biology in general.

All human understanding or comprehension of complex objects and events in this world is connected with our perception of form (*Gestalt*). And this holds equally of all our activities of comparison, arrangement, and classification, no matter whether we describe these forms with the concepts of anatomy, ethology, biochemistry, systematics, or ecology.

Morphe is derived from the Greek, and means "Gestalt," form or appearance (as well as beauty). Thus, morphology is the doctrine of Gestalt, of the bodily structure as well as of the development of organisms in particular. And what would the theory of evolution be? The theory of the development of organisms. Thus, it is paradoxical to ask whether the doctrine of development plays a role in the theory of the development of organisms.

And yet the question is not merely paradoxical. Behind it there is really more: the paradox of the epistemological position of biology, as a very concrete dilemma. That is the dilemma of knowledge. I want, therefore, to show (1) what morphology ought to be; (2) what has happened to it, in order to present (3) a solution of the problem of its method, and (4) its object. For our intellect not only represents this nature, it is also its product, and to be understood only as such – once more biologically.

What is morphology?

From the start, the place of morphology is characterized by the fact that we must first define our subject matter; and this obtains even

Translated by Marjorie Grene.

though all biology has arisen on the foundation of morphology (as I understand it). In the United States, I tried asking what morphology is, and there were nearly as many answers as respondents. A fair number held that morphology describes "the form of external appearance" or "the organization of the body." But even those who knew that the concept goes back to Goethe viewed it as obsolete, almost as a "dirty word," which one would never dare to use in a grant proposal. It would be better to speak of "structural design" or "pattern of features" if one wanted one's grant approved. Ernst Mayr said to me, when I called his attention to this decline, "Morphology is nothing but German idealistic philosophy." And that is the very worst allegation one could make about a science.

But even in the German-speaking world, where there is a stronger morphological tradition, we speak of "functional" or "comparative morphology," which also shows that we no longer know how this discipline was conceived, or what its most essential task was supposed to be. So what is morphology?

Morphology is the doctrine of Gestalt (even here we find no English equivalent),[1] above all in the biosciences, although disciplines of biochemistry and of the human sciences also use this concept. *Morphology contains the methodology of scientific comparison, namely, the distinction between essential similarities (homologies) and accidental similarities (analogies)*. It is supposed to establish the foundation for the formation and weighting of anatomical and systematic concepts. Thus, morphology grounds the cognitive procedure of comparative anatomy, embryology, and ethology, as well as of systematics and phylogeny. Its method is, in principle, hermeneutical.

It does not thereby contain the explanatory procedure for the phenomena in these disciplines. But we shall see that it also exerts an influence on the theory of explanation of homologies, of systemic categories (the way organisms are assigned to taxa),[2] and of the phenomena of cladogenesis. And it does this because an isomorphism exists between the processes of knowledge and of origin.

In view of morphology's having as inclusive and fundamental a domain as I ascribe to it, one may well ask, how we can understand that biology makes such excellent progress although this allegedly so fundamental area is in decay and, in part, already wholly liquidated.

To this there are two answers.

1. Biology is advancing without hindrance only in a one-sided, that

---

[1] "Gestalt" is now used as an English word (Translator).

[2] German usage does not distinguish between "category" and "taxon." Where possible, I have tried to render the text so as to take into account this distinction (Translator).

is, a reductive, fashion. This analytical reduction of systems or resolution into their parts destroys most systems and runs the danger of reducing itself to biochemistry and thus destroying itself.

2. The approach to morphological method is innate, it is a part of our hereditary cognitive apparatus, which operates preconsciously. Brunswik (1934, 1955) called it "ratiomorphic," that is, an intellect-resembling apparatus. Lorenz (1941, 1973) speaks of our innate forms of visualization (*Anschauungsformen*: usually translated "forms of intuition");[3] this is a product of adaptation, experienced as our unreflective "common sense." And it functions even without scientific theory and without rational control.

In contrast, the decline of morphology is to be ascribed chiefly to the attempt to rationalize the ratiomorphic, preconscious, and basically correct approach, to substitute for it rational reflection. And because the underlying mechanism was unknown, both the substitutes that were developed led investigators astray. One method commits idealistic–rationalistic, the other materialist ("mechanistic")–empiricist errors.

In order to adopt this, my point of view, we need two new theories: the "system theory of evolution" and "evolutionary epistemology."[4] I shall discuss both, but first I must glance at the history of morphology.

The rise and fall of morphology

The development of morphology is connected with that of evolutionary theory, and both coincide with the rise of the Enlightenment and of positivism. A relation between the two pairs is not detectable at the start; but later the influence of positivism and of scientism (as it is depicted by F. v. Hayek [1979]) on evolutionary theory and even more on morphology becomes decisive. Granted, something of morphology can be traced back to Aristotle. But it is only questions of knowledge that have retained their influence from those days. I shall deal with these in the next section.

The key figures in the development of morphology are Lamarck and, a generation later, Cuvier and Geoffroy St. Hilaire: all in Paris from 1780 to 1830. We owe to Lamarck (1809) the idea of evolution,

---

[3] This is the standard English translation. Not only is it thoroughly misleading, but to compound the difficulty, Lorenz seems to use the term to refer to the Kantian categories as well as to the forms of intuition – a very grave error from the point of view of Kantian scholarship (Translator).

[4] It appears to me that Riedl understands here by "evolutionary epistemology" only a minimal component of what most of its adherents intend by the term (Translator).

the concept of a continuous development of organisms; to Geoffroy and Cuvier the discussion of synthetic versus analytic method. Lamarck's principle of evolution is brushed harshly aside by Cuvier and replaced by his catastrophe theory, which is closely allied to the conservative notion of the Deluge. Constant creation de novo once more stands opposed to evolution.

In 1790 Goethe published his first essay on the metamorphosis of plants, which, as he laments in 1807 "was fated to receive a cold, almost unfriendly reception" (Goethe, 1928, p. 10). In 1795 the "First Draft" of the introduction to osteology appears. Here too, he observes, "insufficiently clarified distinctions (*Vorstellungsarten*) are used. Either they took the matter [of Gestalt] too trivially and clung to the mere appearance, or they sought aid in final causes, and so strayed even further from the idea [notion, concept] of a living being."[5] "The similarity of animals [i.e. mammals] to one another and to man are obvious," yet "a norm is lacking, in relation to which one could test the various points; and in consequence a series of principles is lacking, to which one would have to subscribe . . . Therefore what happens here is a suggestion of an anatomical type, a common picture, in which the forms [*Gestalten*] of all animals (e.g., the species of mammals) would be potentially contained." But "even from the general idea [notion, theory] of a type it follows, that no individual animal could be the pattern of all . . . One ought rather to proceed as follows: Experience [*sic*] must first teach us the parts that are common in all animals [mammals], and how these parts differ. The idea [notion, abstraction] must hold sway over the whole and derive the common picture in a genetic [coherent] fashion. If such a type has been suggested even as a trial [as hypothesis], then we can very well use the usual kinds of comparison to test it" (Goethe, 1928, pp. 274–6).

What is proposed, therefore, is an empirical method reciprocally determined: first of the common *Bauplan* of a group of organisms from its special representatives; and then the testing of their affiliation by their difference from the tentatively proposed general pattern. This is a relation of exchange between theory and consequent correction, in the expectation that the corrected *Bauplan* can be more and more univocally determined and demarcated from its representatives. This is the method by which, even if unconsciously, all our comparative concepts arise. And it is the method of the "hermeneutical circle," in which, almost a hundred years later, Dilthey, without knowledge of this passage in Goethe, divided the method of human science (*Geisteswissenschaft*) from the scientism of the natural sciences.

---

[5] The parenthetical insertions are the author's.

But even the tracing of the individual parts or characters, which compose the type, and which were soon to be called "homologues" by Oken, had been introduced by Goethe. We will recognize the essential part, Goethe tells us, "according to the place it assumes in the organism," according to its destination (*Bestimmung*), and we will "determine the form that it can have according to its destination, and that it must have in general" (Goethe, 1928, p. 292). In doing this "we are presupposing a certain consequence in nature, we trust her to behave in all individual cases according to a certain rule" (Goethe, 1928, p. 293).

And finally it is recognized that type stands within type. "Classes, genera, species and individuals are related as instances to law; they are contained in it, but they contain it and do not give it" (Goethe, 1928; p. 325).

According to Goethe, the type would be explicable out of an "esoteric principle." And now the misunderstandings begin. The *Naturphilosophie* of German idealism, which was soon to begin, sought in Goethe its guarantor. It interprets its term "idea," not as an abstraction of the general from the particular, to be secured empirically and tested in its instances, but as something given in advance. It interprets the type concept as an a priori, Platonic idea and reads "esoteric" as "mysterious" or "transcendent." No one noticed, or wanted to notice, that Goethe started from the pair of concepts "esoteric/exoteric," by which he understood "internal" as against "external" causes. Today we would translate "esoteric" as system-immanent, as this is demanded also as the cause of homologues, *Baupläne*, and system-groups.

By Goethe's successors, however, all this was rationalized in the name of idealism. And the empiricists read Goethe's texts even less frequently. But they pointed a finger at the idealists and announced: Idealistic morphology is not a science. To this extent Ernst Mayr was right. Yet we were destined to go on misunderstanding Goethe idealistically.

As a matter of fact, in the time to come, many natural historians felt themselves to be idealistic morphologists, others empiricists, without any deeper effect on the development of systematics. The hereditary human understanding of the appraisal of similarities, that ordinary, garden-variety common sense, guided their judgment so correctly, that Alfred Russel Wallace and, contemporaneously, Charles Darwin were able to discern the origin of species in accordance with hundreds of thousands of correctly grouped species and system-groups.

In this context it is crucial to keep in view that this achievement was possible exclusively because of the fact that all the hundreds of thousands of species then known had been systematically correctly ordered; and that this ordering was possible without biologists' having understood their own method. The knowledge of evolution as a whole is founded on the correct understanding of Gestalt, on the correct weighting of essential similarities, on morphological method, on an achievement of our inherited preconscious intellect.

Not even a unified theory of explanation was available to assist in this achievement. Lamarck's evolutionary perspective had hardly been noticed. Even Goethe, who follows attentively the events in Paris and translates and comments extensively in the very same year (1830) on the learned debate between Cuvier and Geoffroy St. Hilaire, appears never to have heard of Lamarck, who had died only the year before. It should be mentioned that, on the other hand, Goethe did know about his contemporary Erasmus Darwin, but took no notice of his evolutionary perspective (Goethe, 1928, p. 405). For Goethe it was indeed the path of the discernment of organic similarity that was important, and not the path of explanation. And to be sure, the discernment, the concept, of a subject matter (*Sache*) must necessarily precede its explanation.

As we know, Charles Darwin's *Origin* was an instant success. The reading public of England, which, with Victorian industrialization, had demonstrated its (often ruthless) efficiency, could now see the rights it arrogated to itself on the ground of that efficiency legitimized as a law of nature. And out of the general euphoria there arose the impression that the explanation of evolution would be already adequately given through the mechanism of natural selection. It was overlooked that Darwin was a Lamarckian (!). He knew that selection could be effective only for the most efficient possible diversification of forms, and that is precisely what Lamarck's theory had proved. It was overlooked (and is still overlooked), that into his old age, Darwin sought for a hereditary mechanism for the Lamarckian principle, and published this in his theory of "pangenesis." Indeed, the biographers of Darwin today are still ashamed of "the awkward, not to say dilettantish, attempt" of the revered master (Hemleben, 1964, p. 122), if they do not conceal it altogether.

Now it would be wrong to expect this explanatory theory of evolution, once established, to displace research in comparative anatomy. On the contrary. So much the more intensively was the Darwinian doctrine filled in with a world of anatomical and embryological facts. Haeckel published his challenging *General Morphology* (1866), and the second half of the nineteenth century became the high point, as yet unexcelled, of morphological research, and thus of the nearly exclusive application of a method, the essence of which was not yet rationally understood. Whatever aspired to the rank of science had to be adjusted to this method. Biology remained unaffected by the scientism that had led physics and then chemistry since Galileo to their great achievements.

This situation changed only about the turn of the century, when with the rediscovery of Mendel's laws, neo-Darwinism developed out of Darwinism, and when with physiology the methods of pragmatic reductivism characteristic of chemistry and physics had their advent in biology. It was only then that reasons were found to quibble about the scientific character of the morphological method: "The morphologist Goebel himself compared it with the products of the creative imagination of poetry" (Remane, 1971, p. 1).

It was only after a generation of uncertainty that two works of an opposing tendency appeared, which independently and also in different ways again approached the solution for the cognitive procedure of morphology. These are the volumes by Hennig (1950) and Remane (1952). Their destiny is as characteristic as their aims. Willi Hennig is concerned with tightening the evaluation of characters in the service of phylogenetic systematics, to produce a simplified logic for practice. Adolf Remane's work stands against the same practical background, but with the intention of resolving the problem of conceptual definition before any logical procedure with the concepts so defined. Thus Remane is concerned with the formation of concepts in morphology and this is the decisive level of this methodology.

At first neither book had much impact. Only when the controversy about numerical taxonomy began in the United States (Sokal and Sneath, 1963) was Hennig translated into English (1966) and began to have some influence, also affecting those who came out of Remane's school. Hennig's schematic proved easier to handle and to discuss. I shall return to "numerical taxonomy."

Remane's absolutely decisive contribution, however, was never translated. The fault must have lain in the mistrust that people in "modern countries" brought to the "old morphology." There may also be personal coincidences. Ernst Mayr and Adolf Remane were both active at the same time at the Berlin Museum, and whereas destiny led the one to leave Germany, it led the other to come to terms with that Germany. Whereas one sought the solution in intraspecific evolution, the other sought it in evolution at the transspecific level.

Thus Remane's work had no influence – neither here nor there. There, in the United States, its diffusion was hindered by the modest foreign-language proficiency of our American colleagues and by the widespread opinion that everything important would be translated in any case (an attitude that still leaves open the question of who shall decide what is important). But even in German-speaking countries, scientism, long since widespread, was sufficient to displace Remane's contribution, as essential as it was, into the sphere of the inessential. The authority of physiology now provided the modern paradigm and punished deviants with social sanctions.

For the specialist, Remane's work is a direct continuation of Goethe's concern for a theory, or metatheory, that would give a rational account of the process of concept formation in the investigation of biological form and its development. Thus it was a carrying forward of the study of morphological methodology. In this connection it is striking that not only Hennig, but also Remane, scarcely mentions Goethe. I know from many conversations with Remane, that he knew his Goethe well. It was the many academic disputes and imputations that made him avoid Goethe. And, in consequence, he emphasized repeatedly "that the notion of type . . . developed logically and necessarily on the basis of correlations and connections established in biology, and does not owe its origins to the direct borrowing of an intellectual contemplation in the sense of Plato's theory of ideas" (Remane, 1971, p. 24).

Yet even this caution and clarity have not helped. If one applies the principle of causality to morphology, such an obvious circularity in reasoning results, that Hassenstein (1951), before as well as after Remane (Hassenstein, 1958; see Remane, 1952), banished morphological method to the human sciences, if not, as unscientific, all the way to the Platonic theory of ideas. Yet Remane had observed: It would not be "surprising if, before the possibility of a scientific explanation for biological types was known, the attempt had been made to build up an intellectual background in idealistic morphology, as it was provided for similar phenomena in the field of the human sciences" (1952, p. 25). He mentions the doctrine of style, linguistics, prehistory.

Has Remane then explained the method of morphology? Not indeed entirely, but he has come much further than Goethe on the track of the solution. It was his work that made the solution evident to me. For my understanding, his work contains two decisive contributions. First, order is brought into the hodgepodge of concepts of original form (*Urform*), ancestral form (*Stammform*), and type (think of the concept of "systemic type," which had come into use out of complete ignorance of the true state of affairs). Second, and perhaps even more important, is his analysis of the criteria of homology. There ancestral form and morphological type become clear to us, there the twofold interplay of the heuristic process of morphological judgment is fore-shadowed.

Remane distinguishes principal and auxiliary criteria of homolog-ization and among the former the criteria of position (*Lage*) and of special quality (structure). Altogether there are five (or six; 1971, p. 58), and we could well ask, why precisely five? Does any equivalent in nature correspond to this division, whether in thought or object?

Indeed, this question has never been asked, for the thought, like the work, was carried on by no one. Instead the attempt was made simply to cut through the Gordian knot of type and homology. Sokal and Sneath (1963) declared the morphological method to be circular, abandoned both concepts, and recommended that characters simply be counted. Hence, the designation "numerical taxonomy." In fact, this method has the simplicity of an assembly line, so that it quickly found many imitators.

Now there is no objection to counting. But the critics asked, "How do you know what comparable characters are?" (See the literature on this discussion in Sneath and Sokal [1973] and in Riedl [1975].) "But," the pheneticists asked the phylogeneticists in turn, "what does *your* computer program for the weighting of characters look like?" And the phylogeneticists could only reply: "that the weighting is always better left to the brain of the experienced systematist" (Mayr, 1969, p. 60). "There you have it," the pheneticists, in turn, concluded, "where there is no program, there is no method. The morphological method is at best an art form, in any case not a science." We end here, where we began.[6]

But this "computer-brain," I was convinced, has indeed "up to now created the whole 'marvel' of insight into the natural relatedness of organisms without . . . knowing its (own) functions well enough. This is the real marvel . . . ; to clarify it would be a matter for the psychologists, to our advantage" (Riedl, 1975, pp. 310–11). But the psychologists were busy elsewhere (although they recognize this task today; Wagner, review, 1981). I have, therefore, concerned myself with this clarification (Riedl, 1979). Evolution itself furnished the key to the solution.

In the meantime the debate about the investigation of relatedness without homology continued and remained stubborn enough, be-cause the group "New Systematics" represented a firmly established antipole. Paradoxically, this group comprised those who supposed

[6] Riedl equates phylogenetic with evolutionary taxonomy. There are numerous presentations of the three modern positions; for a recent discussion, see E. Mayr, Biological classification: toward a synthesis of opposing methodologies (*Science* 214:510–17, 1981) (Translator).

the phenomena of transspecific evolution solved along with those of the intraspecific process; those, in other words, who wanted to consign classical morphology to the junk-yard of Platonic idealism. (Both sides avoided the concept of type, not realizing that the homologues of any system-group always add up to a type.)

In any case, the "numerical taxonomists" conceded, and recognized homology (Sneath and Sokal, 1973), but went wrong again, insofar as they believed that would be a problem of secondary significance. "We still lack a consistent system of operational homology. It appears that phenetic analysis can be carried out in spite of this drawback, but nevertheless it is philosophically [better: epistemologically] embarrassing to rest [sic] character coding on such shaky foundations" (p. 248).

Here the authors are twice right. For no enumeration can be better than the determination of what is to be enumerated. For one thing, the summing up of a jumble of this and that has never produced any deeper knowledge. For another, we know why the pheneticists got on without a concept of homology. As long as it was left in peace, their preconscious cognitive apparatus determined the homologies. We shall speak of that in the next section.

But what solution of this cognitive problem, which, as foundation, doubtless determines all further attainable degrees of certainty, was offered by these authors? One was to rely in the determination of homologies on "the production of 'self-evident' results" (p. 434), and on common sense. Again we have ended up where we began. The problem of knowledge has been moving in a circle for 2,000 years.

Whenever the attempt was made to look into the method by rational means, the outcome was a slide into the nonsense of philosophical extremes. At one end, the result was the extreme rationalism of the idealistic theory of forms; at the other end, extreme empiricism, a materialistic mechanism, a tabula rasa point of view on knowledge. And both attempts at solution only obscured the possible way out; with the vanishing prospect of solution, they brought morphology into uncertainty, led it to its downfall, and even wholly destroyed it.

We may now consider the attempted solution of the theory of ideas as obsolete. But the mechanistic attempt at solution is snugly embedded in the established paradigm of scientism, and this, in turn, is supported by the pragmatics of neopositivism, which, consciously or unconsciously, still governs the philosophy of all modern natural science.

Is morphology then exhausted? No, certainly not. For a biology without the determination of similarity and relationship, without attention to historicity and descent, is no biology at all. Whatever it produces has meaning only on this foundation. This foundation, further, is still relegated to the unknown mechanisms of preconscious reason, Brunswik's "ratiomorphic apparatus." So morphology cannot achieve greater certainty than what the preconscious offers.

The foundation of morphological method

The reason why the cognitive procedure of morphology could not be given a foundation lies beyond morphology. It has to do with the fact that our whole cognitive apparatus cannot be grounded in itself. Cognition is possible only through precognition and this, in turn, through pre-precognition. One ends up in three kinds of dead end: either the recognition of a circular argument, or of an infinite (and hence irresoluble) regress, or with the abandonment of the undertaking. Albert (1968) aptly calls this the "trilemma" of knowledge.

The same outcome is expressed in the rationalism–empiricism controversy. The rationalists say: "Experience always presupposes preexperience." The empiricists explain: "Experience is possible only through experience." As we know, Immanuel Kant presented with special exactitude those preconditions that must be presupposed for the achievement of any human knowledge: preconditions that, as the presupposition of all reason, are consequently not to be grounded in reason alone. These are the familiar Kantian a prioris.

To put it simply, it is a question of our expectations or "forms of intuition" of space and time, probability, comparability, of causes and qualities.[7]

Konrad Lorenz was the first to recognize (1941) that in those a prioris it was a question of genetically anchored products of adaptations "that our forms of intuition fit this world for the same reason that the fish's fin fits the water even before it slips out of the egg." Consequently, we have systematically examined the hierarchical structure of our forms of intuition (Riedl, 1979, 1980*a,b*) and found that all the a prioris of the individual are to be understood as a posteriori products of ancestral learning. Thus, both the rationalists and empiricists are right. However, the achievement of knowledge is precisely not only a matter for the associating individual, but for evolution as such.

Therewith the problem of isomorphism is also solved – the question: How do our forms of thought fit this nature? They are shown

---

[7] Clearly, "probability" and "comparability" are not Kantian categories. Nor, again, can the "a prioris" spoken of in the next paragraphs be *only An-schauungsformen* (Translator).

to be products of adaptation to the regularities of this world, relevant and available to the organism.

I have described the most fundamental of these innate forms of intuition as four "hypotheses" of our expectation (Riedl, 1979). And of these the first two are of interest for the foundation of morphological method, for the formation of concepts (for the comprehension of the predictable, of order in this world).

The first, "the hypothesis of the apparently true," includes the expectation "that much of what is experienced can be predicted with probability under corresponding conditions, and thus be confirmed through its recurrence" (Riedl, 1979, p. 53). It constitutes the presupposition for separating the predictable, lawlike, and orderly from the accidental. It is of high selective value, because chances of survival are connected with the extent of correct prediction.

Indeed, the whole algorithm of the cognitive procedure possible for us depends on the fact that actual perception can be confirmed repeatedly as an expectation of nature; as a reinforcement of the expectation of the sequel, such measure of confirmation, in turn, determines the probability of expected degrees of certainty (detailed in Oeser, 1976; Riedl, 1975, 1979). This principle mirrors the high redundancy of nature.

But in the same degree as the repeated confirmation of a prediction allows the probability of a possible explanation through chance to vanish, in that same degree this holds also for the richness of the characters and the number of prognoses that permit a repeated expectation. (One can also say: The number of alternatives one believes can be expected. Turning up heads three times may be ascribed to chance. Drawing the jack three times out of a pack of thirty-two scat cards will already convince us that it is a question of purpose. The probability of chance is related as $(\frac{1}{2})^3 = 138$ to $(1/32)^3 = 1/32,768$.)

With these considerations I am grounding, to begin with, the polarity of Remane's principal and auxiliary criteria of homology. In the case of the principal criteria it is a matter of the effect of the richness of characters in the possible prognoses; in that of the auxiliary criteria, of the effect of repetition. The two can be multiplied in the manner of simultaneous and successive coincidences. (That ten simultaneously tossed coins should all show heads is exactly as improbable as to expect one coin in ten tosses to show heads every time: namely, $(\frac{1}{2})^{10} = 1/1024$; but that ten coins tossed ten times should always show heads would give rather $(\frac{1}{2})^{10} \cdot (\frac{1}{2})^{10} = 1/1,048,576$).

The second, "the hypothesis of the comparable" (*vom Vergleichbaren*) contains the expectation "that the dissimilar in the perception of things should be evened out, and that similar entities, even though they are clearly not the same, would prove to be comparable also in many characters not yet perceived" (Riedl, 1979, p. 93). In the present context it is interesting that the characters of things can be anticipated in a hierarchical arrangement; for example, teeth occur only in jaws and jaws only in skulls. One can also say that concepts get their content through their lower concepts and their "sense" (*Sinn*) only within the whole series of their higher concepts. If we took the concept of apple, for instance, out of its higher systems of tree fruits, fruits, reproductive organs, plants, organisms, it might be an Adam's apple or the apple of your eye. It would not be an apple in the above sense.

It is wholly in this manner that I understand the two sides of Remane's principal criteria of homology (Riedl, 1975). The probability that it is a question of a tooth or an apple depends not only on the object's structure (special quality), but in accordance with nature [*sic*], also on its situation (*Lage*), on the fact that it is found in a jaw or on a fruit tree. In the same way the type, like the systematic category (taxon?), for instance of primates, consists of the genera, species, and individuals that make them up; and they have "meaning" (*Sinn*) only within the mammals, vertebrates, chordates, and metazoa.

From this perspective, too, the problem of weighting evoked by the debate between "numerical taxonomy" and the "new systematics" is long since solved.[8] The most important characters are those that coincide most clearly with the boundaries of a "field of similarities" (of a systematic category, a taxon, a concept). In the reciprocal procedure already familiar to us, whereby we distinguish one level from the neighboring levels, the likeliest boundary of a "field of similarity" is that in which the greatest number of character boundaries coincide. The reciprocal, hermeneutical manner of concept formation is innate (Riedl, 1979; Kaspar, 1980). The numerical taxonomists were wrong in believing in a circular argument. The "new systematists" were right in trusting common sense.

In the hierarchical structure of the levels of all natural objects (quanta, atoms, molecules, biomolecules, ultrastructures, cells, tissues, organs, individuals, societies, cultures) this two-sidedness of predictability and determination holds sway (Figure 1). Every level contains the structures of its lower levels and is contained in the determinations (or determinants) of situation of its upper levels. This relation holds even within levels. Thus, for example, at the level of organs, our vertebral column is determined according to the structure of its regions as well as of its dorsal position in the skeleton; the region of the cervical section of the backbone, in turn, is determined by the

[8] See Note 7 (Translator).

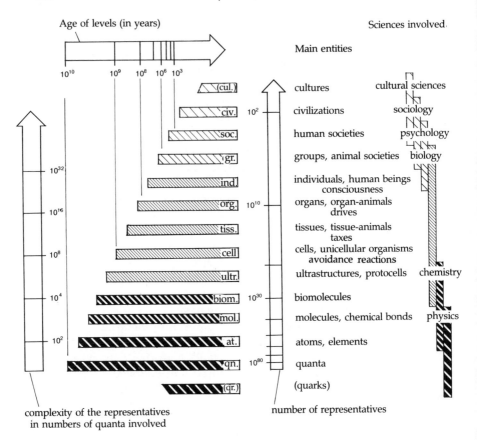

Figure 1. The system of levels of the real world, simplified to only one dozen levels with about equal distances of the growing organization. The pyramid is derived from the age of each level and the complexity of its representatives. [After Riedl, 1978/79.]

structure of its seven vertebrae as well as by their position in the vertebral column, and so on.

The same thing is said when we note that we explain the construction, the material of a system, as well as the source of its motive power, from its parts or lower levels, but that we explain its ordering and form, as well as its function and goal or meaning (*Sinn*) from the higher levels to which it belongs (Riedl, 1975, 1976). In this connection it becomes clear that the four causes of Aristotle are returning here: the *causa materialis, efficiens, formalis,* and *finalis* (material, moving or efficient, formal, and final cause). And it turns out (as Aristotle, in part, already knew) that these four forms of cause, applied to the hierarchical structure of the real world, are related to one another in a double symmetry. First, material and efficient cause take effect from all lower levels to every higher level; form and final cause (conditions of function, choice, and selection) work from all higher levels to all lower levels. Second, as far as our forms of intuition go, forces (power) and goals run without conceptual alteration through all levels. But the material and formal conditions (conditions of choice or selection) are altered in our conceptualization from level to level (Figure 2; Riedl, 1976, 1978/79, 1979).

Now if explanations are available from two directions in the hierarchical structure of this world, these correspond, in turn, to the two alternative explanations of the world: the materialistic and the idealistic. If materialism tries to understand the world from its forces (power) and materials – to reduce it to these – idealism, which would better be called "idea-ism," attempts to explain above all meaning and ends. The natural and human sciences (scientistic and hermeneutical, respectively) have split apart according to this polarity.

But, in fact, the understanding of a system is possible only through a synthesis of explanations from both directions. This was already known (in different fashions) to Aristotle, Goethe, Dilthey, and Remane, without their noticing that they had in view the same conditions for the system of this world.

At this point we should recall once more that the approach to such a contemplation of the world is hereditary, given a priori to the human individual, as the product of a posteriori learning, adaptation and selection of our ancestral history, anchored as foreknowledge and aid to decision making in our genes, these pregiven "forms of intuition" (Lorenz, 1973) corresponding in their prejudices and predilections to the achievement of our "ratiomorphic apparatus" (Brunswik, 1934, 1955), that is, our unreflective common sense (Riedl, 1979).

As products of selection, these are achievements of learning and cognition of the genome of yesterday and the day before. And in correspondence to the slow rate at which genetic knowledge is acquired, they are adaptations to the relevant environment at the latest of early men, partly of early primates, even of early mammals. And although in this learning process in evolution it is always a question of the extraction of natural laws from the environments of the organisms concerned (and of their incorporation as instructions for organization and activity), these are, nevertheless, results of learning in the far simpler environment of that world of our very remote ancestors.

Neither the real world, nor even the "being-so-and-so" of things, is mirrored in these forms of intuition, but just whatever simplifi-

220   *Rupert Riedl*

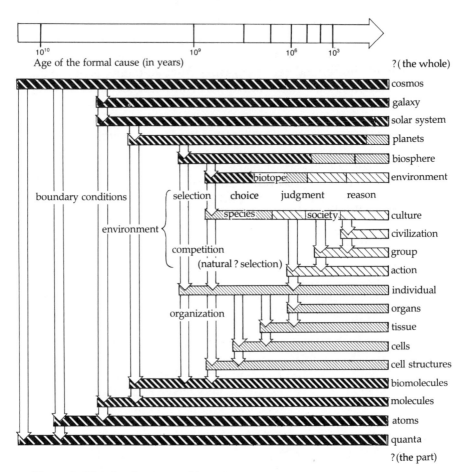

Figure 2. The development of formal causes in connection with the evolution of the system of levels of the real world (cf. Figure 1), organized according to age and complexity of the levels. Three items are of interest: (1) Each differentiation develops between the "part" (components) and the "whole" (entity); (2) the stepwise differentiation of the formal cause, which (3) has different expressions, ranging from "boundary conditions" to "reason" [After Riedl, 1978/79.]

cation of them was permitted, and this too only with respect to the modest segment of the environment's role for survival. This makes comprehensible why the system of causes is conceivable for us only in segments of four qualities that appear incapable of being united, rather as the four-dimensional space–time continuum is conceivable for us only in two qualities that appear incapable of being united: in a one-dimensional time and a space understood as being three-dimensional. And just as Einstein taught us our error,[9] so we must master the synthesis of the causal systems – and this despite and against our hereditary, unalterable, and rationally no longer teachable forms of intuition.

The most significant obstacle that this simplification in the grasp of cause presents us with consists in the expectation that causal contexts will be understood in linear form (if A, then B; if B, then C, etc). This must be the reason why materialists looked for first causes, idealists for the last cosmic goals (and that here, too, they sought to find the extrapolation of this notion in "the unmoved mover" – today we say "the big bang" – or in God's final purpose, the *causae exemplares*). And because this linear form also suggests the idea of a primal original cause (*Ur-Ursache*), a first cause of all causes, our cosmic explanations are split into materialistic and idealistic camps, which still divide our culture (Snow, 1963; Riedl, 1976, 1978/79, 1981).

Correspondingly, we now understand why previous attempts to give a rational account of our hereditary, ratiomorphically directed morphological method became embroiled in disunity and contradiction; on the one side, materially and scientistically and, on the other, idealistically and finalistically. We also understand why the method achieved so much without having been decoded rationally. And with this third, synthetic attempt I hope I have moved closer to the ground of the cognitive procedure of morphology.

It is of interest historically that this "evolutionary view of knowledge" has repeatedly been taken up only to run dry and be forgotten without establishing a tradition, until now, for the first time, it is attracting general attention. I had (1975) not noticed Lorenz's approach (1941, 1973). Lorenz was active without noticing Popper (1972). And we had not observed that Ernst Mach and even more explicitly Ludwig Boltzmann had long since put forward the evolutionary thesis (see Kreuzer, 1981, p. 121). But again the corresponding approaches of Herbert Spencer and David Hume had been equally forgotten.[10] Evolutionary theory developed without influence from evolutionary epistemology. That Hume had no influence on Lamarck and none on Goethe is easy to understand, but neither did Spencer influence Haeckel's morphology nor did Boltzmann influence Remane's. Only when morphology and epistemology meet in evolutionary theory does a new form of "reciprocal illumination" begin.

[9] About space and time (Translator).

[10] Hume as evolutionary epistemologist? Professor Riedl replies that he was thinking of the fact that for Hume causal judgment is a *Seelenbedürfnis*. As such, however, it is, of course, purely subjective; it derives from an impression of *reflexion* and has no source in the external world (Translator).

The foundation of the object of morphology

My theory, which asserts that our patterns of thought should be understood as the product of selection in accordance with patterns of nature (Riedl, 1979), has been borrowed with surprising speed. But if this is correct, then the question is also legitimate: Where did those natural patterns come from according to which our thought patterns had been formed? But this I had known years earlier. Differently from Lorenz (1973) and Popper (1972), who came from ethology and from philosophy of science, respectively, to an evolutionary theory of knowledge, I came to it from evolutionary theory itself.

A new study of the phenomena of transspecific evolution showed me that they all lack explanation: homology, type and the systemic categories, taken together, the whole hierarchy of the "natural system" of organisms, not to mention a host of such dynamic phenomena as systemic mutations, spontaneous stabisms, Haeckel's law, and so on. I must ignore all the latter here; I have dealt with them extensively elsewhere (Riedl, 1975, 1977, 1978). For the sake of brevity, I shall confine myself to the static phenomena.

I found that all the phenomena of transspecific evolution could receive a common explanation through a "system theory of evolution," and that these systemic conditions for evolution led to four fundamental patterns of order in which these phenomena are contained: norm, interdependence, hierarchy, and historicity (*Tradierung*). Again, for reasons of space, I shall confine myself to the case of hierarchy.

You will recall that the phenomenon of type remains unexplained. Goethe's "esoteric principle" was misinterpreted as "mystic" or "transcendent." The idealists turned the type into a Platonic idea; the scientistically minded denied its existence. The fate of homologies, which compose each type, is similar. Remane (1971) seeks the ground of their discernment; the grounds of their explanation remain open. It turns out that without homology no genealogical research is possible, so now the grounds of both discernment and explanation lie open.

It is different with the systemic categories.[11] Their definitions express nothing but simplified descriptions of the static components of the type. But no one has worried about the fact that here, too, neither the process of their recognition nor that of their explanation has been demonstrated in correspondence with nature. I conjecture that, to some extent, one was so used to concepts like fish, bird, deer, that

---

[11] Or the names of the taxa subsumed under them? I am not sure if one can disentangle "category" and "taxon" in the present paragraph (Translator).

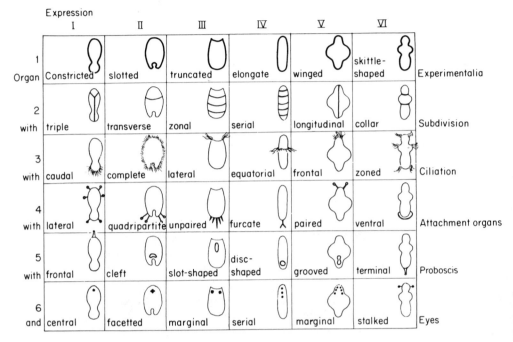

Figure 3. The 36 features of an imaginary group of animals. The six organs have Arabic numbers and their character-states ("expressions") have Roman numbers. A diagnosis of a group of organisms, for example, which only showed character-state I, would be as follows: "grooved experimentalia with tripartition, caudal ciliation, lateral attachment organs, rostral snout, and a central eye." [From Riedl, 1978, p. 118.]

the want of an explanation was not even noticed. Indeed, I am afraid that even today I startle systematists themselves as well as epistemologists or philosophers of science with this statement.

Let us now ask exactly: What is required for a system of similarities to assume a hierarchical pattern? The answer is simple. It is required that characters be fixed according to a hierarchical sequence (transformed from a variable and adaptive phase to one that no longer changes).

Let us verify this. We ask, what would happen if we allowed all characters the same chance of alteration (for example in the six organs and expressions I to IV in Figure 3). We leave it to one die to decide which adaptive channel will be occupied and to a second to decide whether the appropriate mutation has appeared. Under these conditions all similarity or comparability disappears. Only the tempo of disappearance depends on the mutation rate that we introduce. But definitive disappearance is always the necessary outcome (see Figure 4).

No systemic concept could be so much as formed. What would remain of similarities in such an evolution would be solely accidental and functional similarities and, hence, analogies: horns, spines, wings, legs: the contrary of essential similarities, or homologies. A system here could describe only more analogous "types of life form" (in the sense of Remane, 1943; Kühnelt, 1953). No homology, no type, no single unity in the sense of consanguinity or of the "natural system" would be perceived or expressed.

But, in fact, systematists formulate for the nearly 2 million species known today a hierarchical system comprising nearly a million names of taxa. So evolution must have proceeded differently.

The structure of the "natural system" can be understood only if one recognizes that with every branching in the field of similarities in the family tree, at least one character was permanently fixed (Figure 5), at least to the degree that it can be indubitably recognized, like the vertebral column, for example (from *Petromyzon* to the frog, boa constrictor, tortoise, or man). The number of these fixed characters in the system is remarkable. It corresponds to the number of differential diagnostic characters, let us say (on the average) ten, times the number of taxa distinguished (about a million): so there are about 10 million: scarcely an exceptional case!

We may test this by a thought experiment, with the question: What kind of system would arise if, statistically, say, after each speciation event, half the branchings acquired a fixed character? Let us allow the characters A to Z and a to f to arise from three dichotomous development steps (Figure 6). The die may now decide with the probability $\frac{1}{2}$ if a character is fixed (in our case, this affects the characters A,B,E,F, etc., up to P; they are marked with a cartouche). The remaining characters (C,D,G, etc., up to O) would again disappear. But since the alphabet means nothing to the systematist (as we must assume) a system results that does indeed contain hierarchical components (Figure 6), but such as do not exist in reality. Through this we characterize the fact that with the scope of the systemic categories[12] (in our example, from genera to classes) the dichotomous portion tends increasingly to disappear.

In the "natural system" this situation is reversed. The dichotomous hierarchy governs the splitting even of the largest units. The manifold branchings, the so-called mass hierarchy (Riedl, 1975) scarcely appears in families, becomes more common in genera, and is dominant

---

[12] See Note 11 (Translator).

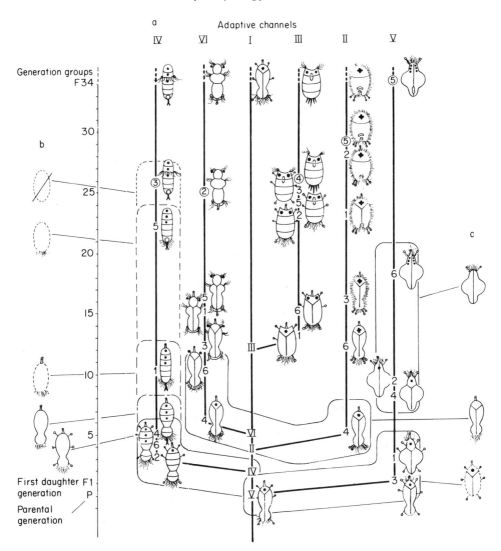

Figure 4. Disappearance of phyletic similarity given equal prospects of mutational change in all features. (a) The occupation and adaptation to adaptive channels II to VI as decided by the dice, is reached by accident in 34 generations. Decisions causing alterations are marked according to generation and channel. (b) As a result, at each step a sixth of the resemblance to the original form is lost. (c) Four agreeing character-states occur in a chain of three, or at most four, of the closest related mutants. After 34 steps no relationship can any longer be recognized. [From Riedl, 1978, p. 120.]

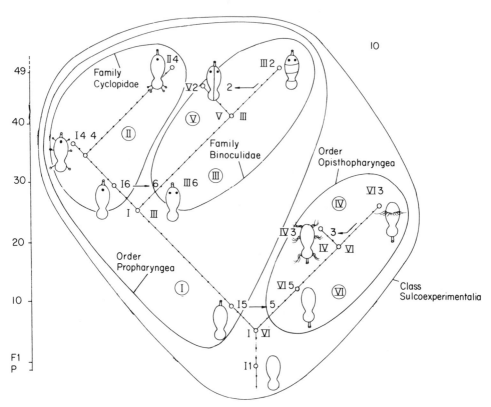

Figure 5. The origin of a dichotomous hierarchy of alternatives by the fixation of alternative character-states of the same organ after the dichotomous branching of the stream of determinacy. Points signify unsuccessful throws of the dice, and circles signify successful ones. The arrows indicate when dice decisions have been carried over to the alternative channel. The artificial groups of animals have been given artificial names. [From Riedl, 1978, p. 123.]

in the species of the genera. But, to the contrary, this is to be attributed to the fact that the expression of the dichotomous fixation is not plain enough (too little developed in its consequences) to have been perceptible and unambiguous. In the advanced stage of development, fixations are, therefore, a presupposition for every systemic category of the "natural system."

Now what reason can be given for this phenomenon of fixation?

At first there still seems to be the possibility of believing in the effect of chance. One might assume that the chance encounter of some *Bauplan* with the series of eventual environmental conditions that con-

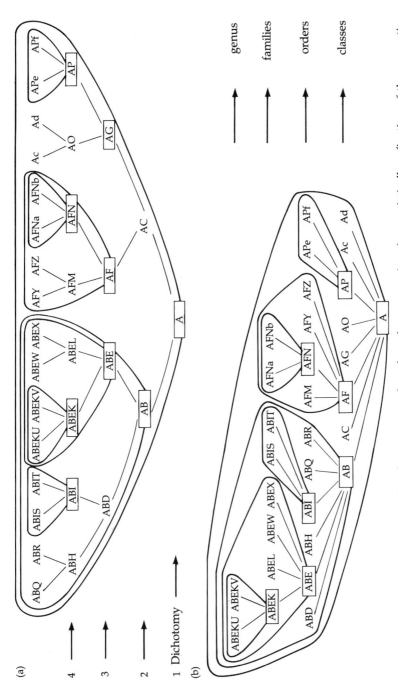

Figure 6. Development of a model of a taxonomy, based on the assumption that statistically a fixation of the respective new characters (A–Z) and (a–f) will occur only at each second bifurcation (frame) (part a). The resulting taxonomic system demonstrates a reduction of the dichotomous hierarchy from genus to classes (part b).

front its descendants will cause many characters to disappear, although others will not. The latter have persisted, since no altered demands of the environment force them to disappear.

A second thought experiment seems to confirm this. If for every dichotomous branching we assume the novel occurrence of just so many new characters that the chance rate of nonnecessary change would affect at least one of these each time, then, in fact, a hierarchically dichotomous system would arise. However, the model fails to answer the question why (as here presupposed) the subsequent divisions and subsequent environments would result in no loss of any character, once established. So in nature it would come once again to the dissolution of systemic groups, as we have already seen (Figure 4).

Thus the retention of a character under manifold change of environmental conditions (consider dolphins, sea-serpents, sea-tortoises) will be no chance event; for with the duration of the phylogenetic process the explanation through chance becomes increasingly improbable.

Decisive now is the observation that the characters that become fixed appear in series and stages, which themselves form a hierarchical pattern. Table 1 illustrates this with the example of a sequence (or series) of hierarchically nested groups of systems from the chordates to the mammals. The four differential-diagnostic characters given in the example show that they presuppose one another in any given series.

The systemic definition of the (class) Amniota, for example, reads: "Tetrapods, the centrum of whose atlas vertebra is usually fused with the axis (epistropheus), with vestigial remains of the notochord, reduced *ductus botalli* and reduced fifth aortic arch, with intraventricular septum, reduction of the *truncus arteriosus* and disappearance of the nephrostomes." But this presupposes that the anatomist already knows that the regions of the vertebral column are already differentiated, that the pulmonary, carotid, and aortic arches are present, that the pulmonary section of the heart with the atrian septum has been formed and the glomeruli have exhibited a regular development. But this, so the systematist knows, are the differential-diagnostic characters of the (division) Tetrapoda. They are the preconditions for the more recent adaptations.

Two insights result from this. First, the fixed characters build on one another, like stories of a building. Let us take the example of the support system. The fixation of the seven cervical vertebrae (Mammalia) presupposes the fixation of their differentiation (Amniota), which presupposes the differentiation of the formation of the regions of the backbone (Tetrapoda), this, in turn, the preparation of the shoulders and pelvis (Gnathostomata), this the articulation of the vertebral column (Vertebrata), and this the presence of the notochord (Chordata). But these are functional units ranked one above another, in such a way that the capacity to function of each differentiated lower character (upper story) presupposes the fixation of all upper characters (lower stories). Whenever an error occurs in the construction of the notochord or the articulation of the vertebrae, the embryo of a mammal is eliminated as a lethal mutant. Thus the fixation is the consequence of inner causes ("esoteric" conditions, as Goethe said) of functional, systemic interconnections of the *Bauplan* under the control of selection, where it is not the environment (in the egg for the notochord? in the maternal uterus for the vertebra?), but the conditions in the system itself that decide on the breakdown of the developmental process.

In this context, it is, indeed, taken for granted that such fixations arise continuously and gradually. To use a further example, we may illustrate this with the development of the mammalian limb. Within the context of genealogy (or similarity) (Figure 7), not only do the spheres of the succeeding taxonomic groups (Figure 7, top) succeed one another[13] but the stages of differentiation of the extremities run parallel (I–IV, Figure 7, bottom). All these stages are passed through very gradually in full adaptive freedom until complete fixation (Figure 8). Where, for example, in the sphere of the vertebrates (from the *Agnatha* on) there is totally free experimentation, even with the number of paired appendages, in the sphere of the tetrapods (from the labyrinthodonts on), this number is largely fixed. But where in the tetrapods there can still be experimentation with the axis and the sequence of bones of the extremity, in the sphere of the mammals (from the protherians on) axis and sequence are fully fixed.

But the functional network through the phenotypes of the adult organism is only a starting point, a first cause of a much deeper systemic interconnection in the development of living things. As we recall (from Table 1), the notochord is a first presupposition for the development of the articulation of the seven vertebrae of the cervical region, and for the formation of the renal pelvis, the construction of a subchordal pronephron. But both notochord and pronephron have long since lost their function in the adult. For 240 million years the notochord has been absorbed into the nuclei pulposi (in the discs); for 350 million years the pronephron has been replaced by the mesonephron (Wolffian body). They now have their function as carriers of genetic information, of developmental instructions, for the organs

[13] Again, *Systemkategorie*, but presumably taxa, not categories (Translator).

Table 1. Dependent series of differentially diagnostic features. The functional dependence of differentially diagnostic features, as illustrated by four functional systems in six sequentially hierarchical levels down to the mammals.

	Supportive system	Vascular system	Heart	Excretory system	No. of species
Chordata (phylum) are bilaterians with	Dorsal notochord situated ventral to the central nervous system	Antero-ventral principal vessel and paired gills	Medio-ventral contractile portion of the principal vessel	When present myomeric subnotochordal kidneys	43,000, beginning 5 $\times$ 10^8 years ago (Cambrian)
Vertebrata (subphylum) are chordates with	Jointed vertebral column and branchial skeleton	Six paired, ventro-lateral, primary aortic arches	heart divided into auricle and ventricle	Paired Wolffian ducts with postero-ventral openings and proximal tubule portions	41,700, beginning 4 $\times$ 10^8 years ago (Ordovician)
Gnathostomata (superclass) are vertebrates with	Jaws (visceral arches) and two pairs of limb anlagen			Pronephros now only functioning in the larva	41,650, beginning 3.5 $\times$ 10^8 years ago (Silurian)

Tetrapoda (division) are gnathostomes with	Primarily 5 vertebral regions, 2 girdles, and stylopodium and zeugopodium	Differentiation of the pulmonary (6), carotid (3) and aortic (4) arches	Pulmonary portion of heart and formation of an interauricular septum	Regularly developed glomeruli	21,100, beginning 2.8×10^8 years ago (Devonian)
Amniota (group of classes) are tetrapods with	The centrum of the atlas vertebra mostly fused with the axis, with atrophied remnants of the notochord	Reduction of the ductus Botalli and disappearance of the 5th aortic arch	Formation of the intraventricular septum and reduction of the truncus arteriosus	Complete loss of the nephrostomes	18,600, beginning 2.4×10^8 years ago (Lower Carbonifercus = Mississippian)
Mammalia (class) are amniotes with	Primarily a fixation of seven cervical vertebrae	Conservation of only the left aortic arch and loss of the renal portal vein	Complete separation of the chambers of the heart (as also with the birds)	Formation of the renal pelvis; complete development of Henle's loop, renal medulla, and cortex	3,700, beginning 2×10^8 years ago (Trias)

Note: In each functional system the features of lower systematic value presuppose the features above them in the column. In the right column note the decrease in age and number of species. Groups after Romer (1959); numbers of representatives after Mayr (1969); age of groups after Müller (1963).

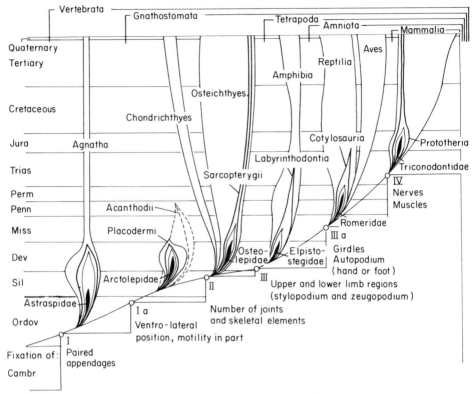

Figure 7. Taxonomic survey of the precursors of the mammalian limbs according to time and estimated morphological distance. Along the supposed ancestral line I have shown the sequence of consecutive groups, with the families in black and three higher hierarchical groups up to the class. Above the diagram the systematic categories of the sequential hierarchy are shown. Beneath it are inserted the features becoming fixated in the fixation stages I–IV. [From Riedl, 1978, p. 152.]

that succeeded them (articulation of the vertebral column and mesonephron). But as such they are no less indispensable. Indeed, they are now wholly irreplaceable, because the success of all subsequent development depends on the information they purvey.

So we find the functional interconnections of the phenotypes anticipated in the functional interconnections of the "epigenetic system" (the system of genic interactions). The interdependence of the phenes must have copied the interdependence of the genes, in the sense that those random shots were favored by selection that, through regulator genes, brought into functional dependence such structural genes as

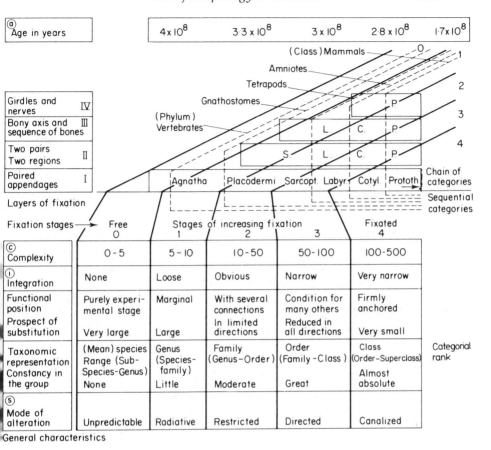

Figure 8. The fixation path of a group of features as illustrated by the mammalian limb. The building up, step by step, of the fixation layers I to IV in passing through fixation stages 0 to 4 of certain limb features, within the systematic groups closest to the ancestors of present-day mammals. In the sequence of these groups the class or subclass names are better known and have been used in the table instead of the family names. Thus the Agnatha correspond to the Astraspidae; the Placoderms to the Arctolepidae; the Sarcopterygians to the Osteolepidae; the Labyrinthodonts to the Elpistostegidae; the Cotylosaurs to the Romeridae; and the Prototheres to the Triconodontidae. [From Riedl, 1978, p. 157.]

coded for functionally dependent phenes. And this in such a way that through such a lucky shot the chances of adaptation, and we can also say, the speed of adaptation, thus essentially rose (Riedl 1975, 1977; Kaspar, 1977; Wagner, 1980).

From our consideration of the static phenomena this context has significance, because we can understand from it how organisms remain fixed, even though practically everything about them, form, function and location, has basically changed. As a further example of this, we may refer to our three auditory ossicles which have not lost their interconnection since they formed the jaw apparatus of the cartilaginous fishes some 300 million years ago.

Thus the systemic conditions for each kind of organization are cause for the object of morphology. Homologies (as constancies despite changing demands) are to be explained in the first instance from the fixations of a hierarchy of functional dependencies, and then also of epigenetic limitations; and type and *Bauplan* are to be explained from the system of homologies.

But there remains a second insight for us to derive here. The nature of the "natural system" as well, the object of the systematist, is to be explained from within this context. For not only did the homologies compose the *Baupläne*; the *Baupläne* themselves are, in turn, described by the diagnoses of the appropriate systemic categories (taxa) [*sic*], and, indeed, according to the fixations and remaining degrees of freedom of the homologies.

And, what is most essential, the hierarchy of homologies corresponds to the hierarchy of systemic groups. The series of hierarchically superimposed homologies – epistropheus (axis), cervical vertebral column, vertebrate column, dorsal supporting organ (development of the notochord) – corresponds to the series of hierarchically arranged systemic groups: mammals, tetrapods, vertebrates, chordates.

The foundation of the object of morphology lies in the inner "esoteric" systemic conditions of organization. These are the material conditions provided by all inferior systems in their interchange with all the conditions of form and selection from all upper systems, between which any new system arises, as an insertion between the former. And these foundations of the object of morphology are the consequences of the field of application of morphology: the foundations of homologies for the anatomist and the foundations of defining characters for the systematist.

The history of the search for a solution of the problem of morphology was soon transformed into a search for the solution of the problem of cladogenesis (of the phenomena of transspecific evolution). Thus one can also say that the problem of morphology has become wholly clear through the problem of cladogenesis. And the search for a solution of both has never been abandoned: against the perspective of positivism, against the doctrine of reductivistic scientism, and against the synthetic theory (of neo-Darwinism), which wholly failed to recognize the problem.

We may recall how close Goethe was to a solution, or that Darwin was looking for it in his theory of "pangenesis." And because there have been pure Darwinians, that is, adherents of the doctrine that the environmental selection of chance alterations alone explained evolution, the opposition has not been abandoned. It begins, even in Darwin's time, with Karl Ernst von Baer (1876) and continues through Plate, Osborn, Schindewolf, and Haldane, to mention only a few (survey in Riedl, 1975, p. 89), to Bertalanffy, Waddington, and L. L. Whyte (1965), all seeking an inner principle, with vitalism (Driesch), idealism (Bergson), nominalism (Sokal, Sneath) in their wake – and also with metascientific approximations like Nicolai Hartmann's *nexus organicus* (1950), seeking a two-sided concept of cause (a causal–finalistic synthesis) for the solution of the problem of living things. But none of them achieved a breakthrough.

A higher-order isomorphism

I am of the opinion that the problem of morphology, like that of cladogenesis, is not to be solved apart from the problem of method. And the problem of method is not to be solved apart from the problem of knowledge. What stood in the way of a solution were our hereditary forms of intuition, which suggest to us linear causal chains, as well as an opposition, or reciprocal exclusion, of causal versus teleological solutions. This was an error, or rather a simplification in the foundations of our cognitive capacity, that escalated, through rational extrapolation, to error and to the fission of our culture (or our world view). But another obstacle to the solution was our ignorance of the structure of our ratiomorphic apparatus, that unreflective common sense, which, without rational intervention, guides the superlative performances of morphologists and systematists. The absence of explicit knowledge of its contribution permitted the impression that there was neither problem nor method.

But now we can take the question of the origin and structure of our cognitive capacity as solved, at least as far as concerns the problem of isomorphism, which has an important role in it. Again, this problem of isomorphism, or uniformity, or correspondence, contains the question: How did it happen that our thought fits this world; how this world can be thought about specifically by our thought.

We solved this problem by means of evolutionary theory. We said that our patterns of thought fit this world because they are to be understood as the products of adaptation and selection with respect to natural patterns. The order of our thoughts is an image of natural order. Our intuitions of space and time present a simplification of the dominant space–time continuum. Our probabilistic expectation is a mirror of the high redundancy of natural objects. Our expectation of succeeding comparisons corresponds to the nonrandom combinability of characters in things and events in this world.

But still more complex structures, the construction of our real world itself, turn out to be anticipated in the expectations of our ratiomorphic guidebook; namely, our commitment to a hierarchically structured world. Indeed, this expectation is reflected in the way we experience Gestalt or structure, as well as in what we expect in superordinated contexts and call an explanation.

We experience the Gestalt and structure of every system, at every hierarchic level, as content and form, structure and function, as organization and meaning (goal) in a two-sided relation. Accordingly, we possess a two-sided symmetric expectation of so-called causes, the expectation of understanding a subject matter from its conditions in terms of matter and energy, and from its conditions of form (selection) and end. In this sense we are born as hermeneuticists.

And this equivocity, this understanding of nature from two contrary explanatory directions is the mirror image (or the product of selection) of a differentiation of this world that is just as equivocal in its origin and organization. There is no doubt that that step of differentiation between parts and whole occurs, whether in cosmic, chemical, organic, or cultural evolution: in organic evolution between the DNA sequences and their species. Certainly the force and structure of a muscle is determined by a series of its materials (its lower levels), its form, function and goal by its situation in the organism (e.g., in the forearm, the wing, the bird, and the functions of the species, and thus by a series of its higher systems).

Thus, on the whole, our cognitive procedure is to be understood as a reflection of the process of origin of this world. Cognitive patterns of order are a product of selection from the ontological. Our creative cognition corresponds to the order of nature even in the dimension of its creative development. Thus there is an isomorphism of a higher order.

I believe that in the last analysis it is this context that we must recognize in order to understand morphology and evolution. For the morphological method of acquiring knowledge is not to be understood without an acquaintance with natural order, because it was through selection in relation to its structures that we won acquaintance with this world. And conversely the "strategy of genesis" cannot be entirely disclosed without the knowledge of our cognitive apparatus, because the latter has developed as the product of evolution. And in relation to our more modest theme, to the question of the role of morphology in evolutionary theory, we can say: Morphology contains the theory by which the internal and systematical conditions of evolution are to be understood, and we describe the conditions of evolution by a theory according to which we understand the a prioris of our morphological preunderstanding.

References

Albert H. 1968. *Traktat über kritische Vernunft*. Tübingen: Mohr.

Baer, K. E. von. 1876. *Studien aus dem Gebiet der Naturwissenschaften*. St. Petersburg: Rottger.

Brunswik, E. 1934. Wahrnehmung und Gegenstandswelt. *Psychologie vom Gegenstand her*. Leipzig-Vienna: Deuticke.

1955. "Ratiomorphic" models of perception and thinking. *Acta Psychol.* 11:108–9.

Goethe, J. W. von. 1790. *Morphologische Schriften*. Weimar: Bohlau.

Goethe, J. W. von. 1928. *Werke*, vol. 36. Berlin: Bong.

Haeckel, E. 1866. *Generelle Morphologie der Organismen*, 2 vols. Berlin: Reimer.

Hartmann, N. 1950. *Philosophie der Natur*. Berlin: De Gruyter.

Hassenstein, B. 1951. Goethes Morphologie als selbstkritische Wissenschaft und die heutige Gültigkeit ihrer Ergebnisse. *Neue Folge Jahrb. Goethe-Gesellschaft* 12:333–57.

1958. Prinzipien der vergleichenden Anatomie bei Geoffroy Saint-Hilaire, Cuvier und Goethe. *Act. Coll. Int. Strasbourg, Publ. Fac. Lettr.* 137:155–68.

Hayek, F. von. 1979. *Missbrauch und Verfall der Vernunft*. Salzburg: Neugebauer.

Hemleben, J. 1964. *Ernst Haeckel in Selbstzeugnissen und Bilddokumenten*. Hamburg: Rowohlt.

Hennig, W. 1950. *Gründzuge einer Theorie der phylogenetischen Systematik*. Berlin: Deutscher Zentralverlag.

1966. *Phylogenetic Systematics*. Urbana: Univ. of Illinois Press.

Kaspar, R. 1977. Der Typus – Idee und Realität. *Acta biotheoret.* 26(3):181–95.

1980. Naturgesetz, Kausalitat und Induktion. Ein Beitrag zur Theoretischen Biologie. *Acta Biotheoret.* 29:127–47.

1981. Das Werden lebendiger Ordnung. *Prax. Naturwiss.* 11:131–41.

Kreuzer, F. 1981. Ich bin – also denke ich. Die evolutionäre Erkenntnistheorie. Vienna: Deuticke.

Kühnelt, W. 1953. Ein Beitrag zur Kenntnis tierischer Lebensformen. *Verhandl. Zool. Bot. Ges. Wien* 93:57–71.

Lamarck. J. 1809. *Zoologische Philosophie* [ed. H. Schmidt]. Translation of *Philosophie zoologique*; Leipzig (1909): Kroner.

Lorenz, K. 1941. Kants Lehre vom Apriorischen im Lichte gegenwärtiger Biologie. *Blätter Deut. Philos.* 15:94–125.

 1973. *Die Rückseite des Spiegels. Versuch einer Naturgeschichte menschlichen Erkennens.* Munich-Zürich: Piper.

Mayr, E. 1969. *Principles of Systematic Zoology.* New York: McGraw-Hill.

Müller, A. 1963. *Lehrbuch der Paläozoologie,* vol. 1. Jena: Fischer.

Oeser, E. 1976. *Wissenschaft und Information. Systematische Grundlagen einer Theorie der Wissenschaftensentwicklung* (3 vols.). Vienna-Munich: Oldenbourg.

Popper, L. 1972. *Objective Knowledge:* An Evolutionary Approach. Oxford: Clarendon.

Remane, A. 1943. Bedeutung der Lebensformtypen fur die Ökologie. *Biol. General.* 17:164–82.

 1952. Die Grundlagen des natürlichen Systems der vergleichenden Anatomie und der Phylogenetik, 2nd ed. [Authorized reprint of 1st ed., 1952, Leipzig: Geest and Protig.] Konigstein-Taunus: Koeltz.

Riedl, R. 1975. *Die Ordnung des Lebendigen. Systembedingungen der Evolution.* Hamburg-Berlin: Parey.

 1976. *Die Strategie der Genesis. Naturgeschichte der realen Welt.* Munich-Zürich: Piper.

 1977. A systems-analytical approach to macro-evolutionary phenomena. *Q. Rev. Biol.* 53:351–70.

 1978. *Order in Living Organisms. A Systems Analysis of Evolution.* New York: Wiley & Sons.

 1978/79. Über die Biologie des Ursachen-Denkens. Ein evolutionistischer, systemtheoretischer Versuch. In *Mannheimer Forum 78/79,* ed. H. V. Ditfurth, pp. 9–70.

 1979. *Biologie der Erkenntnis. Die stammesgeschichtlichen Grundlagen der Vernunft.* Berlin-Hamburg: Parey.

 1980a. Die Entwicklung des Begriffs vom taxonomischen Merkmal oder das Problem der Morphologie. *Zool. Jb. Anat.* 103:155–68.

 1980b. *Homologien; ihre Gründe und Erkenntnisgründe. Verh. Deutschen Zool. Ges.,* pp. 164–76. Stuttgart-New York: Fischer.

 1981. Die Folgen des Ursachen-Denkens. In *Die erfundene Wirklichkeit,* Hrsg. P. Watzlawik. Munich: Piper.

Sneath, P., and R. Sokal. 1973. *Numerical Taxonomy. The Principle and Practice of Numerical Classification.* San Francisco: Freeman.

Snow, C. 1963. *The Two Cultures: And a Second Look.* Cambridge: Cambridge Univ. Press.

Sokal, R., and P. Sneath. 1963. *Principles of Numerical Taxonomy.* San Francisco: Freeman.

Wagner, G. 1980. Empirical information about the mechanism of typogenetic evolution. *Naturwissenschaften* 67:258.

 1981. "Biologie der Erkenntnis," Review in: *Psychol. Beiträge Jahrgang 23,* Heft:1.

Whyte, L. 1965. *Internal Factors in Evolution.* New York: Braziller.

PART IV. SOME CONTEMPORARY ISSUES:
THE SYNTHESIS RECONSIDERED

9

Paleobiology at the crossroads: a critique of some modern paleobiological research programs

ANTONI HOFFMAN

Introduction

Paleobiologists have long been concerned to justify their own existence as scientists. To be sure, their descriptions and distributional and functional analyses of fossils, as well as their reconstructions of ancient life environments have always found readers, but almost exclusively among other paleobiologists who dealt with the same or very similar problems. Public opinion could show some interest for hominid fossils and for such curiosities as dinosaurs, but the bulk of paleobiology has been a closed and self-perpetuating system at the margin of science.

The most obvious remedy to this frustrating situation is provided by the traditional paleontological, or, better, paleontographical, task of deciphering and reconstructing the course of biotic evolution. In spite of the criticism raised by the radical phylogenetic systematists and vicariance biogeographers, there can be little doubt that the fossil record may provide some crucial information to this end, if only by documenting extinct taxa.

However, such justification might seem satisfactory only to those adhering to the idiographic approach to science, whereas such a radically positivistic attitude toward science is nowadays rather outdated.

The main sources of my inspiration are the evolutionary ecologists from Cracow (Adam Łomnicki and his colleagues), the philosophers of science from Warsaw (Stefan Amsterdamski and his colleagues), and the paleobiological community of Warsaw. Several ideas considered in this essay stem from my discussions and/or correspondence with Peter Bretsky, Stephen Gould, John Maynard Smith, David Raup, Thomas Schopf, and Nils Stenseth. I have found in Tübingen an ideal environment to think about the pitfalls and the promise of paleobiological research, which is in response to the stimuli coming from Wolf Reif and Adolf Seilacher. Needless to say, the full responsibility for all the failings of this essay is my own.

Financial support by the Alexander von Humboldt-Stiftung, Bonn, is most gratefully acknowledged.

There is no science without theories, without conjectures and refutations. There is no science unless the question *how* is followed by *why*. It is probably for this reason that paleobiologists have been so frequently and fervently engaged in modern disputes on theoretical and evolutionary biology, and also why they have so desperately searched for methods and/or approaches that might give them a chance to contribute, either theoretically or empirically, to those discussions. As an example one might cite a considerable number of paleobiological studies pretending to have documented mortality patterns and life-history strategies, character displacement, and other morphological responses of populations, patterns of speciation and genetical evolution, patterns of colonization and ecological succession, and ecological equilibrium of ecosystems.

Most of those studies, however, have failed. There now seems to be a consensus among paleobiologists that biological processes operating on the ecological time scale have all too often been projected uncritically onto the fossil record. Consequently, paleobiologists have found themselves confronted with the question, why be a paleobiologist, and with an even more urgent one, what to do as a paleobiologist. This then is their fundamental dilemma.

No doubt, geological, or evolutionary, time is the only dimension of biological processes that is accessible to paleobiologists but not to biologists. It should, therefore, present the major constituent of any uniquely paleobiological field of research. Paleobiologists, indeed, emphasize more and more commonly the reality of changes in structure of the biosphere in geological time, and the irreducibility of macroevolutionary processes to population–genetical and evolutionary–ecological principles. This emphasis forms the core of the majority of modern paleobiological research programs.

My aim in this essay is to evaluate critically the theoretical background of those research programs and to look for alternative programs, alternative solutions to the fundamental dilemma of modern paleobiologists. But before I attempt this, the basic premise of my entire argument must be made explicit. It is the conservative belief that new and more complex principles, concepts, or theories should not be accepted in science unless their necessity, or the insufficiency of the old and simpler ones, is demonstrated. No new complex theory should be introduced for so long as the processes envisaged in terms of the old theory can successfully and parsimoniously account for the observed patterns. This is a pragmatic reductionist approach. I do not claim that everything in nature can be reduced to a single basic level of reality. I demand only that prior to looking for processes that operate at different levels of reality, their distinctness from each other must be shown, or at least their irreducibility to each other at the present state of knowledge. In the context of the following discussion this means that I always start with neo-Darwinian principles. Only if these fail, will I move beyond neo-Darwinism, toward neo-Lamarckism, or Goldschmidtian mutationism, or holism, or anything else I find appropriate.

Community paleoecology

Community paleoecology has been among the most popular research programs intended to solve the fundamental dilemma of modern paleobiologists. Even Stephen J. Gould, who has recently become so critical about this program and its derivatives (Gould, 1980a), wrote a few years earlier the article "Palaeontology plus Ecology as Palaeobiology" where by the term "ecology" he meant various theories dealing with r-K selection, niche patterns, ecological succession, and community stability and evolution (Gould, 1976). I myself shared the hope and enthusiasm of the most fervent adherents of this research concept (Hoffman and Szubzda, 1976; Hoffman 1977, 1978a; Hoffman, Pisera, and Studencki, 1978), which may explain why I perceive its achievements so critically (Hoffman, 1979) and also why I am so desperately attempting to appreciate its actual promise (Hoffman 1981a).

The popularity of community paleoecology appears entirely consistent with the apparent increase in popularity of community or ecosystem ecology, which is probably related to the widening interests of scientists, nonscientists, and governments in pollution problems, environmental protection, wildlife management, and so forth. The well-known susceptibility of man-made agricultural monocultures to pests, and the disastrous effects exerted on whole ecosystems by environmental modifications by humans, such as the introduction of rabbits to Australia, the extermination of flies in China or leopards in cedar forests of the Atlas Mountains, or the enrichment of streams and rivers in nitrogen, demonstrate clearly that in order to cope successfully with problems of that sort one has to understand ecological interrelationships within natural communities. However, community paleoecologists are obviously unable to contribute much to that understanding. The fossil record is simply inadequate to give insight into the nature of processes operating within communities. Ecological succession, for instance, may be recorded in geological sequences only under exceptional depositional and diagenetic conditions, in some, but by far not in all, fossil reefs (Crame, 1980). In fact, even the classic writers of community paleoecology did not restrict the scope of their empirical analyses and theoretical considerations to ecological time (Olson, 1952, 1966; Shotwell, 1964; Valentine, 1968, 1972, 1973; Bretsky, 1969; Boucot 1975, 1978). They regarded community paleoecology as something more than, or at least different from, merely community ecology projected into the geological past.

It is not easy to tell what precisely has been meant by the term "community paleoecology." This can be done by pointing out the most important problems that scientists commonly recognize as being appropriate for solution by community paleoecologists. This approach is certainly less precise but at the same time more illuminating than any normative definition. There are two major topics discussed by community paleoecologists. One is the community approach to the reconstruction of ancient environments. This is actually a part of geological facies analysis. The major paleobiological problem addressed by community paleoecologists is community evolution, that is to say, modes of control of changes in ecological communities through evolutionary time.

Much inconsistency has been introduced in discussion on community evolution by use of the same term for the designation of two distinct processes. One process occurs in ecological time (Dunbar, 1960; Margalef, 1968; Odum, 1969), whereas the other refers to evolutionary time (Dunbar, 1960; Olson, 1966; Valentine, 1968; Bretsky, 1969). Changes occurring in ecological time actually grade into changes occurring in evolutionary time, but the two processes can be distinguished at least conceptually (Hoffman, 1980). When considering evolutionary time, one deals not as much with single communities as with community types, because individual communities themselves disappear as soon as their particular biotopes vanish or undergo major changes. The term "community type" is meant here to refer to a group of communities making part of a single bioprovince, resembling each other very closely in their ecological structure, and limited by the same environmental factors. The taxonomic composition may vary among particular communities but the replacement species are expected to be ecologically very close to each other. This concept thus refers to the permanence of ecological structure, given a constant environmental framework.

Gould (1977) has proposed that three main questions have organized the scope of paleontological research since the very inception of evolutionary paleontology. The same problems are also among the most pervasive topics of community paleoecology, although the questions are obviously expressed in another theoretical language. Community paleoecologists ask: (1) whether or not there is any consistent trend in community evolution; (2) whether community evolution is induced mostly by environmental or by purely biological factors; and

(3) whether it is gradual or spasmodic. The focus on these points may arise from the widespread hope that a high-level order can be found somewhere in the biosphere, imposed upon, or emerging from, the apparent disorder of evolution at the level of populations. Some biologists look for that order in the succession of *Baupläne* (Teilhard de Chardin, 1963; Waddington, 1963), others in the development of the ecological organization of the biosphere.

The methods of community paleoecology involve, first of all, reconstruction of paleoecosystems and comparisons among paleoecosystems that are chronologically disparate but functionally analogous, that is, occupying entirely analogous sets of physical environments. The reliability of results, however, is strongly reduced by various biases and limitations inherent in the very nature of the fossil record. First, a lot of organisms have no fossilizable parts whatsoever, whereas skeletal remains of the others are commonly subject to postmortem destruction, fragmentation, transportation, sorting, and mixing. Because of these geological biases and information losses, paleoecological data represent merely a small and unpredictably biased sample of the original community of living organisms. Second, except for very rare, unusually favorable geological conditions, fossil assemblages originate generally through gradual accumulation of organic remains. Therefore they may or may not derive from a single persistent community, because microhabitats and their associated communities are distributed patchily and often rapidly wandering in ecological time. Third, there is always a considerable variability in maximum individual size and longevity as well as in growth rate and mortality pattern among species belonging to a single community. Hence, dominance hierarchy of a fossil assemblage does not reflect the dominance hierarchy of the original biological community. Fourth, autecological characteristics of organisms and synecological interrelationships among community members are evidenced only exceptionally in the fossil record. Each paleoecological inference about community structure must, therefore, be based upon fairly speculative premises about the ecology and behavior of particular taxa, and, at best, upon functional morphological arguments supplemented with what Dodd and Stanton (1981) have called "taxonomic uniformitarianism." The latter methodological concept allows for ecological data on extant organisms' being directly transferred to their extinct relatives. Validity of such an extrapolation obviously declines with increasing taxonomic rank and geologic age. Finally, paleoecosystem reconstruction is hampered considerably by the fact that a large majority of important ecological characteristics of natural environments (primary productivity, short-term predictability, etc.) cannot be determined from the fossil record, and even those accessible to geological investigation (water salinity and turbulence, substrate nature, etc.) can be estimated as a rule only very vaguely. This makes comparisons among paleoecosystems questionable because one can only rarely get certainty that the compared paleoecosystems are really comparable, that is, that they did occupy the same set of physical environments.

All these biases and limitations of paleontological data can hardly be overcome. There are some paleoecological situations where they can be kept at a reasonable minimum. This is the case with some Cenozoic mollusk-dominated shallow-marine benthic communities and long-term persistent reefs that might provide a scanty but fairly firm empirical basis for case studies on community evolution. Thus far, however, community paleoecologists have devoted much more effort and time to Paleozoic and Mesozoic paleoecosystems (Hoffman, 1979; Boucot, 1981; Gray, Boucot, and Berry, 1981; and references therein) and consequently, their concepts of community evolution are precarious at best.

However, the major problem with community evolution is not so much in the scarcity of adequate empirical data as in the weakness of the theoretical background. The idea of community evolution goes back at least to Maxwell Dunbar and his natural selection at the level of ecosystems (Dunbar, 1960). Its basic premise is that the recurrent species associations that comprise ecological communities or biocoenoses represent a distinct level of biotic organization achieved through ecological integration and coevolution among the species. Under this assumption, communities are claimed to be real biological units each of which is defined by its particular taxonomic composition and ecological structure. They are also expected to be dynamic in both ecological and evolutionary time, that is, to change their taxonomic composition and ecological structure not only in response to environmental disturbances, but also as the result of their own intrinsic mechanisms. Community dynamics is here meant as an inherent trend to maximize a certain selection value (diversity, stability, resilience, efficiency, persistence, etc.). The term "selection value" obviously does not necessarily refer to any competition between communities. It reflects merely a hierarchy of community states, by analogy to the theory of irreversible thermodynamics and its macromolecular derivatives.

This set of premises makes up what I have termed the "community paradigm" (Hoffman, 1979). It is entirely consistent with system ontology as developed by Bertalanffy (1968; see also Koestler and Smythies, 1969) and consequently, system-theoretical concepts and methods should provide the tools to analyze the principles of community evolution. In fact, there is no doubt that communities exist. Co-occurring organisms and populations obviously interact. Organisms feed upon, are eaten by, compete with, and behave in response to, other organisms. Commonly, communities are very distinct because of their nature (e.g., reef communities), or the characteristics of environmental gradients that bound them (e.g., pond communities). They may also be stable in both ecological and evolutionary time. However, none of these observations proves that communities are anything more than a descriptive convention, a mode of delimiting the subject of scientific research, an epiphenomenon of geobiological reality. If the community paradigm is to resist Occam's Razor, the selection value assigned to communities must be considered as irreducible to, or non-deducible from, characteristics of the component populations, and even more so with those of individuals. This implies that there exist some ecological and evolutionary processes acting directly upon communities that affect, through mediation of communities, the component populations and individuals. These processes are determined by laws that cannot be applied to the lower levels of biotic organization. To be sure, the community paradigm does not follow Smuts, Alexander, and Lloyd Morgan, the classics of true biological holism. It does not imply any sort of emergent evolution as claimed by the latter. What it implies is merely an ontological emergence of the community level of biotic organization.

The community paradigm has tended to become part of ecological folk wisdom. The holistic, system-derived distinction between community nature and the attributes of its component populations is among the central, even though commonly implicit, concepts of modern ecology (Peterson, 1975). The widespread trend in ecological theory is to analyze the structural characteristics of a community and to infer from those data the patterns of community behavior through time, while disregarding the actual behavior of the component populations. To quote from May (1976): "The approach is to abandon description of the behavior of individual species, and to focus instead on overall aspects of community structure . . . There are many examples where the world appears chaotic and vagarious at the level of individual species, but nonetheless constant and predictable at the level of community organization." At the extreme, this approach restricts the scope of ecological analysis to mere topology of a food web and its constraints upon the community behavior (Levins, 1975).

I claim that the community paradigm is untenable. Three lines of argument can be developed to support this view (Hoffman, 1979). First, the degree of community integration seems generally insufficient to provide the driving force for structural developments predicted by the system–ecological theory, even though existence of some integration can hardly be denied. This is so because, as argued by Christiansen and Fenchel (1977) from purely theoretical considerations, most community members can be interrelated only weakly at equilibrium state. Multispecies competitive situations usually are unstable, whereas elimination of one or more species from an exceptionally stable competitive system does not disturb the system. As a matter of fact, coevolution among competitors seems to be an extremely rare phenomenon (Connell, 1980). In spite of intensive search for corroborating evidence, only a single case (Harger, 1972) has been documented convincingly. Of course, coevolution does work in predator–prey situations and the like, but these alone do not allow for several trophic levels and cannot account for any real ecosystem. Complex communities involving many trophic levels and many competitors cannot be integrated to any significant degree. Their composition and structure are always more or less flexible, which greatly undermines the community paradigm.

Second, it is assumed within the conceptual framework of the community paradigm that ecological communities tend to maximize their selection value until they achieve an equilibrium state that reflects the optimum partitioning of habitat and resources among the component species; the optimum partitioning results from long-term interspecific competition and coevolution among many community members. This idea requires the assumption of an almost universal occurrence of communities in equilibrium as a necessary prerequisite. At unstable stages of community development, severe interspecific competition would more probably result in local extermination of poorer competitors than in a coevolutionary process; whereas at equilibrium, local extermination is ruled out a priori because it would lead to a decrease in the selection value of the community. However, environmental disturbances, biotic as well as abiotic, occur much too frequently to allow that ecological communities universally or even commonly achieve an equilibrium state. Contrary to widespread opinion, this is what may account for the high diversity of various communities. As pointed out by Connell (1978), there is virtually no habitat partitioning among the keystone species in tropical rain forests and coral reefs. This is confirmed by the empirical observation that if interspecific competition in those ecosystems is allowed to go on for long undisturbed, a few species exclude most others from the habitat and finally the equilibrium community is close to monospecific. The apparent incompatibility of this observation with the ecological theory disappears when one discards the assumption that most tropical rain forests and coral reefs are in equilibrium achieved through coevolution among species.

Third, the community paradigm implies that a significant role in evolution is assigned to group selection. To claim that some ecological and/or evolutionary processes exist that act directly at the community level but are inapplicable at the lower levels of biotic organization, is equivalent to the assertion that some community properties induce a natural selection for certain features that may cause a decrease in individual fitness if it is outweighed by a parallel increase in selection value of the community. Otherwise, all community properties would be perfectly explicable by reference to the component individuals. However, there is no overall structural property of a community that would be able to impose any constraints directly upon the individual selection because the community structure itself is, by definition, expressed in terms of population sizes. Mediation through population regulation is, therefore, necessarily needed by the community paradigm. One may thus conclude that the community paradigm implies universality of population regulation, which is induced by a trend to maximize the selection value of a community even if it is disadvantageous to the individuals. This is consistent with the theory of group selection (Wynne-Edwards, 1962).

This issue is crucial for the validity of the community paradigm because one can hardly estimate community integration in quantitative terms and therefore the problem of community organization cannot be solved empirically. The concept of group selection was proposed to account for those biological phenomena that appeared inexplicable by reference to the traditional population–genetical concept of natural selection. These phenomena included ritualized intraspecific aggression, altruistic behavior of animals, and self-regulation of population size. Group selection has been largely refuted by modern biology because all these phenomena can be explained also by the theories of kin selection (Hamilton, 1964; Charnov, 1977), reciprocal altruism (Trivers, 1971), evolutionarily stable strategy (Maynard Smith and Price, 1973), or referred to the effects of intrapopulation variability on the population behavior (Łomnicki, 1978). All these theories are very plausible and entirely compatible with the population–genetic theory. The problem is, however, that, except for kin selection, none of them permits a rigorous, quantitative, empirical test. This situation leaves much space for tedious discussion (see Wade, 1978).

As a matter of fact, D. S. Wilson (1980) hopes to reconcile the reality of community organization and the action of group selection with the principle of individual selfishness. He starts from an analogy to human communities. A large majority of our activities as bakers, shoe-makers, teachers, scientists, or novelists are intended to directly ben-efit the community as a whole rather than ourselves. This is, however, not to say that we behave altruistically, because we are paid for our jobs and so we indirectly gain ourselves, too. Wilson's point is that this liberal view of human societies holds for ecological communities as well. He believes that the individual behavior of organisms is al-ways selfish; but because of the very nature of biotic interactions, this selfishness turns out to be also good for the species and the com-munity. Were it so, group selection could operate and communities should be considered as real biological units. However, what Wilson (1980) did actually demonstrate is only that selfish interests of indi-viduals may contribute to maintenance of predator–prey relation-ships and the like, whereas competitive interactions are out of the scope of his analysis. This is not a sufficient justification for group selection.

The refutation of group selection also refutes the community par-adigm. As a matter of fact, some leading evolutionary theorists still hope to derive the reality of community organization from population genetics as emended by the Red Queen hypothesis, without any ref-erence to group selection (Maynard Smith, 1976; Stenseth, 1979; Sten-seth, Maynard Smith, and Haigh, Ms). However, as long as they have not succeeded the community paradigm cannot be accepted; and even if they finally succeed, the objections raised by Connell (1978, 1980) and others still remain.

I conclude that ecological communities and community types are to be considered as epiphenomena rather than as evolving entities. They arise, change, and disappear due to the evolution of species that associate or disassociate as a result of their particular adaptations and biogeographical distributions in changing environments. There is no intrinsic, biologic mechanism inducing community dynamics in either ecological, or evolutionary time. In the terms of Hull (1980), communities and community types are not individuals, that is, spa-tiotemporally localized entities with reasonably sharp beginnings and endings in time. Consequently, there is no real process in nature to be termed community evolution. Its description and analysis cannot provide any valid problem for a paleobiological research program.

Evolution of organic diversity

Perhaps the most spectacular and widely appreciated, specifically pa-leobiological research program is the one concerned with the evo-lution of the structure of the biosphere or its large segments (e.g., the marine realm). The problem can be addressed purely descriptively, in qualitative terms (Valentine, 1972), but such explanations are generally untestable and, hence, unsatisfactory. Therefore, a number of quantitative approaches to the problem have also been developed. What they share with each other, and also with the community paradigm, is the idea that organic diversity, as measured by the number of species and their frequency distribution, provides the diagnostic characteristic of the structure of the biosphere. Within this conceptual framework, the actual changes in organic diversity through evolutionary time are estimated by means of various sets of paleontological data, mathematically modeled, and explained by an evolutionary scenario at the level of the biosphere. The latter can be presented either in deterministic (Valentine, 1969, 1970, 1973, 1977; Bambach, 1977) or in stochastic terms (Raup *et al.*, 1973; Schopf, 1974; Raup 1976*a*, 1976*b*, 1978; Sepkoski, 1978, 1979). This is indicative of a specific ontological option whereby taxa are considered to be, at the analyzed level, particles indiscernible from one another (Schopf, 1979). This option reflects what I have termed the "uniformistic," as opposed to individualistic, approach to the history of life (Hoffman, 1981*b*).

The uniformistic approach is fully justified in physics and chemistry because individual electrons, atoms, and molecules are indeed particles, each of which is identifiable *via* its spatiotemporal position but not *via* its unique characteristics. This is not only an axiom, but also an empirical statement, if only because of the uncertainty principle. The order of the physical microcosm therefore depends upon universal laws, but not upon the behavior of individual particles. Evolutionary biology, in turn, and, by implication, paleobiology, is an individualistic science unless all biological phenomena are conceived as being entirely reducible to, or deducible from, quantum physics. In fact, uniformistic or particle biology implies that there is no fundamental incompatibility, distinction, or discontinuity in the very nature of animate and inanimate matter; whereas individualistic biology claims that life is an irreducible phenomenon. In the last analysis, this is an ontological issue that defies scientific resolution. One is reminded, for instance, of the contrasting views of this problem advocated by Jacob (1970) and Monod (1970), both of which derived from exactly the same empirical experience and theoretical background. I believe that we are, at the ontological level, on the horns of a dilemma that cannot be solved through any scientific method. To be sure, my own option is for the individualistic view of life.

At a more pragmatic level, however, this option may be supported by several arguments. In our everyday practice as well as in research

we perceive individual organisms attributable to various organic species, and not unidentifiable living beings. On a larger scale, we are
dealing with individual organic species attributable to various supraspecific taxa and various ecological categories. The apparent order
of the biosphere is a by-product of the ecological and evolutionary
behavior of individual organisms, populations, and species. Behavior
of particular individuals or taxa may or may not be constrained by
any universal laws or principles valid at higher levels of biotic organization. One is reminded of Van Valen's (1973, 1976) Red Queen
hypothesis, Ghiselin's (1974) view of sex, and Dawkins' (1976) selfish
gene metaphor and their obvious individualistic implications. In the
context of the evolution of organic diversity, this is indicated also by
the well-established correlation of the evolutionary longevity of particular species to their various ecological characteristics. There is a
quantity of empirical data that evidences a strong correlation between
long-term evolutionary persistence of species and their cosmopoli-
tanism (Jackson, 1974; Vermeij, 1978; Hansen, 1980; Jablonski, 1980;
Koch, 1980; Hoffman and Szubzda-Studencka, 1982), the latter being
dependent to a considerable extent upon reproductive pattern and
dispersal strategy. This correlation shows that individual adaptive
strategies determine the apparent order discovered by Van Valen
(1973) in taxonomic survivorship curves. However, this correlation
tells nothing about the longevity of any particular species, or about
the longevity-frequency-distribution in any particular fauna unless
we know all their individual ecological characteristics and individual
geological settings.

The point being made here is that one can hardly recognize whether
a general statement about the order of the biosphere is actually a
universal law or just an inductive generalization that does not permit
any prediction. This is, I believe, a fundamental, ontological difference between physics and biology. In physics, indeterminacy makes
up the very nature of reality. Therefore so-called statistical or weak
determinism may indeed furnish an appropriate model for the physical microcosm. However, it is incompatible with the neo-Darwinian
idea of overwhelming causality in evolutionary biology, because a
probability distribution can never be deduced from purely deterministic causal laws. One may always ask why a probability distribution
is as it is, and there is no answer to that question except by a *regressus
ad infinitum* (Krylov, 1979).

Within the uniformistic conceptual framework for evolutionary biology, species can be conceived only operationally as being indistinguishable particles (Schopf, 1979; Hoffman, 1981*b*). Consequently,

biological processes and phenomena can be merely described, but not explained. This may be important for practical reasons, as in Schopf's (1979) prime example, demography. But what is the ultimate, biological sense of a proposition that empirical data on changes in, say, test size in a foraminifer species through evolutionary time fit the random-walk model? This hypothesis is representative of the uniformistic approach, for it assumes that despite their differential phenotypic appearance (implied genotypic constitution) and environmental setting, all the particular populations in that foraminifer lineage can be justifiably considered identical; otherwise, the probability distribution could not have been constant. Is this application of random-walk model equivalent to a claim that the investigated process was beyond control by natural selection and hence, that the foraminiferal test size was adaptively neutral? Or to a claim that the selective forces were too multifaceted to be identified, or that the selective agent itself was oscillating randomly? If the uniformistic background were intended ontologically, the first alternative would be the obvious answer. However, having deprived this approach of its ontological status, one must admit that such questions are beyond its reach (see Raup, 1977*a*).

Incidentally, the uniformistic approach is equally questionable in theoretical ecology, from which it has been directly imported to paleobiology (Gould, 1976). The theories of island biogeography (MacArthur and Wilson, 1967; Simberloff, 1978) and ecosystem stability (May, 1973; Levins, 1975) are among the most influential examples of uniformistic ecologic models. However, nowadays the individualistic approach is gaining more attention also in the field of ecology. The individualistic tradition in community ecology, whereby organic populations are considered to be something other than merely a member in the trophic web of an ecosystem, has actually a very long history (Gleason, 1926; Connell, 1972). This view is strongly supported also by Simberloff (1974), who has rejected Heatwole and Levins' (1972) application of the community paradigm to island biogeography. On the other hand, Łomnicki (1978, 1980) has recently demonstrated the far-reaching, but thus far neglected, consequences of the concept that individuals in an animal population must not be regarded as indistinguishable particles.

Let us, however, agree, for the sake of the following discussion, that a uniformistic approach to the evolution of organic diversity is justifiable. Certainly, the method that is most widely applied to analyze the patterns in evolution of organic diversity is *stochastic modeling*. The argument is that one should search for the residuum left after subtraction of the result of stochastic modeling from the real world, because apart from that residuum nothing requires a causal explanation (Raup *et al.*, 1973; Raup, 1977*a*, 1977*b*).

This approach is highly objectionable in practice because the real organic diversity of the biosphere or its major realms can only be measured in terms of supraspecific, commonly suprageneric, taxa. This is caused by various taxonomic and preservational biases. Supraspecific taxa, however, are incompatible with one another among major organic groups, and their purely artificial nature may obscure the underlying patterns at the specific level (Valentine, 1980), whereas the latter are the only interesting ones because there is no real biological process above the species level. Therefore, the most common criticism against this approach has been that the fossil record is so biased that the actual course of the evolution of organic diversity cannot be reliably reconstructed even for the marine biosphere; consequently, the residuum left after stochastic modeling may have very little in common with the real world.

Sepkoski *et al.* (1981) have recently undertaken an attempt to overcome this criticism by evaluating the significance of the fossil record to this end. They analyzed correlations among five major and essentially independent estimates of marine diversity in evolutionary time. These estimates are based upon:

1. The number of ichnospecies observed in marine sequences of variable geological age (Seilacher, 1974, 1977).
2. The number of invertebrate species per million years as derived from a sample of the data published in the "Zoological Record" (Raup, 1976*a*).
3. The species richness of some hundred benthic fossil assemblages of variable geological age (Bambach, 1977).
4. The number of genera and subgenera in particular geological periods as derived from the "Treatise on Invertebrate Paleontology" (Raup, 1978).
5. The number of marine metazoan families in particular stages of the Phanerozoic as compiled after 300 paleontological monographs (Sepkoski, 1981).

All these estimates are strongly intercorrelated, which is partly the result of the effect of their correlations on evolutionary time. However, their intercorrelation remains statistically significant even after removal of the latter effect. This shows that the underlying pattern of organic diversity in evolutionary time is real. This general pattern can be summarized as follows: (1) low diversity in the Cambrian; (2) higher but not persistently increasing diversity through the remainder of the Paleozoic; (3) low diversity in the Triassic; and (4) increasing diversity through the Mesozoic and Cenozoic.

Unfortunately, these results do not solve the whole problem. First of all, the data concern almost exclusively shelly fauna. It is unknown whether the nonshelled fauna follows the same pattern, or an independent one, or whether it has evolved in complementarity to the shelly fauna under constraints of a global equilibrium. Second, the observed pattern differs considerably from what has been recorded above the familial level (Valentine, 1969; Sepkoski, 1979), whereas there is no explanation for a real discontinuity between the familial and ordinal levels of the artificial taxonomic hierarchy. Third, the results are too vague to permit an answer to any important question about the evolution of organic diversity. Was an equilibrium diversity achieved and maintained in the Paleozoic? Can we speak about multiple equilibria in the history of the biosphere, as invoked previously by Bambach (1977) and Sepkoski (1979)? Was diversity smaller in the Triassic than in the Permian? Until we answer these and similar questions, the results of the stochastic modeling of organic diversity in evolutionary time cannot be interpreted reliably. Empirical data are, therefore, badly needed. However, as admitted by Sepkoski *et al.* (1981), the increased taxonomic and stratigraphic knowledge since the oldest estimate of diversity was made by Phillips (1860; see Rudwick, 1972) does not seem to increase appreciably the accuracy of this general pattern.

The approach that applies stochastic modeling to analysis of the evolution of organic diversity is objectionable also in theory. There is no process in nature that cannot be simulated by a more or less simple, or plausible, or even elegant, stochastic process. The question is only: How likely is such an outcome of simulation? Therefore, one is unable to apply stochastic modeling to test real events for randomness unless one has an a priori theory describing the structure of the process to be investigated. First of all, the probability distribution must be assumed. One needs considerable theoretical background to substantiate a preference for one rather than another stochastic process as a model for the investigated events. Why claim, for instance, that a simple random-walk process with equal probabilities of change in either direction is more adequate to explain the phyletic change in size of foraminiferal tests (Raup, 1977a) or ammonite shells (Raup and Crick, 1981), than a model assuming that a stasis or decrease in size is more likely to occur than an increase? The latter assumption about the probability distribution may actually be more plausible when the initial size is close enough to the upper limit of the size range set up by some mechanical, physiological, and/or ecological constraints. Yet the likelihood of obtaining a congruence between the outcome of simulation and the actual events depends upon the assumed probability distribution.

Concerning the evolution of organic diversity, it is crucial to ask whether or not it is justifiable to assume, as did Raup *et al.* (1973), Gould *et al.* (1977), Sepkoski (1978, 1979), and Carr and Kitchell (1980), that the probabilities of species extinction and branching are equal. The underlying theory is that a global, or at least continental, evolutionary equilibrium in organic diversity is commonly achieved and maintained, due to intrinsic, biological controls, over long periods of time (Webb, 1969; E. O. Wilson, 1969; Rosenzweig, 1975). It assumes that the per species speciation rate decreases and the per species extinction rate increases with increasing diversity of the biosphere, because the average population size declines and interspecific competition becomes more and more intense. These assumptions are explicitly derived from the theory of island biogeography, which is obviously restricted in its applicability to the ecological time scale. It is questionable whether that theory can be justifiably extrapolated over evolutionary time (Hoffman, 1981*b*). It is indeed reasonable to expect that per species extinction rate increases with increasing diversity. However, the relationship of per species speciation rate to diversity is unclear. On the ecological time scale, the immigration rate from the mainland to an island decreases with increasing diversity simply because the likelihood that constantly immigrating individuals represent a new species decreases as the island diversity approaches the mainland diversity. This argument is nonapplicable in terms of speciation. As a matter of fact, all ecological and biogeographical factors that supposedly tend to decrease the per species speciation rate due to increased species packing and decreased average population size (Rosenzweig, 1975; Sepkoski, 1978) may well be counterbalanced by an increase in speciation potential by genetic drift and competitive-speciation mechanisms (see Rosenzweig, 1978; Rosenzweig and Taylor, 1980). Perhaps, then, there is no equilibrium point whatsoever (see Flessa and Levinton, 1975). Even if one takes for granted that an evolutionary equilibrium diversity exists as the upper limit to the total number of species, the equilibrium point may be at a level too high ever to have been reached in the course of the evolution of the biosphere. The realized limits to biotic diversity may well be set by abiotic factors, if only because of the frequency of environmental disturbances (Valentine, 1973; Whittaker, 1977). The empirical evidence presented thus far in support of long-term persistence of evolutionary equilibrium diversities in the Phanerozoic (Bretsky and Bretsky, 1976;

Webb, 1976; Bambach, 1977; Mark and Flessa, 1977; Sepkoski, 1979; Sepkoski *et al.*, 1981) is inconclusive. Furthermore, it has been recently demonstrated by Stanley *et al.* (1981) that even if the existence of a global evolutionary equilibrium and its actual damping effect on diversity are accepted, the likelihood of obtaining a given result of stochastic modeling depends, to a large extent, upon the level of that equilibrium diversity and the probability of speciation and extinction events. These basic data, however, can be only assumed for, but not measured in, the fossil record. Consequently, one can never get certainty that the patterns to be subtracted from an estimate of the actual evolution of organic diversity are likely or unlikely to arise by biological processes.

I conclude that the theoretical background of the paleobiological research program concerned with the evolution of organic diversity is unacceptable; and even if it were acceptable, the method that is most commonly applied to this end, that is, stochastic modeling in order to test the course of that evolution for randomness, is inadequate.

Macroevolution

It might seem absurd to expect that a revolution in evolutionary biology be triggered by paleobiological research. This is, however, the case, or at least this is claimed to be the case. The rarity of well-documented gradual phyletic changes in the fossil record and the commonness of missing links between various *Baupläne* or even their varieties have been explained by the theory of "punctuated equilibria" (Eldredge and Gould, 1972; Gould and Eldredge, 1977; Eldredge, 1979). Thereby gaps in the fossil record of evolutionary lineages have been interpreted as reflecting the biological reality, rather than as inevitable deficiencies of geological data. Consequently, it has been proclaimed that the neo-Darwinian synthetic theory of evolution has collapsed and a new evolutionary theory has been demanded to account for macroevolutionary events, or the origin of evolutionary novelties (Gould, 1980*b*). To this end, the prevalence of neutral mutations and the overwhelming structural integration of organisms have been invoked and hence, Goldschmidtian "hopeful monsters," developmental constraints, and epigenetic mechanisms have been referred to in order to account for macroevolutionary change (Gould 1980*b*, 1980*c*, 1982; Rachootin and Thomson, 1981; Alberch, 1982; Maderson *et al.*, 1982). On the other hand, the emphasis on the reality of morphological gaps between ancestral and descendant species also precludes an explication of large-scale evolutionary trends by long-term

persistence of selective pressure. Therefore, the theory of species selection has been put forth. It claims that there are major features of
the biosphere that cannot be explained by the evolution of populations through natural selection, but only by the differential rates of
speciation and extinction in various lineages (Stanley, 1975, 1979;
Gould and Eldredge, 1977; Vrba, 1980; Gilinsky, 1981).

Briefly, the theory of punctuated equilibria has given a stimulus to
decouple macroevolution from microevolution, which is meant in this
context as the processes envisaged by the neo-Darwinian theory of
evolution of populations. It has become the most fashionable, but
also the most controversial, paleobiological research program to analyze the former, perhaps because the latter is beyond the reach of
paleobiological methodology. The controversies are partly due to the
fact that the terminology used is far from being unequivocal. However, there are also real empirical and theoretical issues. The limitations of this research program with respect to species selection are
discussed lucidly by Maynard Smith (1982). Its empirical background
and its actual consequences for the origin of evolutionary novelties
are equally limited (Levinton and Simon, 1980; Schopf, 1981a; Hoffman, 1981c, 1982).

The model of punctuated equilibria claims that the large majority
of morphological changes in evolution take place very rapidly and in
populations that are so small that their discovery in the fossil record
is virtually impossible, whereas the change along the continuous segments of evolutionary lineage is negligible. Therefore, gaps in the
fossil record of evolutionary lineages reflect some real biological phenomena, even though these may not be true jumps or discontinuities.
The opposite model of phyletic gradualism claims that morphological
changes in evolution are accomplished mostly by a gradual shift of
the range of intraspecific variation; this shift does not necessarily have
to be unidirectional or to proceed at a constant rate but it must eventually exceed the limits of a single species. These two models are not
mutually exclusive. A continuous spectrum of modes of evolution
may occur even within a single evolutionary lineage. Morphological
gaps between the continuous segments of a lineage may be either
larger or smaller than the net morphological change along those segments, or they may not exist at all. The latter case would be entirely
consistent with the model of phyletic gradualism, the former one with
the model of punctuated equilibria, but the intermediate case cannot
be ruled out a priori.

In order to evaluate the significance of macroevolutionary consequences of the theory of punctuated equilibria, it has been considered
crucial to estimate the relative frequency of the two modes of evo-

lution. In fact, if punctuated equilibria were the only, or even merely a predominant, mode of evolution, the macroevolutionary theory demanded by Gould (1980*b*) would really be indispensable. Stasis and punctuations might be attributed to the very nature of the epigenetic system, because cybernetic analysis of the latter would predict its discontinuous behavior. It might be expected to shift rapidly from one domain of attraction to another (see Ho and Saunders, 1979; Alberch 1982; surprisingly enough, catastrophe theory has been only marginally considered in this context; Dodson and Hallam, 1977). Punctuated patterns of morphological evolution might then be regarded as reflecting a jerkiness in macroevolutionary processes that lead from one *Bauplan* to another. Consequently, macroevolution would be indeed decoupled from microevolutionary processes, as the latter operate upon conspecific populations within the limits of a single domain of attraction available to each specific morphogenetic and epigenetic system. Macroevolutionary patterns would appear explicable solely in morphological or morphogenetical, but not in genetical terms.

However, this may not be the case. The debate on the relative frequency of the punctuated versus the gradual mode of evolution has already lasted almost a decade and there is no way to settle it through empirical arguments. With all the notorious deficiencies of the fossil record taken into account, each position can be defended endlessly with the use of more or less plausible ad hoc hypotheses. First of all, it is extremely difficult, if not impossible, to provide a positive evidence for punctuated equilibria, if only because one can hardly prove that punctuation events observed in one area reflect more than immigrations from another area (unknown, inaccessible, or even nonpreserved) where the lineage did evolve gradually. This is well exemplified by the record of the evolution of Early Jurassic dactylioceratid ammonites (Howarth, 1973). There is a vertical sequence of distinctive dactylioceratid populations in Great Britain, with no intermediates in between. This pattern seems to be recurrent throughout Europe and perhaps even beyond, and it might be interpreted as a case for punctuated equilibria. However, there are time gaps between the particular chronospecies and there is no way to determine whether the intermediates are lacking because of their extreme rarity, or whether each of the British dactylioceratid populations simply immigrated from somewhere else as the gradualists would claim.

Certainly, an indirect and purely qualitative support for punctuated equilibria can be found in the commonness of morphological stasis in evolutionary time, especially when taken in conjunction with the well-known rarity of unquestionable ancestor–descendant relationships among fossil species (see Gould and Eldredge, 1977; Hallam, 1978; Stanley, 1979; Hoffman and Szubzda-Studencka, 1982). On the other hand, however, quite a large number of examples can be cited in support of the model of phyletic gradualism (Hoffman, 1982). They concern protists as well as metazoans (brachiopods, gastropods, cephalopods, conodontophorids, graptolites), and involve some real morphological changes and not only an increase in size, which has been commonly accepted as an exception to the preponderance of punctuated equilibria. These data greatly undermine the supposed universality, or even numerical predominance, of the punctuated mode of evolution.

There are punctuated as well as gradual patterns of morphological evolution. There are also intermediate cases (Cisne *et al.*, 1980). It thus becomes crucial to ask why some species-level lineages undergo gradual morphologic evolution, whereas others show stasis and punctuations. This variation in the mode of evolution might be assigned to differential characteristics of the changes of habitats and ecospace through time. This would, however, imply a strict adherence either to the billiard-ball model of organisms, or to the Panglossian paradigm of Gould and Lewontin (1979). If the evolutionary behavior of species were considered as controlled entirely by extrinsic, environmental factors, organisms would appear as nothing but puppets driven with perfect efficiency from outside; or else they would appear as optimal machines able to rearrange immediately their form and function in response to new environmental challenges. There is little doubt that evolution is stimulated, or even required, by environmental challenges. Nevertheless, the extrinsic controls are mediated by some biological mechanisms. Organisms may be justifiably regarded as striving constantly for optimality within the constraints of their phylogenetic heritage and constructional possibilities, but not as being optimal under any given environmental regime (Lewontin, 1978).

The tempo and mode of evolutionary response of species to environmental challenges can be expected to vary from one species to another, depending upon their biological and ecological characteristics. This is expressed by the punctuated versus the gradual mode of morphological evolution. An explanation for this variation has been attempted in terms of developmental homeostasis of individuals, genetic homeostasis of populations, and ecologic homeostasis of communities (Eldredge and Gould, 1972; Gould and Eldredge, 1977; Hoffman, 1978*b*), but all those explanations have failed (Hoffman, 1982). Actually, they have been doomed because they referred to the very process of speciation and its biological controls, whereas what we really have evidenced in the fossil record as being punctuated or gradual are only patterns in morphological change. The mutual identity of the patterns of morphological and genetical change has been assumed by the most optimistic paleobiologists, but it seems very unlikely that it does exist (Schopf, 1981a, 1981b).

Various organisms are able to develop into the "normal" adult form under a very wide variety of more or less unusual environmental conditions. This ability obviously is of considerable selective advantage. It is termed "developmental canalization," or "homeorhesis" (Waddington 1957, 1975). This fundamental property of individual development is brought about by the action of the epigenetic system which is responsible for the fact that only certain genetic potentialities of the organism are revealed during ontogeny. The opposite to developmental canalization is developmental plasticity. This is the ability of a single genotype to produce highly variable phenotypes under variable environmental regimes. There is a large body of data demonstrating that species are widely variable in the stabilizing power of their homeorhetic mechanisms, which ranges from strong canalization to considerable plasticity (see Waddington, 1975; Stearns, 1981).

Given this absence of an unequivocal, one-to-one relationship between the genotype and its resultant phenotype, the hypothesis can be put forth that punctuations in morphological evolution occur in species displaying considerable amounts of developmental canalization, whereas gradual evolution is confined to species with a high degree of plasticity; however, morphological stasis may be brought about by canalization as well as by plasticity (Stearns, 1980, 1982; Hoffman, 1981c). To test this hypothesis, the relationship of the mode of morphological evolution to the pattern of morphological variation must be investigated, because the latter can be expected to provide an approximate measure of the stabilizing power of homeorhetic mechanisms. In fact, there are some data to support this hypothesis (Hoffman, 1982). They point out that the lineages that present the best documented examples of gradual morphological evolution all show very large amounts of morphological variation, which may indicate much developmental plasticity. Parallel with a gradual shift in morphology, however, the range of variation commonly undergoes a considerable reduction. On the other hand, morphological punctuations coincide with a significant increase in intraspecific variability in the few well-documented cases of punctuated evolution. These data suggest that a previously established developmental canalization must be broken down in order to permit achievement of a new morphology.

However, this is not to say that a new macroevolutionary theory is necessary to account for those punctuational events because natural selection is irrelevant to, or negligible in the context of, macroevolutionary change. Certainly, evolutionary novelties may arise very rapidly, by single genetical or developmental changes, if a functional threshold is being crossed (Jaanusson, 1981). This can be exemplified by the evolution of the jaw of bolyerine snakes (Frazzetta, 1970), the uniserial rhabdosome of monograptid graptolites (Jaanusson, 1973), and the characteristics of the plethodontid salamander *Aneides* (Larson *et al.*, 1981). All these features have apparently arisen in single steps, without any adaptive intermediate stages, although the associated morphological modifications have occurred gradually. A similar pattern has been envisaged previously for the evolution of cichlid fishes (Liem, 1974) and birds (Bock, 1979). However, the underlying mechanisms may be entirely consistent with the concept of developmental canalization and plasticity being adaptive and genetically determined (Waddington, 1957, 1975; Stearns, 1982). According to that concept, developmental canalization of a character may be adaptively advantageous under a certain environmental regime, but it is broken down under another regime which may permit achievement of a key innovation; whereas developmental plasticity may provide the optimal adaptive strategy under varying environmental conditions.

The punctuated and the gradual modes of morphological evolution appear then as mere epiphenomena of the underlying genetic processes that are controlled by natural selection and constrained by a variety of historical, mechanical, and physiological factors, but not by principles of morphological transformation. Macroevolution turns out to be reducible to microevolutionary processes. Or at least, the theory of punctuated equilibria cannot serve as an argument for the decoupling of macroevolution from microevolution. Consequently, however, there is no reason to expect that the mechanisms of the origin of evolutionary novelties are within the reach of paleobiological analysis. The research program intended to analyze macroevolutionary change should focus first of all upon relationships of canalization and plasticity to various ecological regimes. It should attempt to reconcile evolutionary ecology and developmental biology. Perhaps this can be done within the conceptual framework of the neo-Darwinian synthetic theory.

Paleobiological strategy of research

Paleobiologists have long attempted to discover the fossil record of microscopic evolutionary and ecological processes (character displacement, ecological succession, speciation, etc.), but what they searched for is generally not to be found. The pendulum has now swung to the other extreme. Paleobiologists claim to have discovered macroscopic biological phenomena (community evolution, evolution of organic diversity, macroevolution) that are perceivable exclusively on the evolutionary time scale and, therefore, inexplicable in terms of microscopic processes as discussed by population geneticists and evolutionary ecologists. If this were so, macroprocesses should be discovered and analyzed, and this general aim might provide paleobiologists with new and fruitful research programs.

I have argued that this is not the case. First of all, the paleontological data one can get are generally insufficient to describe the patterns of community evolution, evolution of organic diversity, and macroevolutionary change. Second, I believe that the recently emphasized macroscopic phenomena are reducible to genetic and ecologic processes under plausible geologic boundary conditions. One can, therefore, use those macroscopic terms to describe the biological reality, but not to explain it. It is a consequence of my option for the individualistic approach to biology that I do not believe in any macroevolutionary law. What can be achieved at the supraspecific levels are merely generalizations. This is why a counterexample can always be found in ecology and evolutionary biology, exactly as is the case with psychology, sociology, and history.

All the recent paleobiological debates have, in my opinion, contributed much to the conceptualization of paleobiological research itself, but nothing to evolutionary biology. This is not to say that paleobiologists cannot contribute to evolutionary biology. I contend only that they have not done so. For the promise of paleobiology lies in its potential ability to provide data critical for the evaluation of contradictory hypotheses that are equally acceptable from the standpoint of the evolutionary biologists. This is the only way that paleobiologists can find between the Scylla of hopelessly copying what biologists do and the Charybdis of the equally hopeless invention of new, but illegitimate, evolutionary principles. To this end, however, paleobiologists should focus their research on patterns rather than on processes. Their strategy should be to search for macroscopic patterns, defying observation on the ecological time scale, and to explain them in terms of microscopic biological processes under acceptable geological boundary conditions. In this context geological evidence gains paramount importance.

The few existing paleobiologic programs to follow this research strategy include first of all the one concerned with the explanation of organic form. Observed patterns involve the potential and the realized phenospace of major organic groups, and the change in the realized phenospace through time. Potential phenospace is investigated by the program of theoretical morphology (Raup, 1966, 1967); realized phenospace and its changes in evolutionary time are studied by classic paleontography and phylogenetic reconstruction which may supply also quantitative data (e.g., McGhee, 1980; Chamberlain, 1981). To permit explanations for those patterns, the program of constructional morphology has been developed (Seilacher, 1970; Thomas, 1979). Results of such research may turn out to be crucial in current debates on the causes for discreteness of the realized phenospace, which is of paramount importance for the whole issue of macroevolution (cf. Alberch, 1982; Maderson *et al.*, 1982).

Another program of this kind might be concerned, for instance, with biological response to long-term environmental changes. It is then to be asked how populations and ecosystems react to systematic deterioration and eventual destruction of their habitat islands (e.g., a reef growing up into progressively shallower water, or a lake being gradually filled in with sediments). One might also attempt to discover what really happens in a long-term persistent ecosystem under constant environmental conditions ("constant" here does not mean stable in ecological time but merely persistent within the same environmental framework; e.g., delta in a regressive geological sequence). It is then to be asked whether or not there is any evolutionary equilibrium in niche pattern, a question which has thus far been satisfactorily answered neither in theory nor in empirical practice; whether newcomers to the ecosystem appear in large clusters, or one independently of another; whether they subsequently coevolve, or merely persist with morphological stability through time (Hoffman, 1981*a*).

The fossil record may be good enough to permit an attempt to solve these problems, at least in some case studies of particularly suitable geological sequences. This would be a major contribution to the present debate on the Red Queen hypothesis of Van Valen (1973, 1976), especially because it seems that the Red Queen hypothesis and its opposite evolutionary model are equally plausible on various theoretical grounds (Stenseth, Maynard Smith, and Haigh, Ms). The Red Queen hypothesis tells us that morphological evolution, speciation, and extinction continue even in the absence of any environmental change. The other model claims that the rate of evolutionary change gradually declines and evolution eventually stops in the absence of changes in the physical environment. The two models cannot be tested directly in the fossil record, but it follows from the Red Queen picture that bursts of rapid evolution affecting many ecologically interrelated species should occur, followed by longer periods of slower coevolution among coexisting species; whereas the other model implies that species appear, evolve, and disappear independently of one another, as a rule. Some preliminary data I have analyzed (Hoffman, 1979) suggest that the former picture may hold true for shallow-marine benthic mollusks, whereas the latter is adequate for pelagic microorganisms. Obviously, much additional data are necessary to get even an approximate solution to this problem.

These two examples of paleobiological research programs (one being already realized, the other merely outlined) are here invoked to show that the strategy of research I am advocating is not only productive in the sense of the empirical studies it stimulates, but also that it may permit a significant contribution to the discovery of general biological principles. Nothing more is needed to solve the fundamental dilemma of modern paleobiologists, to justify the existence of paleobiology as a science.

References

Alberch, P. 1982. Developmental constraints in evolutionary processes. In *Evolution and Development*, ed. J. T. Bonner, pp. 313–32. Heidelberg: Springer.

Bambach, R. K. 1977. Species richness in marine benthic habitats through the Phanerozoic. *Paleobiology* 3:152–67.

Bertalanffy, L. von. 1968. *General Systems Theory*. New York: Braziller.

Bock, W. J. 1979. The synthetic explanation of macroevolutionary change, a reductionist approach. *Bull. Carnegie Mus. Nat. Hist.* 13:20–69.

Boucot, A. J. 1975. *Evolution and Extinction Rate Control*. Amsterdam: Elsevier.

1978. Community evolution and rates of cladogenesis. *Evol. Biol.* 11:545–655.

1981. *Principles of Benthic Marine Paleoecology*. New York: Academic Press.

Bretsky, P. W. 1969. Evolution of Palaeozoic benthic marine invertebrate communities. *Palaeogeogr. Palaeoclimatol. Palaeoecol.* 6:45–59.

Bretsky, P. W., and S. S. Bretsky. 1976. The maintenance of evolutionary equilibrium in Late Ordovician benthic marine invertebrate faunas. *Lethaia* 9:223–33.

Carr, T. R., and J. A. Kitchell. 1980. *Dynamics of taxonomic diversity. Paleobiology* 6:427–33.

Chamberlain, J. A. 1981. Hydromechanical design of fossil cephalopods. In M. R. House and J. R. Senior, eds. *The Ammonoidea, Syst. Assoc. Spec.* 18:289–336. London: Academic Press.

Charnov, E. L. 1977. An elementary treatment of the genetic theory of kin selection. *J. Theor. Biol.* 66:1–52.

Christiansen, F. B., and T. M. Fenchel. 1977. *Theories of Populations in Biological Communities*. Heidelberg: Springer.

Cisne, J. L., G. O. Chandlee, B. D. Rabe, and J. A. Cohen. 1980. Geographic variation and episodic evolution in an Ordovician trilobite. *Science* 209:925–7.

Connell, J. H. 1972. Community interactions on marine rocky intertidal shores. *Ann. Rev. Ecol. Syst.* 3:169–92.

1978. Diversity in tropical rain forests and coral reefs. *Science* 199:1302–10.

1980. Diversity and the coevolution of competitors, or the ghost of competition past. *Oikos* 35:131–8.

Crame, J. A. 1980. Succession and diversity in the Pleistocene coral reefs of the Kenya coast. *Palaeontology* 23:1–37.

Dawkins, R. 1976. *The Selfish Gene.* Oxford: Oxford Univ. Press.

Dodd, J. R., and R. J. Stanton. 1981. *Paleoecology, Concepts and Applications.* New York: Wiley.

Dodson, M. M., and A. Hallam. 1977. Allopatric speciation and the fold catastrophe. *Am. Natur.* 111:415–33.

Dunbar, M. J. 1960. The evolution of stability in marine environments. Natural selection at the level of the ecosystem. *Am. Natur.* 94:129–36.

Eldredge, N. 1979. Alternative approaches to evolutionary theory. *Bull. Carnegie Mus. Nat. Hist.* 13:7–19.

Eldredge, N., and S. J. Gould. 1972. Punctuated Equilibria: an Alternative to Phyletic Gradualism. In *Models in Paleobiology*, ed. T. J. M. Schopf, pp. 82–115. San Francisco: Freeman.

Flessa, K. W., and J. S. Levinton. 1975. Phanerozoic diversity patterns: Tests for randomness. *J. Geol.* 83:239–48.

Frazzetta, T. 1970. From hopeful monsters to bolyerine snakes. *Am. Natur.* 104:55–72.

Ghiselin, M. T. 1974. *The Economy of Nature and the Evolution of Sex.* Berkeley: Univ. of California Press.

Gilinsky, N. L. 1981. Stabilizing species selection in the Archaeogastropoda. *Paleobiology* 7:315–31.

Gleason, H. A. 1926. The individualistic concept of the plant association. *Bull. Torrey Bot. Club* 53:331–68.

Gould, S. J. 1976. Paleontology plus Ecology as Paleobiology. In *Theoretical Ecology*, ed. R. M. May, pp. 218–36. Philadelphia: Saunders.

1977. Eternal Metaphors of Paleontology. In *Patterns of Evolution*, ed. A. Hallam, pp. 1–26. Amsterdam: Elsevier.

1980a. The promise of paleobiology as a nomothetic, evolutionary discipline. *Paleobiology* 6:96–118.

1980b. Is a new and general theory of evolution emerging? *Paleobiology* 6:119–30.

1980c. The evolutionary biology of constraint. *Daedalus (Spring, 1980)*:39–52.

1982. Change in Developmental Timing as a Mechanism of Macroevolution. In *Evolution and Development*, ed. J. T. Bonner, pp. 333–46. Heidelberg: Springer.

Gould, S. J., and N. Eldredge. 1977. Punctuated equilibria: the tempo and mode of evolution reconsidered. *Paleobiology* 3:115–51.

Gould, S. J., and R. C. Lewontin. 1979. The spandrels of San Marco and the Panglossian paradigm: a critique of the adaptationist program. *Proc. Roy. Soc. Lond. B* 205:581–98.

Gould, S. J., D. M. Raup, J. J. Sepkoski, T. J. M. Schopf, and D. S. Simberloff. 1977. The shape of evolution: a comparison of real and random clades. *Paleobiology* 3:23–40.

Gray, J., A. J. Boucot, and W. B. N. Berry, eds. 1981. *Communities of the Past.* Stroudsburg: Hutchinson Ross.

Hallam, A. 1978. How rare is phyletic gradualism and what is its evolutionary significance? Evidence from Jurassic bivalves. *Paleobiology* 4:16–25.

Hamilton, W. D. 1964. The genetical evolution of social behaviour. *J. Theor. Biol.* 7:1–52.

Hansen, T. A. 1980. Influence of larval dispersal and geographic distribution on species longevity in neogastropods. *Paleobiology* 6:193–207.

Harger, J. R. E. 1972. Competitive coexistence among intertidal invertebrates. *Am. Sci.* 60:600–7.

Heatwole, H., and R. Levins. 1972. Trophic structure stability and faunal change during recolonization. *Ecology* 53:531–4.

Ho, M. W., and P. T. Saunders. 1979. Beyond neo-Darwinism – An epigenetic approach to evolution. *J. Theor. Biol.* 78:573–91.

Hoffman, A. 1977. Synecology of macrobenthic assemblages of the Korytnica Clays (Middle Miocene; Holy Cross Mountains, Poland). *Acta Geol. Polon.* 27:227–80.

   1978a. System concepts and the evolution of benthic communities. *Lethaia* 11:181–5.

   1978b. Punctuated-equilibria evolutionary model and paleoecology. *Ann. Soc. Geol. Polon.* 48:327–31.

   1979. Community paleoecology as an epiphenomenal science. *Paleobiology* 5:357–9.

   1980. System-analytic conceptual framework for community paleoecology. *Ann. Soc. Geol. Polon.* 50:161–72.

   1981a. Growing points in community paleoecology. In SFB 53 "Palökologie," Tübingen, Bericht 1979–1981; *N. J. Geol. Paläont. Abh.*

   1981b. Stochastic versus deterministic approach to paleontology: The question of scaling or metaphysics? *N. J. Geol. Paläont. Abh.* 162:80–96.

   1981c. Biological controls of the punctuated versus gradual mode of species evolution. In Int. Symp. "Concept and Method in Paleontology," Contr. Papers, ed. J. Martinell, pp. 57–63. Barcelona: Univ. de Barcelona.

   1982. Punctuated versus gradual mode of evolution: A reconsideration. *Evol. Biol.* 15:411–36.

Hoffman, A., A. Pisera, and W. Studencki. 1978. Reconstruction of a Miocene kelp-associated macrobenthic ecosystem. *Acta Geol. Polon.* 28:377–87.

Hoffman, A., and B. Szubzda. 1976. Paleoecology of some molluscan assemblages of the Badenian (Miocene) marine sandy facies of Poland. *Palaeogeogr. Palaeoclimatol. Palaeoecol.* 20:307–32.

Hoffman, A., and B. Szubzda-Studencka. 1982. Bivalve species duration and ecologic characteristics in the Badenian (Miocene) marine sandy facies of Poland. *N. J. Geol. Paläont. Abh. 163*:122–35.

Howarth, M. K. 1973. The stratigraphy and ammonite fauna of the Upper Liassic Grey Shales of the Yorkshire coast. *Bull. Brit. Mus. Nat. Hist. (Geol.) 24*:237–77.

Hull, D. L. 1980. Individuality and selection. *Ann. Rev. Ecol. Syst. 11*:311–32.

Jaanusson, V. 1973. Morphological discontinuities in the evolution of graptolite colonies. In *Animal Colonies*, ed. R. S. Boardman, A. H. Cheetham, and W. A. Oliver, pp. 515–21. Stroudsburg: Dowden, Hutchinson & Ross.

1981. Functional thresholds in evolutionary progress. *Lethaia 14*:251–60.

Jablonski, D. 1980. Apparent versus real biotic effects of transgressions and regressions. *Paleobiology 6*:397–407.

Jackson, J. B. C. 1974. Biogeographic consequences of eurytopy and stenotopy among marine bivalves and their evolutionary significance. *Am. Natur. 108*:541–60.

Jacob, F. 1970. *La Logique du Vivant*. Paris: Gallimard.

Koch, C. F. 1980. Bivalve species duration, areal extent and population size in a Cretaceous sea. *Paleobiology 6*:184–92.

Koestler, A., and J. R. Smythies, eds. 1969. *Beyond Reductionism*. London: Hutchinson.

Krylov, N. S. 1979. *Works on the Foundations of Statistical Physics*. Princeton: Princeton Univ. Press.

Larson, A., D. B. Wake, L. R. Maxson, and R. Highton. 1981. A molecular phylogenetic perspective on the origin of morphological novelties in the salamanders of the tribe Plethodontini (Amphibia, Plethodontidae). *Evolution 35*:405–22.

Levins, R. 1975. Evolution in Communities Near Equilibrium. In *Ecology and Evolution of Communities*, ed. M. L. Cody and J. M. Diamond, pp. 16–50. Cambridge, Mass.: Harvard Univ. Press.

Levinton, J. S., and C. M. Simon. 1980. A critique of the punctuated equilibria model and implications for the detection of speciation in the fossil record. *Syst. Zool. 29*:130–42.

Lewontin, R. C. 1978. Adaptation. *Sci. Am. 239*:213–30.

Liem, K. F. 1974. Evolutionary strategies and morphological innovations: cichlid pharyngeal jaws. *Syst. Zool. 22*:425–441.

Łomnicki, A. 1978. Individual differences between animals and the natural regulation of their numbers. *J. Anim. Ecol. 47*:461–75.

1980. Regulation of population density due to individual differences and patchy environment. *Oikos 35*:185–93.

MacArthur, R. H., and E. O. Wilson, 1967. *The Theory of Island Bio-geography*. Princeton: Princeton Univ. Press.

Maderson, P. F. A., P. Alberch, B. C. Goodwin, S. J. Gould, A. Hoffman, J. D. Murray, D. M. Raup, A. de Ricqles, A. Seilacher, G. P. Wagner, and D. B. Wake. 1982. The Role of Development in Macroevolutionary Change. In *Evolution and Development*, ed. J. T. Bonner, pp. 279–312. Heidelberg: Springer.

Margalef, R. 1968. *Perspectives in Ecological Theory*. Chicago: Univ. of Chicago Press.

Mark, G. A., and K. W. Flessa. 1977. A test for evolutionary equilibria: Phanerozoic brachiopods and Cenozoic mammals. *Paleobiology* 3:17–22.

May, R. M. 1973. *Stability and Complexity in Model Ecosystems*. Princeton: Princeton Univ. Press.

1976. Patterns in multi-species communities. In *Theoretical Ecology*, ed. R. M. May, pp. 142–62. Philadelphia: Saunders.

Maynard Smith, J. 1976. A comment on the Red Queen. *Am. Nat.* 110:325–30.

Maynard Smith, J., and G. R. Price. 1973. The logic of animal conflict. *Nature* 246(5927):15–18.

McGhee, G. R. 1980. Shell form in the biconvex articulate Brachiopoda: a geometric analysis. *Paleobiology* 6:57–76.

Monod, J. 1970. *Le Hasard et la Necessité*. Paris: Seuil.

Odum, E. P. 1969. The strategy of ecosystem development. *Science* 164:262–70.

Olson, E. C. 1952. The evolution of a Permian vertebrate chronofauna. *Evolution* 6:181–96.

1966. Community evolution and the origin of mammals. *Ecology* 47:291–302.

Peterson, C. H. 1975. Stability of species and of community for the benthos of two lagoons. *Ecology* 56:958–65.

Rachootin, S., and K. S. Thomson. 1981. Epigenetics, Paleontology, and Evolution. In G. G. E. Scudder and J. L. Reveal, ed. *Evolution Today*, Proc. Second Int. Congr. Syst. Evol. Biol., Pittsburgh (Carnegie-Mellon): Hunt Institute, pp. 181–93.

Raup, D. M. 1966. Geometric analysis of shell coiling: general problems. *J. Paleont.* 40:1178–90.

1967. Geometric analysis of shell coiling: coiling in ammonoids. *J. Paleont.* 41:43–65.

1976a. Species diversity in the Phanerozoic: a tabulation. *Paleobiology* 2:279–88.

1976b. Species diversity in the Phanerozoic: an interpretation. *Paleobiology* 2:288–97.

1977a. Stochastic models in evolutionary paleontology. In *Patterns of Evolution*, ed. A. Hallam, pp. 59–78. Amsterdam: Elsevier.

1977b. Probabilistic models in evolutionary paleobiology. *Am. Sci.* 65:50–7.

1978. Cohort analysis of generic survivorship. *Paleobiology* 4:1–15.

Raup, D. M., and R. E. Crick. 1981. Evolution of single characters in the Jurassic ammonite *Kosmoceras*. *Paleobiology* 7:200–15.

Raup, D. M., S. J. Gould, T. J. M. Schopf, and D. S. Simberloff. 1973. Stochastic models of phylogeny and the evolution of diversity. *J. Geol.* 81:525–42.

Rosenzweig, M. L. 1975. On continental steady states of species diversity.

In *Ecology and Evolution of Communities*, ed. M. L. Cody and J. M. Diamond, pp. 121–40. Cambridge, Mass.: Harvard Univ. Press.

1978. Competitive speciation. *Biol. J. Linn. Soc. 10*:275–89.

Rosenzweig, M. L., and J. A. Taylor. 1980. Speciation and diversity in Ordovician invertebrates: filling niches quickly and carefully. *Oikos 35*:236–43.

Rudwick, M. J. S. 1972. *The Meaning of Fossils*. London: Macdonald.

Schopf, T. J. M. 1974. Permo-Triassic extinctions: relation to seafloor spreading. *J. Geol. 82*:129–43.

1979. Evolving paleontological views on deterministic and stochastic approaches. *Paleobiology 5*:337–52.

1981*a*. Punctuated equilibrium. *Paleobiology 7*:158–66.

1981*b*. Evidence from findings of molecular biology with regard to the rapidity of genomic change: Implications for species durations. In *Paleobotany, Paleoecology, and Evolution, Vol. 1*, ed. K. J. Niklas, pp. 135–92. New York: Praeger.

Seilacher, A. 1970. Arbeitskonzept zur Konstruktionsmorphologie. *Lethaia 3*:393–6.

1974. Flysch trace fossils: Evolution of behavioural diversity in the deepsea. *N. J. Geol. Palaont. Mh. 1974*, 233–45.

1977. Evolution of trace fossil communities. In *Patterns of Evolution*, ed. A. Hallam, pp. 359–76. Amsterdam: Elsevier.

Sepkoski, J. J. 1978. A kinetic model of Phanerozoic taxonomic diversity. I. Analysis of marine orders. *Paleobiology 4*:223–51.

1979. A kinetic model of Phanerozoic taxonomic diversity. II. Early Phanerozoic families and multiple equilibria. *Paleobiology 5*:222–51.

1981. A factor analytic description of the Phanerozoic marine fossil record. *Paleobiology 7*:36–53.

Sepkoski, J. J., R. K. Bambach, D. M. Rapu, and J. W. Valentine. 1981. Phanerozoic marine diversity and the fossil record. *Nature 293*:435–7.

Shotwell, J. A. 1964. Community succession in mammals of the late Tertiary. In *Approaches to Paleoecology*, ed. J. Imbrie and N. D. Newell, pp. 135–50. New York: Wiley.

Simberloff, D. S. 1974. Equilibrium theory of island biogeography and ecology. *Ann. Rev. Ecol. Syst. 5*:161–82.

1978. Using island biogeographic distributions to determine if colonization is stochastic. *Am. Natur. 112*:713–26.

Stanley, S. M. 1975. A theory of evolution above the species level. *Proc. Natl. Acad. Sci. USA 72*:646–50.

1979. *Macroevolution. Pattern and Process*. San Francisco: Freeman.

Stanley, S. M., P. W. Signor, S. Lidgard, and A. F. Karr. 1981. Natural clades differ from "random" clades: Simulations and analyses. *Paleobiology 7*:115–27.

Stearns, S. C. 1980. A new view of life-history evolution. *Oikos 35*:266–81.

1982. The role of development in the evolution of life histories. In *Evolution and Development*, ed. J. T. Bonner, pp. 237–58. Heidelberg: Springer.

Stenseth, N. C. 1979. Where have all the species gone? On the nature of extinction and the Red Queen hypothesis. *Oikos* 33:196–227.

Stenseth, N. C., J. Maynard Smith, and J. Haigh. Coevolution in ecosystems: Red Queen evolution or stasis? (Ms.)

Teilhard de Chardin, P. 1963. *La place de l'homme dans la nature.* Paris: Flammarion.

Thomas, R. D. K. 1979. Morphology, Constructional. In *Encyclopedia of Paleontology*, ed. R. W. Fairbridge and D. Jablonski, pp. 482–87. Stroudsburg: Dowden, Hutchinson & Ross.

Trivers, R. L. 1971. The evolution of reciprocal altruism. *Q. Rev. Biol.* 46:35–57.

Valentine, J. W. 1968. The evolution of ecological units above the population level. *J. Paleont.* 42:253–67.

1969. Patterns of taxonomic and ecological structure of the shelf benthos during Phanerozoic time. *Palaeontology* 12:684–709.

1970. How many marine invertebrate fossil species? A new approximation. *J. Paleont.* 44:410–15.

1972. Conceptual Models of Ecosystem Evolution. In *Models in Paleobiology*, ed. T. J. M. Schopf, pp. 192–215. San Francisco: Freeman.

1973. *Evolutionary Paleoecology of the Marine Biosphere.* Englewood Cliffs: Prentice-Hall.

1977. General Patterns of Metazoan Evolution. In *Patterns of Evolution*, ed. A. Hallam, pp. 27–57. Amsterdam: Elsevier.

1980. Determinants of diversity in higher taxonomic categories. *Paleobiology* 6:444–50.

Van Valen, L. 1973. A new evolutionary law. *Evol. Theory* 1:1–30.

1976. Energy and evolution. *Evol. Theory* 1:179–229.

Vermeij, G. J. 1978. *Biogeography and Adaptation: Patterns of Marine Life.* Cambridge, Mass.: Harvard Univ. Press.

Vrba, E. S. 1980. Evolution, species, and fossils: How does life evolve? *South Afr. J. Sci.* 76:61–84.

Waddington, C. H. 1957. *The Strategy of the Genes.* London: Allen & Unwin.

1963. *The Ethical Animal.* Chicago: Univ. of Chicago Press.

1975. *The Evolution of an Evolutionist.* Ithaca: Cornell Univ. Press.

Wade, M. J. 1978. A critical review of the models of group selection. *Q. Rev. Biol.* 53:101–14.

Webb, S. D. 1969. Extinction-origination equilibria in late Cenozoic land mammals of North America. *Evolution* 23:688–702.

1976. Mammalian faunal dynamics of the great American interchange. *Paleobiology* 2:220–34.

Whittaker, R. H. 1977: Evolution of species diversity in land communities. *Evol. Biol.* 10:1–67.

Wilson, D. S. 1980. *The Natural Selection of Populations and Communities.* Menlo Park: Benjamin/Cummings.

Wilson, E. O. 1969. The species equilibrium. *Brookhaven Symp. Biol.* 22:38–47.

Wynne-Edwards, V. C. 1962. *Animal Dispersion in Relation to Social Behaviour.* Edinburgh: Oliver & Boyd.

10

Current controversies in evolutionary biology

JOHN MAYNARD SMITH

Evolutionary biologists are arguing about many things – how and why sex evolved, whether some DNA is "selfish," how eukaryotes arose, why some animals live socially, and so on. These problems are, in the main, debated within the shared assumptions of "neo-Darwinism," or "the modern synthesis." Recently, however, a group of paleontologists, of whom Gould, Eldredge, and Stanley have been the most prominent, have announced that the modern synthesis is soon to be swept away, to be replaced by the new paradigm of stasis and punctuation.

In science, a theory is not abandoned unless an alternative theory already exists, ready to replace it. My object in this essay is to identify this alternative, and to explain why I do not find it particularly persuasive. The main aim will be to clarify ideas.

The punctuationist position consists of a minor and a major claim. The minor claim is that the typical pattern of the evolution of species, as revealed by the fossil record, is one of long periods of stasis during which little significant change occurs, interrupted by brief periods of rapid change associated with the splitting of species into two. The major claim is that it is a consequence of this observation, together with a study of development, that the large-scale features of evolution are not the result of the accumulation of changes occurring in populations because of natural selection, together with the processes of speciation as understood by the proponents of the modern synthesis. In brief, macroevolution can be uncoupled from microevolution.

I shall say little about the minor claim. Ultimately, it is a matter of empirical test. I have only two comments to make. The first is that the answer may be hard to come by, as is so often the case in evolutionary biology – compare, for example, the continuing debate about the neutral mutation theory. One difficulty is that a sudden transition from form A to form B in one place may (or may not) conceal the fact that A changed gradually to B in some other place, and that B subsequently migrated to replace A in the area studied. It may, therefore, prove to be easier to establish (or to dismiss) the reality of stasis than to study the nature of the transitions. My second comment is that it will be of little use to analyze the durations in the fossil record of particular named forms, as Stanley (1979) attempts, because this is to study the habits of taxonomists rather than the evolution of organisms. There is no alternative to a statistical study of populations.

At present, I am unconvinced that stasis and punctuation are typical, although I am satisfied that they occur and was persuaded long ago by Simpson (1944) that evolution can proceed at very variable rates. The problem for a population geneticist concerns stasis rather than change. A change comparable to that between species which was completed in 1,000 generations would be rapid to a paleontologist but slow to a population geneticist: consider the changes produced by artificial selection in dogs in little more than 1,000 generations.

For stasis, two explanations have been proposed. One is that the form of a species cannot readily be changed by selection because of constraints during development. This idea lies at the core of the new theory, and I will, therefore, postpone discussion of it until later. The other is that a species does not change because of stabilizing selection; that is, changed individuals are of lower fitness. This raises the question of why the selective optimum should remain unchanged for millions of years. It certainly seems incompatible with Van Valen's (1973) "Red Queen" picture of evolution, according to which each species is evolving as fast as it can to adapt to changes in the others. Nils Stenseth and I have been trying for some time to formulate Van Valen's ideas more precisely. It seems to us that a model combining both ecological and evolutionary time scales can lead to one of two pictures, between which only the fossil record can decide. One is the Red Queen hypothesis, with evolution, extinction, and speciation continuing even in a uniform physical environment. The other is one in which evolutionary change gradually slows down and stops in the absence of changes in the physical environment. We, therefore, have some interest in the outcome of the argument about stasis in the fossil record.

Before leaving the question of stasis, there is one other crucial point to be discussed. This is the difficulty of reconciling the assertion that species display stasis in time with the fact that they manifestly do not do so in space. The essential point can be made by a quotation from Mayr (1963): "We find that in every actively evolving genus there are populations that are hardly different from each other, others that are as different as subspecies, others that have almost reached species level, and finally still others that are full species." In this continuum of degrees of difference, to what is the punctuational event supposed to correspond?

At this point, I cannot resist digressing for a moment to discuss a curious twist in the microhistory of science. In the debate about macroevolution, an alliance has been formed between punctuationalists and "cladists" – the followers of a taxonomic methodology first proposed by Hennig (1966). There is, certainly, no necessary connection between the two points of view; Gould, for example, has said that he is not a cladist. For some time I was unable to see any link between the two positions. Hennig himself was almost too committed to the modern synthesis. However, a recent reading of Hennig, and of Hull (1970), has suggested a connection.

Hennig was concerned that a hierarchical classification (into species, genera, families, etc.) should accurately reflect phylogeny. He went further, and insisted that a classification should reveal to anyone looking at it the phylogenetic hypothesis held by the classifier. If all the species classified are contemporary, this raises no difficulty of principle. If some species are possible ancestors of others, ambiguities arise. Hennig pointed out that these ambiguities are overcome if certain rules are obeyed when naming fossils. The essential rules are that a lineage must *not* change its name except when it splits into two, and that when it does split, both daughter species *must* be given new names. Thus a lineage that changes without splitting can have only one name, no matter how great the change; a species that "buds off" another without itself changing must, nevertheless, change its name.

For Hennig, these were rules of convenience, needed if classifications were to yield unambiguous phylogenies. For punctuationists, they have become assertions about the world – species do not change except when they split. An odd reason for an alliance! Matters have since become still more confused. Hennig's central belief was that classification should reflect phylogeny. His methods – in particular, the distinction between "plesiomorphic" (primitive) and "apomorphic" (derived) traits – make sense only if one supposes that the objects being classified have arisen by a process of branching evolution. Yet there exists today a school of lapsed cladists who continue to use Hennig's methods, while denying that evolution has any bearing on taxonomy. It is as if a lapsed Catholic were to continue to attend Mass.

This discussion of cladism has been a digression: I now turn to the major claim, the so-called uncoupling of macroevolution from the processes of selection in populations. This is best understood in terms of two proposals – "hopeful monsters" and "species selection" – which have so far been offered as alternative mechanisms of evolutionary change.

The idea of "hopeful monsters" goes back to Goldschmidt (1940). Recently, Gould and others have suggested that his ideas, although rejected at the time, may be due for a revival. I want to argue that there are two rather different concepts involved here – which I will refer to as "hopeful monsters" and "systemic mutations" – of which only the latter is clearly incompatible with the modern synthesis. By a "hopeful monster," I shall mean here no more than an individual carrying a genetic mutation of large phenotypic effect. For the term "systemic mutation," I accept Goldschmidt's meaning of a complete repatterning of the genome, by means of a number of rearrangements of the chromosomal material, giving rise in a single step to a new species or higher taxonomic group. It is this latter concept which was, in my view, rightly rejected; I can see no reason for wishing to revive it. In effect, it accounts for novelty by postulating a miracle.

What of a hopeful monster in the simpler sense of a mutant individual so different from typical members of the species that it is able to adopt some new habit or survive in some new environment? Darwinists have had two rather different reasons for expecting evolution to proceed by small steps. The first concerns the "perfection of animals" (Cain, 1964). Even if not perfect, the degree of adaptive fit between organisms and their ways of life can be very striking. A detailed adaptive fit, if produced by natural selection, requires that there is a large amount of finely graded variation for selection to act on. I believe this is correct, but it does not follow that *all* changes should be small – it does not rule out hopeful monsters. It means only that a hopeful monster will need further fine tuning by selection of smaller variants before its descendants achieve detailed adaptation to the new way of life.

A second argument for gradual change, owed to Fisher (1930), would rule out hopeful monsters if it were accepted. Fisher argued that existing organisms are close to a selective optimum, and, hence, that a large change is less likely to improve adaptation than a small one. A large, effectively random change in a complex integrated mechanism is almost certain to have disastrous consequences. This may not be as true of organisms as it is, say, of motorcars. In a paper read at the I.C.S.E.B. at Vancouver in 1980, Thomson argued that development is so buffered that a single large change may be compensated for by many secondary changes in the ontogeny of a single individual, without waiting for further genetic change (Rachootin and Thomson, 1981). To illustrate his point, he referred to a goat described by Slijper (1946) that was born without front legs and consequently adopted a bipedal gait. The anatomy of the backbone, the orientations and lengths of its processes, and the sizes and insertions of muscles and tendons, all changed in ways that could be interpreted as adaptations to a bipedal gait. Presumably, these changes were brought about because bone grows along lines of compression, tendons along lines of tension, muscles hypertrophy if they are used, and so on.

For these kinds of reasons, Thomson argued that a monster may occasionally be hopeful and survive to become the ancestor of organisms with new ways of life. As it happens, I agree with him, at least this far. Indeed, the argument just outlined – including, oddly enough, the use of Slijper's goat to illustrate the point – was included in my book, *The Theory of Evolution*, published in 1958. That book was written to explain the ideas of the modern synthesis to a nonprofessional readership. I did not see then, and do not see now, any contradiction between neo-Darwinism and the idea of hopeful monsters, at least in the sense of a mutant of large phenotypic effect. The essential point is that the fate of hopeful monsters, like that of other mutants, depends on the operation of natural selection in populations.

The preceding discussion, however, is hypothetical. It amounts to saying that one cannot rule out hopeful monsters a priori. But do they happen? As Lande pointed out at the Chicago macroevolution meeting in 1980, the genetic evidence suggests that the differences between morphologically distinct species and varieties is polygenic and does not involve one mutation of large effect (see Lande, 1980). The conclusion is based on an analysis of F_1, F_2 and backcross hybrids; more studies of this kind would be welcome. We are now familiar with the idea that gradual changes in the parameters of a dynamic system can, at critical points, lead to sudden and discontinuous changes in system behavior. It seems certain that gradual changes in genetic constitution can lead to discontinuous changes in phenotype. The only question at issue is how often large changes have contributed to evolutionary change.

The occasional incorporation in evolution of mutations with large morphological effects is an interesting, if unproved, hypothesis. Even if true, it would hardly qualify as a paradigm shift. I now turn to the second theoretical proposal of the punctuationists – "species selection." The idea is as follows. When a new species arises by splitting, it acquires new characteristics. The direction of change is assumed to be random relative to any large-scale evolutionary trends and not to be caused by within-population selection. (This is known as "Wright's rule," in accordance with Wright's earlier views on the random nature of speciation, as discussed by Provine, Chapter 3, this volume). However, species will differ in their likelihood of extinction, or of further splitting, and these likelihoods will depend on their characteristics. Hence selection between species will cause trends in species characteristics.

This theory is isomorphic with evolution in a population of parthenogens, with speciation replacing birth, extinction replacing death, and the acquisition of new, random characteristics at the moment of speciation replacing mutation. Logically, some such process must occur – although there is no reason to assume the truth of "Wright's rule." I have argued elsewhere, that species selection is relevant to the maintenance of sexual reproduction, although not necessarily the only, or even the most important, process involved. But is it responsible for macroevolutionary trends?

There is one category of evolutionary event that must often have occurred, and which might be taken as an example of species selection, but, in my view, misleadingly so. This is the competitive replacement of one group of species by another; examples are the replacement of the multituberculates by the rodents, or of the creodonts by the modern carnivores. In both cases, it is at least plausible that the replacement occurred because individuals of the new taxon were competitively superior to individuals of the old one. If so, should one call it "species selection"? In a sense, this is a purely semantic issue. However, the importance of replacements of this kind has long been recognized and is not a contribution of the punctuationists. Furthermore, such replacements in no way uncouple macroevolution from microevolution, provided, of course, that the characteristics by virtue of which, for example, the rodents replaced the multituberculates arose in the first instance by natural selection. Hence, taxonomic replacements of this kind, however important, cannot in the present context be taken as examples of species selection.

The difficulty in regarding species selection as an important evolutionary force can be seen best by discussing a concrete example. Consider the trends in the evolution of the mammal-like reptiles. In the skull, there were a group of changes associated with the evolution of chewing – reduction of tooth replacement to one, differentiation of the teeth, opening up of the dermal roof of the skull, and the development of a bony secondary palate. The adaptive significance of the reduction in the number of bones in the lower jaw, the acquisition of a new jaw articulation, and the incorporation of quadrate and articular into the middle ear, is less clear to me. Postcranially, there is a whole series of changes in the backbone, limb girdles, and limbs associated with a gait that involves flexing the backbone in a vertical rather than a horizontal plane (referring back to Slijper's goat, many of these changes are not a kind that could have arisen as secondary ontogenetic responses to a single primary change). Changes in the soft parts, associated with homoiothermy, viviparity, a double circulation, and so forth, are not recorded in the fossil record, but were presumably occurring.

I find it hard to believe that anyone seriously thinks that these changes were the result of species selection. There are two difficulties, one quantitative and the other logical. The quantitative difficulty is that the amount of change that can be produced by selection is limited by the number of births and deaths (i.e., speciations and extinctions) that occur. I cannot be precise about this, because I do not know how many species extinctions occurred, or how many independent changes in character states were incorporated, but I doubt whether a quantitatively plausible account could be given.

The graver difficulty, however, is the logical one. Species do not chew or gallop or keep warm or bear their young alive: Individual animals do these things. Hence, if I am right in thinking that the secondary palate evolved because it enables an animal to chew and breathe at the same time, there is no way in which "species selection" could be responsible for its evolution, *except* insofar as species survive if and only if the individuals who compose them survive.

The point may become clearer if I revert for a moment to the evolution of sex. Species evolve, but individual animals do not. Hence sexual reproduction affects a property of a species – namely, the capacity to evolve. It is, therefore, reasonable to speak of sexual reproduction evolving by species selection (as R. A. Fisher in effect did: Whether or not he was right is still a matter of debate, but at least his argument was not illogical). The claim would be correct if:

1. Mutations from sexual to asexual reproduction are commoner than the reverse, which is almost certainly true.
2. Within populations, asexual reproduction tends to replace sexual.
3. Sexually reproducing species evolve faster and are, therefore, less likely to become extinct than parthenogenetic populations.

In the case of the secondary palate, however, the claim that it evolved by species selection would have to mean something like the following. There existed a species, A, all of whose members had a partially developed palate. There arose from A a second species, B, reproductively isolated from A, all of whose members had a better developed palate, and a third species, C, whose members had a less well-developed palate. In competition, the B individuals outcompeted the A and C individuals, because they could chew better, so that only species B survived.

Note that, in this scheme, the reproductive isolation and the changed palate would have to arise simultaneously. If all that happened was that, in the original species A, some individuals arose with better palates, we are back with good old-fashioned natural selection in populations.

Would it make any difference if we abandoned my assumption that a secondary palate evolved because it enables animals to chew and breathe simultaneously? Maybe the palate influences individual survival or reproduction in some other way, in which case the essence of my argument is unchanged. Perhaps it does not affect survival or reproduction at all, but is associated with something that does; if so, my argument is still unchanged. Finally, and implausibly, it may be that a secondary palate does not affect survival and is not correlated with anything that does. If so, the species selection argument collapses altogether, because, if the palate affects species survival at all, it must do so via some effect on the individual that has the palate. If there are no such effects, we are left, to coin a term, with "species drift." To explain the major features of evolution by species drift would surely be the ultimate absurdity.

In the light of this discussion, we can now recognize four different processes which might be called "species selection":

1. Selection operating on emergent properties of the species itself, affecting chances of extinction and/or speciation. Examples are capacity to evolve rapidly (influenced, for example, by sexual reproduction and level of genetic recombination), and likelihood of speciation (influenced, for example, by dispersal behavior). The use of the term here seems appropriate. However, most evolutionary trends (e.g., in the secondary palate) could not be explained in this way. Even traits that could, in principle, be influenced by this type of species selection may, in practice, be largely determined by individual selection. For example, recombination rate is almost certainly the result of individual selection (Maynard Smith, 1978, 1980); the essential point is that there is extensive intraspecific variation in the trait, and dispersal patterns may have been influenced by the individually disadvantageous effects of inbreeding (Greenwood, 1980).

2. Selection acting on traits affecting individual survival and/or reproduction, but that arise suddenly at the time of speciation, simultaneously with reproductive isolation, and not as the result of within-population selection. If such a process occurs, I have no objection to calling it species selection. Even though the relevant traits are properties of individuals and not of species, the adaptedness of such traits

would arise because of the survival and splitting of some species and the extinction of others. My objection to invoking this type of species selection is that I do not think species arise in this way (except in speciation by polyploidy); if they did, I do not think the number of speciation and extinction events would be large enough to account for the extent of adaptation actually observed.

3. The replacement of one species, or group of species, by another, because individuals of the successful species have selectively superior traits that evolved, in the first place, by individual selection within a population (for example, the replacement of multituberculates by rodents, supposing that the traits responsible for the success of the rodents did arise in the first instance by within-population selection). In my view, it would be misleading to describe such events as cases of species selection.

4. The radiation of a taxon into a new adaptive zone by virtue of traits that evolved, in the first instance, by within-population selection. Again, as for category (3), I think it would be misleading to use the term species selection.

It is important that proponents of species selection should make clear which, if any, of these meanings they intend. Discussions at the conference with Gould suggest that he intends meaning (1), and perhaps (4), but not (2) or (3); however, he will no doubt clarify his own position. I suspect that Stanley intends meaning (2), because it corresponds to his view of the nature of speciation.

If, as seems to me appropriate, species selection is confined to meanings (1) and (2), it follows that (with the exception of a few traits, such as sexual reproduction or powers of dispersal, which confer properties on the species as such as well as on the individuals that compose it) the concept applies only if new species arise suddenly, with new characteristics, and are at once isolated reproductively from the ancestral species. Only if this is true does the concept have any meaning.

The hypothesis of species selection, therefore, arises from a typological view of species: The view that only certain forms are possible, and that intermediates are ruled out. There is nothing ridiculous a priori about this view. A typological view of chemistry would be correct. The laws of physics permit the existence of only certain kinds of atom; there are no intermediates between helium and hydrogen. What are the equivalent laws that are supposed to be responsible for the fixity of species? Two different but related proposals have been made; that there are "developmental constraints" and that evolutionary change can occur only in small isolated populations. I will discuss these points in turn.

The concept of a developmental constraint is best explained by an example. Raup (1966) has shown that the shapes of gastropod shells can be described by a single mathematical function, with three parameters which can vary. This is a consequence of the way the shell grows (e.g., that it grows by addition at an edge and not by intercalation). In other words, because of the way the shell grows, only certain shapes are possible; there cannot be gastropods with square shells.

That developmental constraints exist is not at issue. But they cannot by themselves account for stasis, or typological species. They limit the *kinds* of change that can occur, but they do not rule out all change. An infinite variety of different shapes is consistent with Raup's rules, and any one could be changed into any other by minute degrees. Of course, developmental constraints may permit a discrete set of possibilities, rather than an infinite gradation. However, such quantized variation occurs within species, without involving reproductive isolation, and there seems to be no difference between the genetic basis of continuous and quantized variation.

Thus developmental constraints limit the kinds of variation that can arise in a given species, but they do not rule out all variation. I suggested earlier that stabilizing selection maintains stasis; it will do so subject to the range of variation that arises. However, developmental constraints alone could not account for stasis, unless it were true that *no* variation is possible, and that is manifestly not so.

It is worth adding that the concept of developmental constraints is by no means new. As an example, my colleague Dr. Brian Charlesworth has drawn my attention to the following remark by H. J. Muller (1949) "the organism cannot be considered as infinitely plastic and certainly not as being equally plastic in all directions, since the directions which the effects of mutations can take are, of course, conditioned by the entire developmental and physiological system resulting from the action of all the other genes already present."

Of developmental constraints, then, one can say that they will inhibit or prevent changes in some directions, but not in all directions. What of the genetic barriers to evolutionary change? It is sometimes asserted that there is some kind of inertia that prevents evolutionary change in large populations. At least as far as our present understanding goes, this assertion is false. It has been supported by references to Lerner's (1954) concept of "genetic homeostasis"; this is an interesting idea, based on the empirical fact that a small population exposed to strong directional selection often reverts partway to its original phenotype when selection is relaxed. Lerner would have been astonished to hear his idea used to support the view that large populations cannot evolve.

There are two points at issue: Can a large population change, and can it split into two without geographical isolation? On the first, I can think of no sensible reason why it cannot, and good reasons why it should be able to sustain evolutionary change for longer than a small population. The second question is more controversial (see, e.g., Mayr, 1963; Endler, 1977). I will not pursue the controversy here, because it is not essential to the argument about species selection.

The idea that large populations cannot evolve seems to be simply false. However, there is one *kind* of change that cannot take place in a large population, but may, with low probability, occur in a small one. This is the passage from one adaptive peak to another through a selectively inferior intermediate. The simplest case is the passage in a diploid, from a population homozygous for one allele to one homozygous for the other, when the heterozygote is of inferior fitness. In the present context, however, the more interesting cases arise when there are epistatic fitness interactions between loci. The simplest case, in a haploid, is the transition from ab to AB when aB and Ab are both of low fitness.

Such events are the subject of one of the oldest controversies in population genetics. Wright thinks that such transitions were important, and Fisher thought that they were not. As a student of Haldane's I can take an impartial view.

It may help at the outset to explain what is *not* at issue. It is not at issue that there are epistatic effects on fitness. Nor is it at issue that the genes present at different loci in a sexually reproducing population are "coadapted." This is self-evidently true – no one supposes that a genome consisting partly of mouse genes and partly of rabbit genes would give rise to a viable animal. The difference of origins need not be so extreme, as shown by the fact that hybrids, in the first or later generations, between closely related species, or even varieties of the same species, often show lowered viability and/or fertility.

The essential point is to understand that the facts of coadaptation, and of hybrid inviability and infertility, are *not* evidence that populations have in the past crossed adaptive valleys. I show, in Figure 1, two simple cases in which a population can evolve from one state to another by a series of favorable gene substitutions, and yet the F_1 hybrids between the terminal states are of lower fitness. I show the simplest cases for which this can be true; as more loci are involved, the likelihood that two populations that have undergone independent evolutionary changes will give inviable hybrids increases.

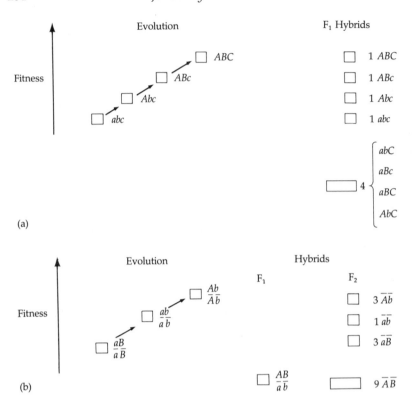

Figure 1. (a) *Haploid case*: Three loci. Suppose it increases fitness to be *A*, and to be *B* only if already *A*, and to be *C* only if already *A* and *B*. (b) *Diploid case*: Suppose *A* is dominant to *a*, and *B* to *b*, and that the fitnesses of the four phenotypes are in the order $A\bar{b} > \bar{a}\bar{b} > \bar{a}B \gg \bar{A}\bar{B}$.

The reason why the Fisher–Wright debate is so difficult to settle should now be clear; we do not know whether all evolutionary change is a hill-climbing process which could occur in a large population, or whether, occasionally, adaptive valleys are crossed. Perhaps the best reason for thinking that valleys are sometimes crossed is that some kinds of structural changes in chromosomes do seem to imply intermediates of lower fertility.

If, provisionally, we suppose that valley-crossing in small populations has been important in evolution, there can still be differences of opinion about its role. Wright imagined a species divided into a large number of demes with little gene flow between them. If one of these demes were, by chance, to cross over an adaptive valley, he thought it could then transform the rest of the species to the new peak, because it would send out more migrants to "infect" the other demes. Thus Wright's model (in its later form) is, in effect, a model of phyletic evolution of a whole species that, because of its demic structure, is able to cross adaptive valleys that would be impossible to a large panmictic population. The virtue of the idea is that the crossing of a valley, itself a very improbable event, would be made more likely because of the very large number of demes, any one of which might achieve the crossing. Its weakness is that most species seem not to be divided into demes that are sufficiently isolated for his mechanism to work. However, whether or not one thinks his mechanism plausible, it is, in any case, a mechanism of phyletic evolution of species of large total number, and hence would lend little support to punctuationist ideas.

The other possible role that has been proposed for events in small populations is Mayr's proposal of "genetic revolutions" occurring in peripheral isolates. Here I face difficulties of interpretation. Thus, suppose that a peripheral population is exposed to environmental conditions different from those of the rest of the species, and that it is sufficiently isolated genetically from the rest of the species to adapt to the new conditions, without being swamped by gene flow from the center. If isolation lasts for long enough, such a population might well evolve into a new species. However, the characteristics of this new species would have arisen by natural selection; they would evolve more readily if the isolate was not too small, because there would then be greater genetic variability for selection to act on. In no sense would these new characteristics be "random" with respect to selection within populations. This type of process must often have happened; I have argued earlier that it should not be called "species selection."

However, an alternative interpretation of Mayr's "genetic revolution" is that the peripheral isolate, by virtue of its small size, acquires new characteristics by genetic drift, initially unrelated to selection. If, at the same time, the isolate acquires reproductive isolation from the parent species, we have, indeed, a process of species origin that meets the criteria needed for "species selection" to be a meaningful concept. A new species has arisen that, simultaneously, acquires reproductive isolation and new characteristics that are not the result of within-population selection.

I doubt very much whether Mayr intended his "genetic revolution" to be interpreted in this way. Thus, writing of Goldschmidt's concept of "systemic mutations," according to which a complete genetic reconstruction gives rise in a single step to a well-adapted individual, Mayr accepts the criticism that this is "equivalent to a belief in miracles." It would be equally miraculous if a series of changes occurring by chance in a small population were to give rise to a new complex adaptation. In fact, it seems clear that when writing of genetic revolutions, Mayr had in mind the possibility that selection would produce different results in a small population, closed to migrants, than it would in a large one. I think he may have exaggerated the difference, but I doubt if he envisaged the origin of new species as being a sudden event, unrelated to the action of natural selection in populations.

References

Cain, A. J. 1964. The perfection of animals. In *Viewpoints in Biology*, ed. J. D. Carthy and C. L. Doddington 3:36–63.

Endler, J. A. 1977. *Geographic Variation, Speciation, and Clines*. Princeton: Princeton Univ. Press.

Fisher, R. A. 1930. *The Genetical Theory of Natural Selection*. Oxford: Clarendon Press.

Goldschmidt, R. B. 1940. *The Material Basis of Evolution*. New Haven: Yale Univ. Press.

Greenwood, P. J. 1980. Mating systems, philopatry and dispersal in birds and mammals. *Anim. Behav.* 28:1140–62.

Hennig, W. 1966. *Phylogenetic Systematics*. Urbana: Univ. of Illinois Press.

Hull, D. L. 1970. Contemporary systematic philosophies. *Ann. Rev. Ecol. Syst.* 1:19–54.

Lande, R. 1980. Review of S. Stanley, *Macroevolution*. *Paleobiology* 6:233.

Lerner, I. M. 1954. *Genetic Homeostasis*. London: Oliver & Boyd.

Maynard Smith, J. 1958. *The Theory of Evolution*. Hammondsworth, Middlesex: Penguin (3rd ed., 1975).

  1978. *The Evolution of Sex*. Cambridge: Cambridge Univ. Press.

  1980. Selection for recombination in a polygenic model. *Genet. Res. (Camb.)* 35:269–77.

Mayr, E. 1963. *Animal Species and Evolution*. Cambridge, Mass.: Harvard Univ. Press.

Muller, H. J. 1949. *Genetics, Palaeontology, and Evolution*. Princeton: Princeton Univ. Press.

Rachootin, S. P., and K. S. Thomson. 1981. Epigenetics, Paleontology and Evolution. In *Evolution Today, Proc. 2 Intl. Congr. Syst. and Evol. Biol.*, ed. G. G. E. Scudder and J. L. Reveal, pp. 181–93.

Raup, D. M. 1966. Geometric analysis of shell coiling: general problems. *J. Palaeontol.* 40:1178–90.

Simpson, G. G. 1944. *Tempo and Mode in Evolution*. New York: Columbia Univ. Press.

Slijper, E. J. 1946. The vertebral column and spinal musculature of mammals. *Verk. Akad. Wet.* (Amsterdam) 42:1.

Stanley, S. M. 1979. *Macroevolution*. San Francisco: Freeman.

Van Valen, L. 1973. A new evolutionary law. *Evol. Theory* 1:1–30.

11

"Adaptation"

RICHARD M. BURIAN

In this essay, I examine three concepts of adaptedness and adaptation found in Darwin and two additional concepts of "Darwinian fitness" employed in current population biology and evolutionary theory. The examination reveals a number of conceptual confusions *within* evolutionary biology and shows that much of the philosophical literature *about* evolutionary theory has been concerned with vulgarizations of that theory. A number of recent controversies in evolutionary theory, all of which focus on the degree of control that natural selection exercises over traits of organisms, are mentioned in passing. These controversies have been plagued by confusions regarding what it is for a trait to be an adaptation, the kinds of evidence required to support such a claim, and the units of analysis appropriate to applications of the concept of (an) adaptation. Removal of these confusions may contribute to the attempt to resolve the underlying biological disagreements (e.g., about the relative importance of various units of selection, the existence of neutral mutations, and "selfish" DNA, etc.); but it will not be sufficient, by itself, to resolve those disagreements. Removal of these confusions will also serve as an important step toward a deeper and more fruitful understanding of the structure and content of evolutionary theory.

Adaptation and adaptedness

In a recent book on *The Development of Darwin's Theory* of evolution by natural selection (Ospovat, 1981), the late Dov Ospovat argues,

This essay has benefited from criticisms by F. Ayala, C. Bajema, R. Brandon, L. Darden, S. Gould, M. Grene, K. Guyot, Ph. Kitcher, E. Mayr, R. Richardson, P. Richerson, A. Rosenberg, G. Simmons, and many others, including discussants at the Reimers Stiftung, Bad Homburg, the Committee on History and Philosophy of Science of the University of Maryland, and the seminar on History and Philosophy of Science, University of California, Davis. I am grateful to all concerned.

in effect, that a major key to understanding Darwin and his theory is provided by grasping the changes in Darwin's concepts of "adaptation" and "adaptedness" and the subsequent changes that Darwin's work helped to bring about in the use of cognate concepts by biologists. Although I am not entirely persuaded by all the details of Ospovat's account of Darwin's concepts of adaptation,[1] I shall employ some of his views, uncritically, as a starting point. The issues about the content and character of the concept of adaptation that I shall be exploring will not be greatly altered (though they may be clarified) by further historical scholarship regarding the path that Darwin followed.

Before proceeding, I must address one terminological point. As will become apparent in the course of this essay, the use of such terms as "fit," "adapted," "fitness," and "adaptation" in the biological and philosophical literatures has been confused and confusing. In order to build my account of the relevant concepts, I have found it useful to fix one aspect of usage by fiat. I shall employ the term "adaptation" in an historical sense, the terms "adapted" and "adaptedness" in an ahistorical sense. Fleetness contributes to the adaptedness of a deer (or makes the deer better adapted) if, and only if, other things being equal, it contributes to the solution of a problem posed to the deer – for example, escaping predation. Fleetness is an adaptation of the deer if, and only if, the deer's fleetness has been molded by a historical process in which relative fleetness of earlier deer helped shape the fleetness of current deer. This usage (which, in its present sharp form, derives from a suggestion by Robert Brandon) is my own, though it reflects an important tradition in the literature. This specialized terminology will facilitate the clarification of a number of the confusions to be discussed subsequently.

Ospovat documents Darwin's struggle to arrive at a satisfactory concept of adaptation for use in the theory of evolution by natural selection. According to Ospovat, Darwin first employed two distinct notions of "absolute" or "perfect" adaptedness (only one of which I shall discuss) in the period from 1838–1854 and then, only after considerable difficulty, arrived at a version of the notion of relative adaptedness employed in the *Origin* sometime in 1857. However Darwin got to it, Ospovat is surely right that the notion of relative adaptedness developed by Darwin is of crucial importance to evolutionary theory. Yet I am prepared to argue (I can give only part of the argument on the present occasion) that the development of the modern neo-Darwinian (or synthetic) theory of evolution has, in fact, minimized the

---

[1] Nor of some of his chronology regarding the development of Darwin's thought.

role of Darwin's concepts of adaptation and adaptedness and obscured their importance by running them together, in a confused and confusing fashion, with a whole battery of other related notions. For instance, two quite un-Darwinian notions, both of which pass under the label of "Darwinian fitness," are common in population biology and evolutionary theory. The widespread use, and occasional misuse, of these notions has, at times, contributed to the illusion that evolutionary theory is viciously circular or tautologous. Distinguishing between these concepts of Darwinian fitness (so-called) and related concepts of adaptation, including Darwin's, is crucial to the proper understanding of evolutionary theory.

Until about five years ago, when a number of new papers bearing on these topics began appearing, the most important and influential attempt to straighten out the conceptual confusions involved in various recent uses and misuses of terms like "adaptation" and "fitness" was G. C. Williams' *Adaptation and Natural Selection* (1966). Williams' main concern was to bring about correct use of the concept (or concepts) of adaptation in contemporary biology – not by means of a logic-chopping analysis, but by showing biologists how practical and theoretical difficulties arose from their misuses of the relevant concepts and by persuasive arguments to the effect that the biological difficulties and disagreements could be handled most effectively by cautious and correct use of the concept of adaptation along the lines he advocated. Williams' book was an immensely useful corrective to the near loss of Darwin's concept in theoretical biology. But now, fifteen years later, it is time to correct certain flaws in Williams' treatment of the concept of adaptation. In particular, his account of adaptation is wedded to quite strong and controversial commitments about the effectiveness of natural selection and about the *units* of selection – that is the *level* at which selection is most effective. I hope to separate the concept of adaptation from these commitments.

I shall proceed indirectly, carrying out only part of this rather large task. I shall try to fill in certain parts of the relevant background, mainly by developing my own account of some of the pertinent concepts of adaptation, fitness, and adaptedness, discussing their interrelations and the confusing consequences of their interplay. In a sequel to this essay,[2] I will show in detail the relevance of these matters to contemporary concerns about the units of selection and exhibit the interplay between controversies concerning group selection, macroevolution, and related topics. In the present essay, my main target is

---

[2] See "On some controversies concerning macroevolution," scheduled for publication in a *Festschrift* in honor of Marjorie Grene, edited by A. Donagan, A. Perovich, and M. Wedin.

to carry the account of the concept of adaptation a few steps beyond Williams.

Before proceeding with this agenda, I shall examine briefly what we should expect of an account of the concept of adaptation. How can we tell a good account from a bad one?

If, as I shall argue, many biologists are careless and inconsistent in their use of "adaptation," "fitness," and cognate terms, we should not make our evaluation of an analysis of the concept of adaptation turn on mere conformity to usage. Rather, it seems to me, we must isolate certain central claims to serve as touchstones against which to test alternative analyses. As the analysis becomes more sophisticated, we may have to reexamine and improve the touchstones, to divide and complicate our account of the concepts involved.

To illustrate what I mean by a touchstone, consider a thought experiment tracing back at least as far as Scriven (1959). In this experiment two identical twins are standing in a forest during a thunderstorm. One is struck by lightning and killed, the other is unhurt; he later marries and has ten children. One may now ask whether one twin is *better adapted* (or *fitter*) than the other. In conformity with *Darwin's* usage, which, I maintain, is standard in this regard, the twins have the same degree of adaptedness because the difference between them is a matter of happenstance and they are, in all biologically relevant properties, virtually identical; they are equally fit or unfit vis-à-vis their environment. Later we will see that use of this little fable as a touchstone shows that one of the most common measures of so-called Darwinian fitness in population biology does not serve as a direct measure of fitness in the sense of adaptation. But I shall save that point, as well as two further touchstones, for the appropriate juncture of the argument.

Darwin on adaptation

Let me start the real work of this essay by turning to Darwin's approach to adaptation. On Ospovat's account of Darwin's early unpublished work bearing on evolution by natural selection, Darwin held, at least from 1838–1850, a theory quite different than the one for which he is known. A remarkable feature of this theory is that it was based on the concept of *limited perfect adaptedness*. According to this theory, organisms are constrained by their structure and constitution – by being built out of organic, not inorganic matter, by being structured as vertebrates or invertebrates, by being warm-blooded or cold-blooded, and so on. Within the limitations of the constraints thus imposed, each species was – or was normally – *perfectly adapted* to its environment. By this was meant that the organisms in question were optimally designed within the applicable constraints to solve the problems posed by the environment; problems of heat, cold, wind, rain, opening up seeds, capturing prey of the sizes and speeds available, and so on.

One of Darwin's central concerns was to explain the diversity of living organisms. At this phase of his development, according to Ospovat, he held that diversity is a secondary consequence of *three* factors:

1. Gradual geological change in general, which altered the environment slowly and imperceptibly but which ultimately confronted organisms with environments to which they were not perfectly adapted.
2. Isolation of one population from another, meaning that different populations of organisms belonging to the same species would face different environments and, hence, different design problems.
3. The laws of variation (what Darwin in the *Origin* called "the mysterious laws of variation and correlation of parts") which, at this time, Darwin thought were such as to call forth variation *only when adaptedness became imperfect,* and then to call forth *directed* variation – variation adjusting the properties of the organism in the direction of perfect adaptedness vis-à-vis the new environmental circumstances.[3]

The role of natural selection at this stage in the development of Darwin's theory, if Ospovat is correct, was dual: First, it guaranteed that there would be pressure on all populations to maintain or to achieve perfect adaptedness; and, second, it preserved the most favorable variations and moved the population mean toward them. Variation provided the *direction* of evolution, whereas the *reduced* variation of perfectly adapted organisms ensured that natural selection would alter the constitution of a population only when the organisms in question were not perfectly adapted.

Whether or not this thumbnail sketch, or Ospovat's detailed account, is fair to Darwin's private speculations, it points to certain features of his concept of adaptedness, of his account of what it is for an organism to be fit with respect to its environment, common to the notebooks and early drafts and to the *Origin*. I shall capitalize on these. To put the essential points informally, an organism is well adapted when its structure and habits enable it to solve optimally the

---

[3] We see here that Darwin already had good reasons for taking the inheritance of acquired characteristics seriously in his earliest speculations on transformation; such inheritance might provide a mechanism that would give variation its direction.

expectable challenges of the environment. An organism is fit or unfit according to its ability to meet these challenges by virtue of its design and its programmed patterns of behavior. Employing engineering criteria, it is possible to compare organisms by type and to determine that organisms of type 1 are adapted (designed) to meet challenges A (or challenges A to Z), whereas organisms of type 2 are not adapted (or not as well adapted) to meet that challenge (or those challenges). Certain features of organisms, for example "organs of extreme perfection" like the vertebrate eye, can be recognized as adaptations; that is, as features existing in some sense or other *because* they are designed for – adapted to – the performance of certain tasks useful or necessary for the survival of the organism.

One reason for citing Ospovat's account of Darwin's early work is that it highlights the possibility of having both absolute and relative concepts of adaptedness. An absolute concept evaluates the design of a product – for example, the eye or the whole organism – in its own right. In order to evaluate the design, it is necessary to specify, tacitly or explicitly, the design problem and the pertinent constraints governing its solution; from there evaluation can proceed by some form of "static" engineering analysis. Because the analysis is static, not dynamic, the absolute concept of adaptedness has no direct bearing on the process by which the design was achieved. Accordingly, it needs supplementation if it is to play a role in the theory of evolution. Darwin's early theory provided that supplementation by reference to "the mysterious laws of variation" that, he thought, directed heritable change toward optimal design whenever an organism was insufficiently adapted. In this theory, natural selection played a relatively secondary role – it was the force that called unadapted organisms to account and it was the preserver of adaptations; that is, of optimally designed features and organisms.

Once Darwin came to realize that variation is ubiquitous and largely undirected with respect to the needs of the organism,[4] he was forced to employ a relative concept of adaptedness, a concept tied much more intimately to the process of natural selection than the absolute

---

[4] P. Richerson (University of California, Davis) has reminded me that one must not lose sight of the fact that Darwin retained some role for directed variation throughout his career. The supposed effects on an organism's descendants of its use or disuse of its organs, amplified in various "neo-Lamarckian" ways in Darwin's later theory of pangenesis, mark the strength of Darwin's commitment to directed variation. In his later work he employed directed variation to speed up evolutionary processes and to overcome the difficulties posed by (theories of) blending inheritance (see Lord Kelvin's and F. Jenkin's criticisms of Darwin's views). For a brief exposition regarding directed variation, see Gould (1971).

one was. All organisms face a multitude of problems bearing on survival and reproduction. If they all vary (at least slightly) in virtually all their features, then typical organisms are *not* perfectly adapted. Some, however, are at an advantage with respect to others – that is, they are better designed to meet the expected or expectable challenges of the environment. These organisms would, therefore, be more likely to survive the insults of the environment. To be sure, their design offers no guarantee that they will do so, but it provides a statistical bias in their favor. To the extent that two further conditions are met, *natural selection is likely to increase adaptedness*. These conditions are: first, that the environment is stable enough that the problems organisms face are reasonably constant or predictable;[5] and second, that the offspring of an organism or of a pair of organisms tends to differ from the population mean in the same ways that its parents do. (This latter condition, the "statistical heritability" of features, is necessary if differential survival in one generation is to affect the characteristics of the organisms of the next generation.)

Although Darwin did not do so in any clear way, it will help us to separate two intertwined concepts pertaining to adaptation. The first is the relative engineering adequacy of a design (given the relevant constraints) as a solution to a particular problem. The second concerns the process by which the design was produced. On this usage, if the variations of a given feature, system, or behavior pattern were causally efficacious in the refinement of that feature or system or behavior pattern by means of natural selection, then that feature counts as an adaptation relative to its alternatives. Note, however, that it will *not* do to say simply "If a feature was produced by natural selection, it is an adaptation"; account must be taken of the "mysterious laws of correlation of growth." Darwin rightly emphasized this point; so far as the process of natural selection is concerned, many features are epiphenomenal, or are produced in spite of being mildly deleterious, or arise independently of their slight usefulness, simply

---

[5] For sufficiently long-lived organisms, the *range* of problems faced should not change greatly from generation to generation. But for short-lived organisms, the environment will vary, both randomly and cyclically, on a scale of many generations; the variation will be life-threatening, but is likely to fall within roughly predictable limits. Indeed, typically, cyclic variations (such as seasonal changes) are associated with cues (such as changes in the number of daylight hours) which are utilized by short-lived organisms (e.g., as a signal to produce an overwintering form). Although the environment has many components which are random *with respect to the organism*, other things being equal those organisms with design features that enable them to utilize or withstand the fairly regular excursions of the environment are better adapted – and more likely to leave descendants – than those that are not.

because their selective consequences are far smaller than those of correlated features. Given what we now know about pleiotropy, multiple functions of organic structures, multigenic determination of features, and so on, we should recognize the broad application of this point. In other words, one is making a substantial claim if one maintains that a feature is an adaptation in the *process* sense. One is claiming not only that the feature was brought about by differential reproduction among alternative forms, but also that the relative advantage of the feature vis-à-vis its alternatives played a significant causal role in its production. This claim is a historical claim about the way in which the feature was produced, subject to all the epistemic difficulties attendant on such historical claims (Gould and Vrba, 1982).

Incidentally, this discussion reveals two more touchstones for analyzing adaptation. The first is that an analysis of the process of forging adaptations must make it possible to distinguish selective processes from random processes, design from drift, and effects of natural selection from effects of mere de facto differential reproduction. Darwin's way of achieving this is, effectively, to distinguish between differential reproduction and natural selection. For Darwin, natural selection is systematic differential reproduction due to the superior engineering fitness of certain available variants – due, that is, to the relatively better design of the favored organisms.

The second touchstone cannot be stated comfortably in Darwinian terminology. It is simply this: claims about relative adaptation concern, in the first instance, a relation between a phenotype (or range of phenotypes) and the environment relative to a field of alternative phenotypes. Thus although genes, gene complexes, genotypes, and so on may be said to be well or ill adapted and may be compared with respect to their degree of adaptation, within *Darwin's* theory such claims ought to be treated as depending on analyses of the various advantages manifested by the (range of) phenotypes that the relevant genes or genomes produce in the appropriate (and appropriately weighted) environmental circumstances. This is true whether one speaks, statically, of the degree of adaptedness (or co-adaptedness) of a gene or gene complex, or, dynamically, of the process by which genetic adaptation is achieved. *Full analysis of adaptation cannot bypass the phenotype.* To be sure, the phenotype alone will not suffice; one must decompose phenotypic variation into heritable and non-heritable components. The former is the raw material on which Darwin's process of natural selection acts in creating adaptation.[6] Some subsequent concepts of fitness (as we will see later) are used in ways that bypass the phenotype; I shall argue that they are weaker than

---

[6] I thank P. Richerson for this last point.

the concept we are articulating currently and that the move of by-passing the phenotype raises difficulties that the present concept does not face.

It is time to codify our results concerning Darwin's concepts of adaptation. (For ease of reference, a short version of this codification is given in Table 1.) So far I have discussed three interrelated concepts. The first of these might be labeled "limited perfect adaptedness," or "absolute engineering fitness." A (type of) feature or a (type of) individual possesses absolute engineering fitness if, and only if, its design manifests an optimal engineering solution to the appropriate (real) challenge or range of challenges posed by the environment. Following Ospovat, I suggested that in developing the theory of the *Origin*, Darwin replaced this concept with one which might be appropriately labeled "relative engineering adaptedness" (or fitness). A (type of) feature or a (type of) individual possesses higher relative engineering fitness than an alternative type if, and only if, its design manifests a better engineering solution within the appropriate (real) design constraints to a specific (real) challenge or range of challenges posed by the environment. Finally, a (type of) feature is an adaptation if, and only if, its design characteristics were produced as a causal consequence of their relative engineering fitness as compared with those of the relevant alternative types, as a solution to a problem or range of problems posed by the environment in the evolutionary history of the organisms in question. One may be able to explain the fact that a (type of) individual is relatively better adapted than alternative (types of) individuals if, and only if, its greater relative engineering fitness is a consequence of specific adaptations. This yields a third sense of fitness which I will call (relative) "selected engineering fitness."[7]

Again, it is important to realize that traits with high engineering fitness need not have been produced by natural selection. The fact that a trait confers advantage with respect to survival or reproductive success does not, by itself, justify any claims about its historical origins. The notorious opportunism of evolution amounts to an ability to turn traits to advantage no matter how they originated, no matter how they were initially produced (see Williams, 1966, p. 12). Accordingly, it is crucial to understand that the theory of the *Origin* requires one to connect the concept of adaptation with that of natural selection. The mere fact that variant forms are differentially fit – that

[7] It is this notion of adaptation that Williams (1966, p. vii) has in mind when he says that "evolutionary adaptation is a special and onerous concept that should not be used unnecessarily, and an effect [i.e., a fitness-increasing use to which a trait is put] should not be called a function [a *designed* fitness-increasing use] unless it is clearly produced by design and not by chance."

Table 1. *Concepts of adaptation*

Label	Definition	Historical knowledge required?
1. Absolute engineering fitness or "limited perfect adaptedness"	A type of feature (or organism) manifests an optimal engineering solution (within real design constraints) to a real environmental challenge.	Only to specify the design constraints and the design problems.
2. Relative engineering fitness	The type of feature (or organism) manifests a better engineering solution (within real design constraints) to a real environmental challenge than specified alternative.	To specify design constraints, design problems, and the range of alternative types.
3. Selected engineering fitness	The characteristics of the type of feature (or organism) are present as a consequence of their higher engineering fitness with respect to real environmental problems (faced in the evolutionary history of the organism) as compared with the historically available alternative types or characteristics.	To specify design constraints, design problems, the range of alternative types, and the causal history behind the prevalence or fixation of relevant characteristics.
4. Realized fitness ("Darwinian fitness"; "tautological fitness")	An organism (or class of organisms sharing some property) has higher realized fitness in environment E than alternative organisms (or classes of organisms) if, and only if, its actual rate of reproductive success is higher than those of the alternatives.	Only to specify actual rates of reproductive success.

Table 1. (*Continued*)

Label	Definition	Historical knowledge required?
5. Expected fitness ("Darwinian fitness"); (propensity for greater reproductive success, de facto, rather than by virtue of design considerations)	A type of organism (or other replicating entity) has higher expected fitness than its competitors in environment E if, and only if, it has an objective propensity to out-reproduce them in E. Usual measure: relative reproductive success within replicate populations in "identical" environments.	Only to specify actual rates of reproductive success of replicate populations and to certify comparability of environments.

is, that they exhibit differences in relative engineering fitness – is not enough to justify employment of the concept of selected engineering fitness in evolutionary theory.[8] This becomes clear when one realizes that Darwin's earlier theory used differences in relative adaptedness (engineering fitness) as the means of restoring (limited) *perfect* adaptedness – that is, absolute engineering fitness – but derived the direction of evolution from the direction of variation. In contrast, the central concept of Darwin's theory in the *Origin* is the concept of features of organisms with high relative engineering fitness that have been designed by the historical processes involved in natural selec-

---

[8] Contrast Gould and Vrba (1982) who argue that many traits have been "coopted," that is, that their current uses are "effects," not "functions." In order to work out the problems for evolutionary theory raised by such a position, it is necessary to distinguish between traits that were originally shaped by natural selection for their current use and those that originated as correlated characters or via selection for a *different* use. Gould and Vrba wish to restrict the trait-descriptor "adaptation" to *traits shaped ab initio for their current use by natural selection* and to employ the neologism "exaptation" for traits coopted for new uses (see their Table 1). I do not like this terminology, but their claim that we need some such distinction and that we should mark it with appropriate terminology is well taken. Indeed, such a distinction is of critical importance if we are to evaluate the Darwinian claim that adapted features in a static sense (characters with high relative design adequacy) are largely adaptations in an evolutionary sense (traits shaped by natural selection *in the first instance* to perform those tasks that they now perform).

tion, rather than by directed variation. This is what later writers like Ernst Mayr mean by the creative power of selection, this is "the onerous concept of design" which G. C. Williams seeks to restore to prominence in evolutionary biology. This is the feature of Darwin's theory that his contemporaries found least persuasive.

Concepts of adaptation in neo-Darwinism

I turn now to the so-called genetical, or synthetic or neo-Darwinian, theory of evolution by natural selection articulated in the 1930s and 1940s, and the regnant orthodoxy since the 1950s. My strategy will be to exhibit two further interconnected concepts of fitness, to demonstrate *that* and *how* they are interconnected, and to show that failure to recognize the complexity of their interactions with each other and with the Darwinian concepts pertaining to adaptation results in a vulgarization of evolutionary theory. On the present occasion I shall proceed somewhat dogmatically, with only slight documentation of the use of the alternative concepts to be discussed.

There are many versions of the synthetic theory and many accounts of its central features.[9] For present purposes, one may venture to say that it placed the Darwinian theory on a new foundation – to wit, on a foundation of Mendelian population genetics. A significant dispute, though not central here, concerns the relative importance of population genetics vs. that of the principle of natural selection. Michael Ruse, for one, sees natural selection as a *consequence* of the principles of population genetics where Mary Williams – and I – see population genetics as the contingent elaboration of the consequences of natural selection for a limited class of organisms [Ruse (1973, especially pp. 48ff); Williams (1973, especially pp. 86–8).][10]

Whatever the issue of that dispute, however, one consequence of the central role of population genetics in the synthetic theory has been a series of shifts in the concepts of fitness, adaptedness, and adaptation, and in the measures of fitness and adaptation employed in evolutionary investigations. These shifts are complicated and have

---

[9] The best single source (though its coverage is partial and biased) is Mayr and Provine (1980). Examples of the controversies over the precise content of the synthetic theory are Gould (1980), Orzack (1981), and Gould (1981).

[10] Ruse has reaffirmed the stance he took in 1973 in a number of recent writings. Williams' views on this topic are never stated as directly as Ruse's. My interpretation of her position rests on a reading of a number of her other works and on the general character of her views in addition to the passage cited. Incidentally, her interpretation of the confusions surrounding the use of the concept of fitness in the synthetic theory (1973, pp. 88–100), in spite of a quite different starting point and style of argumentation, is quite close to the one put forward in the present essay.

resulted in considerable confusion both on the part of biologists and on the part of those analyzing evolutionary theory. In the present essay I will attempt to disentangle only a few of the central confusions.

Both the following sense of "fitness" have often been labeled "Darwinian fitness" in the recent literature. The first of the two applies in the first instance to individual organisms. According to this sense of fitness, an individual (or a class of individuals) is more fit (or better adapted) than its competitors, not if it has a higher *expectation* of survival and reproduction in virtue of its design, but simply if it, *in fact*, enjoys relatively greater reproductive success. Degree of adaptation, in this sense, is an empirical property of the organism (or of the class) in question, but one which can only be known post hoc.

It is ironic that this concept has come to be labeled "Darwinian fitness." Darwin almost certainly meant the phrase "survival of the fittest" to stand for the tendency of organisms that are better engineered to be reproductively successful. Because the use of the present concept turns the phrase "survival of the fittest" into the tautology, "the reproductively successful are reproductively successful," one traditional label for this kind of so-called "Darwinian fitness" is "tautological fitness." I prefer the less tendentious label *realized fitness*, especially since *realized fitness is an empirical property*, namely, *relative reproductive success*.

It should be clear that this sense of fitness falls victim to Scriven's thought experiment concerning the pair of twins. Although relative (actual) reproductive success is, I repeat, an empirical property of the relevant individuals, it simply does not measure degree of fitness or adaptedness in any of the senses relevant to evolutionary theory – all of which have something to do with *systematic* or *designed* fit with the environment, with causally mediated propensities to reproductive success.[11]

There is, however, a much more interesting concept, also sometimes called "Darwinian fitness," that plays a prominent role in the literature. I am grateful to Francisco Ayala for forcing me to recognize its importance. Although it is frequently used, it has only recently

[11] P. Richerson suggests that realized fitness (relative rate of actual reproductive success) applies only to organisms exhibiting *heritable* differences. If the concept is so restricted, the present counterexample is avoided, for there is no heritable difference between identical twins. But the point of the example remains unaffected if we replace our twins with full or half siblings – or even unrelated individuals. One-time rates of reproduction, whether of individuals or of individuals belonging to a certain group or kind, are subject to chance effects (lightning bolts and sampling errors) and extraordinary occurrences. For this reason they provide uncertain evidence of the propensities or tendencies underlying the actual reproductive rates.

been articulated with some clarity in the literature (Brandon, 1978; Mills and Beatty, 1979),[12] so I will discuss it with some care.

It is possible to recognize, and even to work up a good quantitative estimate of the propensity of a type of organism to out-reproduce its competitors in a certain environment or range of environments even though one can offer no historical, causal, or design-based analysis of that propensity. Consider an idealized experiment in which, say, 100 replicate cultures of *Drosophila melanogaster* are begun, each with 50 percent flies of type A and 50 percent flies of type B. If 50 of these cultures are raised at high temperatures and 50 at low temperatures in otherwise standard environments and, after 20 or 100 generations, a roughly stable equilibrium is reached such that at high temperature roughly 85 ± 10 percent of the flies in each culture are of type A and at low temperature roughly 65 ± 10 percent of the flies are of type B, the natural conclusion would be that type A flies are better adapted to (have greater engineering fitness in) the warm environment, whereas type B flies are better adapted to the cool environment.

Unfortunately, as K. Guyot and R. Lewontin have reminded me, life is not so simple. In our hypothetical experiment, type A flies out-reproduce type B flies *when the ratio of the two types in the population is 50:50*. But when the population is at equilibrium (85A:15B), their relative reproductive rates are equal. Therefore, the selection coefficients are frequency dependent; the constitution of the population alters the likelihood of relative reproductive success for the two types. If type A flies were always likely to out-reproduce type B flies in a stable warm environment, over a long enough time period type A flies would entirely replace type B. This shows how crucial it is to take the population of conspecific organisms into account in describing the environment.

For the purposes of the present calculations it does not matter whether the flies in question are identified by their phenotype or their genotype, although it does matter that at least some of the differences between them must be heritable. Either way, in the new sense of fitness, organisms of type A are relatively fitter than organisms of other types in environment E if, and only if, in E, organisms of type A have an objective propensity to out-reproduce those of the alternate types. Usually the best experimental measure of such a propensity is a statistically significant difference in the reproductive success of rep-

---

[12] But for important criticisms of the concept of fitness described by Brandon and Mills and Beatty and in this essay (criticisms whose fundamental point I reject), see Rosenberg (1982) and (Ms.). I thank Professor Rosenberg for supplying me with advance copies of his papers.

licate populations.[13] Although this notion is now commonly called "Darwinian fitness," a more useful label found occasionally in the literature is "expected fitness." To repeat, the (relative) expected fitness of a type of organism as compared with specific competitors in a specified environment is its propensity to manifest a certain (relative) rate of reproductive success as compared with those competitors.[14] In a secondary sense, a trait of the organism may be said to make a certain contribution to its fitness according to the change in fitness – that is, the change in the propensity to reproductive success – that is correlated with its presence as opposed to its absence or to the presence of some specified alternative(s), as appropriate.

Usually the best experimental measure of such a propensity is a statistically significant difference in the reproductive success of the relevant types of organism in replicate populations. In a good experimental study, the replicability of the results allows a powerful statistical argument to the effect that there is an objective tendency at stake rather than a random or unique phenomenon. Thus in the case discussed, in the warm environment with 50 percent of both types of flies present, type A flies have greater expected fitness than type B flies; with 85 percent of type A flies present, both types have equal fitness. Such an argument removes the central objection to realized fitness: If we use the various track records of replicate populations in a particular environment as a measure of the relative fitnesses of various types of organism in that environment, it quickly becomes obvious that in any *single* run organisms of the relatively fittest type may not out-reproduce their competitors; indeed, there are occasionally cases in which none of the fittest organisms survive. Thus the claim that the fittest did or will survive in any particular run, even one lasting hundreds of generations, is no tautology. Nor is it a tautology that the survivors in any natural population are, or were, the fittest.

[13] There are serious technical difficulties with such measures. Important discussions of some of these may be found in Prout (1969) and (1971) and Lewontin (1974). Many of the problems covered by these authors are particularly acute when one deals with natural populations. Some of the problems can be circumvented in properly designed laboratory experiments – at the price of relativizing their direct testimony about fitnesses to laboratory, but not to natural environments. The difficulty, often serious, remains that traits of considerable importance in laboratory populations and environments may not be of comparable importance in natural populations and environments.

[14] Actually, for technical reasons discussed in Mills and Beatty (1979), pp. 274ff., it is preferable to employ a *distribution* of propensities to leave various numbers of offspring. For present purposes, however, we gain considerable clarity by abstracting from these complexities, which does not affect the central points being made here.

Careful attention to the distinction between realized and expected fitness is crucial to a proper understanding of the literature of the synthetic theory in at least two ways. The first of these is the elimination of serious conceptual confusions from the classic texts. The confusions that result from conflating the two senses of fitness just introduced have given needless support to those who charge the synthetic theory with mistaking tautologies for empirical claims (and worse). I shall not document these confusions at length (though innumerable examples can be found in the dreary literature affirming – or denying – that evolutionary theory is tautological), but I will illustrate the matter by reference to two passages written by founding architects of the synthetic theory.

In Dobzhansky (1970) there is a discussion of the various differences among members of a population which "influence the contribution that the carriers of a given genotype make to the gene pool. This contribution [i.e., the *actual* contribution], relative to the contribution of other genotypes in the same population, is a measure of the Darwinian fitness of a given genotype" (p. 101). In the terminology of the present essay, Dobzhansky seems to be claiming that realized fitness is a measure of expected fitness. In a somewhat stronger passage, Simpson (1949) claims the following: "It must, however, be noted that the modern concept of natural selection . . . is not quite the same as Darwin's. He recognized the fact that natural selection operated by differential reproduction, but he did not equate the two. In the modern theory natural selection is differential reproduction, plus the complex interplay in such reproduction of heredity, genetic variation, and all the other factors that affect selection and determine its results" (p. 268).[15]

Now a full reading of the books from which these quotations are taken, as well as of other major works by their authors, reveals that, in spite of verbal confusions, *in fact* (most of the time) they employ the concept here labeled "expected fitness" rather than that of realized (or tautological) fitness. Thus pages 219–229 of Simpson (1949) constitutes an extended argument that natural selection is a "systematic" and "orienting" process, that "the correlation between those having more offspring, and therefore really favored by natural selection, and those best adapted, or best adapting to change [i.e., those that have high expected fitness] is neither perfect nor invariable, [but] only approximate and usual" (p. 221); that the *statistical* bias of "selection . . . [as] a process of differential reproduction" (p. 224) favors "well-integrated" organisms (p. 224) and "favorable or adaptive combinations" of genes (p. 225). Similarly, in the third edition (1951) of

[15] I thank G. Simmons (University of California, Davis) for calling this passage to my attention.

his path-breaking *Genetics and the Origin of Species,* Dobzhansky had defined the Darwinian fitness of a genotype as "the relative *capacity* of the carriers of a given genotype to transmit their genes to the gene pool of the following generations" (p. 78, my emphasis). And in a passage of his (1970) preceding the one quoted above, he had accepted Lerner's definition of selection as *"non-random* differential reproduction of genotypes" (p. 97, my emphasis). Indeed, considerable portions of the four chapters on selection (and fitness) in the later book are devoted to illustrating the statistically systematic character of the process and properties in question. The removal of the conceptual and verbal confusions in such passages, though an unexciting task, contributes significantly to an articulation of the synthetic theory able to withstand misguided criticisms.

A rather deeper contribution of the articulation and clean separation of realized from expected fitness is an improved understanding of the difficulty of applying the synthetic theory to natural populations. One piece of this puzzle stems from the peculiar way in which application of the concept of expected fitness plays an intermediate role between application of the concepts of engineering and realized fitness. The interplay among these three concepts is subtle and interesting. On the one hand, expected fitness resembles engineering fitness (relative adaptedness) in designating a *propensity,* shared by the members of a kind. On the other hand, it resembles realized fitness in that it does not entail or involve any kind of design analysis; the propensity in question is simply to have thus and such a relative reproductive rate in the relevant environments and relative to the specified competitors.[16] Like realized fitness, expected fitness is con-

---

[16] F. Ayala (University of California, Davis) has suggested the following example to illustrate that increased expected fitness may, in specifiable circumstances, be associated with *reduced* engineering fitness. Two species (say, of beetles) compete in a population cage. One is regularly able to displace and eliminate the other. However when a certain mutant (manifesting, say, red eyes) appears in the population of the fitter species, it displaces the wild type, and then is displaced and eliminated by the erstwhile less fit competitor. (This is precisely what happens in experimental populations of *Tribolium castaneum* and *T. confusum* as briefly described in Ayala and Valentine [1979]. The original report is in Dawson [1969].) Ayala's (unconfirmed) hypothesis is that the red-eyed mutant emits a systemic poison to which its conspecific competitor is highly sensitive, it itself is moderately sensitive, and the competing species is only slightly sensitive. Should this hypothesis be correct, the increased expected fitness within the conspecific population of the mutation causing red eyes is associated with a decrease in the engineering fitness (and degree of adaptedness) of its carrier, for the organism has been weakened by the mutation. Thus increase in expected fitness need not – and (I am confident) in real life at least sometimes does not – correlate with improvement of design.

cerned with track record, not design; like engineering fitness it deals with a propensity (i.e., a propensity to have a certain track record) rather than actual success or failure on a case-by-case basis.

When one deals with natural populations, it is nearly always impossible to replicate the environmental conditions (including presence of, and comparable opportunity for, interacting with other organisms) and the genetic or phenotypic constitution of the original population. Because replicate cultures or populations in "the same" environment are unavailable, it is extremely hard to obtain adequate measures of propensities to reproductive success from field studies. Yet a biologist who knows his or her organisms well is often (rightly or wrongly) confident that the reproductive outcome of a particular case study reflects the design of the organisms (engineering fitness), and will, thus, be very tempted to treat the actual reproductive success rates of those organisms (realized fitness) as a reflection or a measure of their objective propensity to success (expected fitness). This is probably why Dobzhansky took the *actual* contribution per genotype to the next generations to be a measure of the "Darwinian" fitness of those genotypes.

The resultant conflation of actual differential reproduction (realized fitness) with a systematic propensity for such differential reproduction (expected fitness) poses serious and often unnoticed epistemological difficulties. This becomes apparent when one notices that if one prevented the use of tacit background knowledge and pretended that one knew *only* the actual survival or reproduction rates, the support for the inference from realized to expected fitnesses would virtually disappear.[17] *Yet almost nothing is known about the reliability of our beliefs about the extent to which "chance" events affect reproductive success from case to case.* Nor do we have a good way to discriminate interactions in which, effectively, organisms with constant fitnesses have been exposed to relevantly changing environments from interactions in which, effectively, the relevant fitnesses have varied in frequency-dependent ways.

This argument shows that field studies (which are automatically fraught with unique occurrences, irreplicability of the relevant pop-

[17] As the preceding footnote shows, both the inference from known aspects of the relative engineering fitness of a group of organisms to relative reproductive success and the converse inference are quite chancy. When one adds that a broad distribution of outcomes in replicate populations is compatible with fairly extreme differences in Darwinian fitnesses (especially when the environment is variable or "patchy"), it is clear that inferences between actual reproductive outcome and reproductive propensities require considerable support.

ulations, and limited repeatability of the biotic and other environmental circumstances) provide a rather slender basis for estimation
of expected fitnesses. When such estimates are challenged, about the
only way to bolster them is by showing that the actual survival and
reproduction rates reflect *known* differences in the design-based abilities of the organisms to overcome the challenges of the environment
or of their competitors. That is, information about reproductive outcomes is supplemented by information regarding engineering fitnesses; a sufficiently powerful analysis of *engineering* fitness supports
the use of actual reproductive outcome as an estimate of *expected* fitness. (So, too, does a consistent pattern over a large number of generations, given adequate environmental stability.) Biologists have, of
course, often argued in support of fitness estimates along these lines,
but only a handful of cases have been worked out in adequate detail
to be generally acknowledged as fully persuasive. Perhaps the best
of these stems from the studies of industrial melanism of the moth
Biston betularia (and some other species), conducted largely by H. B.
D. Kettlewell and his colleagues, and described in virtually all modern
textbooks of evolution. The repetitious centrality of the example of
B. betularia in the standard texts provides a crude and informal measure of the difficulty of demonstrating *in convincing detail* that the actual
changes in the composition of natural populations reflect propensities
(expected fitnesses) or are truly consequences of the design features
(engineering fitnesses) of the organisms in question. And it is of rather
considerable interest that in the case of *B. betularia* (and in many other
interesting but less fully elaborated cases) the primary support for
the inference to expected fitnesses consists of showing that the actual
outcomes conform to the expectations generated by an analysis of
engineering fitnesses.

This analysis of some of the conceptual pitfalls encumbering the
application of contemporary concepts of adaptation enables us to articulate and support one of the major concerns in Williams (1966).
Williams sought to restore the importance of design considerations
in evolutionary theory, emphasizing the importance of "historical"
analysis of the problems faced by the relevant populations in their
evolutionary past; it requires a (partly) historical analysis of the problems faced by organisms that belonged to the relevant populations
to forge the connections between current design (evolved function)
and the process and outcome of natural selection. In my idealized
Drosophila experiment, in the case in which type A flies completely
supplanted type B flies in warm environments and vice versa in cold
environments, one can draw the conclusion that flies of type A are
better adapted to the cold environment. Yet whereas it is almost cer-

tainly true in such a case that (if one holds other environmental factors constant) flies of type A have relatively higher expected and engineering fitnesses in the warm environment, it in no way follows that the recognized distinctive features of type A flies constitute an adaptation to warmth and the distinctive features of type B flies, an adaptation to cold. For example, the problems faced by each type of fly may revolve around temperature-sensitive disturbances of reproductive physiology that are only accidentally correlated with our ability to differentiate As from Bs, or on some other more-or-less bizarre possibility. Such issues can be resolved (if at all) only by a (difficult) study of the evolutionary history of each organism. Typically, thanks to our ignorance, we substitute plausible scenarios (what Gould calls "just-so stories") for such history. Needless to say, the evidential worth of such scenarios is subject to challenge.

In the absence of firm knowledge of the evolutionary history of our flies and in light of the multiple selective effects which *may* have operated on them in natural, as opposed to laboratory, environments, it remains an open question whether the greater expected fitness (and supposedly greater engineering fitness) of the type A flies in warm environments was shaped by differential reproduction operating on the favorable response of the ancestors of the type A flies to warmth (in which case they are exhibiting selected engineering fitness), whether what is involved is really frequency-dependent selection, whether their advantage is an example of "the mysterious laws of correlation" together with the operation of selection on *other* consequences of the type A constitution, or whether their advantage is a consequence of some sort of random change, for example, a series of point mutations that, taken singly, were selectively neutral *in the natural environments of the flies*, although they are highly consequential in the chosen laboratory environments.[18] To repeat Williams's formulation from the précis of his book: "Evolutionary adaptation is a special and onerous concept that should not be used unnecessarily, and an effect should not be called a function unless it is clearly produced by design" (p. vii).

"Fitness," "adaptation," and recent evolutionary controversy

Having completed my direct discussion of the principal uses of the terms "adaptation," "adaptedness," and "fitness" in evolutionary

---

[18] This discussion of the *Drosophila* experiment is easily summarized in Gould and Vrba's terminology: the outcome of the experiment leaves open the question whether the greater expected fitness (and the putatively greater engineering fitness) of type A flies in a warm environment is an adaptation or an exaptation.

theory, I shall devote this final section of my essay to an exploration of certain ramifications of my analysis and of the value of keeping the various concepts of fitness properly separated. I shall make four major points.

First, in light of my discussion, it is easy to see that many recent controversies over the status of evolutionary theory turn on a vulgarization of that theory. As soon as evolutionists employ the concept of *realized fitness* in lieu of the concept of an objective propensity toward reproductive success, they lay themselves open to the charge that their theory has no empirical content. As soon as natural selection is characterized as differential reproduction rather than as differential reproduction *in consequence of systematic or of design differences*, one is barred, by one's mistaken choice of definitions, from distinguishing between selective and random or accidental cases of differential reproduction. The subtle difference between using realized fitness as a contingent measure of expected (or engineering) fitness and of replacing the latter concepts by the former is, therefore, of considerable importance. One must be on guard against confusing the two moves.

The use of actual reproductive outcomes as rough measures of (but not as substitutes for) expected or engineering fitnesses in the study of natural populations can indeed be justified in appropriate circumstances. What biologists must recognize is that *in order to justify the use of de facto reproductive success in actual cases as a measure of evolutionary fitnesses, they must provide good grounds for believing that (in the instances in question) the actual reproductive successes are manifestations of the relevant propensities.* This requirement is comparable to the need to prove an existence and uniqueness theorem when one defines rational numbers in terms of sets of ordered pairs of integers[19]; the procedure is legitimate when the justificatory requirement is met, otherwise not. Conventional wisdom in evolutionary theory holds that that condition is often met – that the organisms within an undisturbed population are usually at, or near, a local maximum of *expected* fitness and are unaffected by unusual or stochastic factors. I do not know (and it is not easy to establish) how often such claims are correct, but when they are, the measurement of approximate expected fitnesses by means of tautological fitness is a sound procedure. Critics of this

---

[19] Thus, one must show that there is a number defined by each and every properly constructed set of ordered pairs (e.g., in some systems, {(1,2), (2,4), (3,6). . . .} defines the number ½) and that each and every such set picks out a unique number. Once this is proved, symbols such as "½" may be used in an unambiguous way in the system in question. Similarly, once it is shown that actual reproductive rate probably reflects the relevant propensities, that rate may be used as an approximate measure of those propensities.

procedure should recognize that, in the appropriate circumstances, *this procedure does not bypass the use of the concepts of both expected and engineering fitness.*

This discussion makes apparent both how easy it is to fall into conceptual confusion here and also that the confusion (in spite of its prevalence in the literature) is avoidable. It is the concepts of expected and engineering fitness of which evolutionary theory makes essential use, and a proper analysis of the structure of the theory should address the role of these concepts rather than that of realized fitness. In the remainder of my discussion I shall proceed accordingly.

Second, in general it is as difficult to justify the step from expected to engineering fitness as it is to justify the step from realized to expected fitness. This is especially apparent when one considers discussions of fitness at the level of genes and gene complexes. Very often, although we know which alleles (or variants of a gene complex) have the highest expected fitness in a given environment, we do not know the precise reasons for these relative fitnesses. Because many, probably most, genes have pleiotropic effects when they are expressed, because their precise expression (if any) depends on the presence or absence of other genes, and because the contribution to fitness of their expressed effects may well vary at different stages of the life cycle, the inference that a particular effect is the gene's primary contribution to fitness is quite substantial.[20] Thus the concept of expected fitness developed previously in this chapter is considerably weaker (and more general) than Darwin's concept of engineering fitness. Accordingly, because it uses the former concept, the synthetic theory's doctrine regarding adaptation is somewhat weaker than Darwin's. On the synthetic theory, the forces involved in natural selection stem from differences in expected (rather than engineering) fitnesses. Because differences in engineering fitness constitute, ipso facto, differences in expected fitness but not conversely (see Note 16), in principle the synthetic theory allows a broader range of causal antecedents

[20] There are exceptions of course – for example, when the effect is a dramatic reduction of viability or loss of function. But even then, matters can be complicated – for example, the familiar case of the gene causing sickle cell anemia whose positive contribution to fitness when in heterozygous condition was not at all easy to locate. And notice that it is really the *genotype* (homozygous normal, heterozygous normal and sickling, homozygous sickling), not the gene that is *causally* relevant to the fitness of the carrier. This raises problems (several of which will be mentioned) treated usefully by Wimsatt (1981, especially sections 2 and 3), and elegantly developed by Sober and Lewontin (1982). Unfortunately, I read this paper too late to shape the text of the present section. (I thank Professors Wimsatt and Sober for sending me their respective preprints.)

to yield adaptations. Whether or not the weaker forces permitted by the synthetic theory, primarily at the genic level, in fact play an important role in the production of adaptations is not an easy question. To my knowledge, it has not been directly addressed in the literature.

At this point it will be useful to review what is involved in claiming that a trait is an adaptation. There are two components to this claim.

1. The trait in question is an optimal – or at least the relatively best – engineering solution of a real problem bearing on (a) survival of the organism; (b) incorporation of energy or other environmental resources into the organism; or (c) incorporation of energy and other resources from the organism and the environment into viable offspring. *All* organisms face problems in all three of these categories, and those which have relatively better (or even optimal) solutions to these problems are those with greater (or optimal) engineering fitness.

2. The design features that yield the high engineering fitness of the trait have been produced by natural selection, that is, in the synthetic theory, by differential survival of organisms with high expected fitness.

One important strand of biological criticism of the synthetic theory turns on a denial of the easy assumption that the distinctive traits of an organism with high expected fitness are, ipso facto, likely to be (or to yield) adaptations in the technical sense just described. This is not a matter that can be resolved by conceptual analysis; its resolution requires extensive and difficult field and laboratory studies. In particular, when expected fitnesses are ascribed to genes, genotypes, or phenotypic markers whose function (if any) is unknown, the inference to engineering fitness is fraught with peril.[21] There is a gap between the genuine expected fitness of such elements and the proper units of analysis in treating of adaptation. *The latter are determined by engineering and causal considerations.* It is this gap that plagues the derivation of engineering from expected fitness; it is only when the units of analysis are at the right level – that is, when the causally relevant

---

[21] This point has both biological and philosophical content. An example of the former is provided by Thornhill (1979), p. 365: "[The] weakest aspect [of the study under review] is its failure to separate the concept of evolved function from the notion of advantage or benefit [i.e., contribution to expected or engineering fitness] . . . The function of a trait is defined as the advantage that characterizes the selective history of the trait; function tells how the trait has contributed to most effective reproduction over evolutionary time. (The elucidation of function is difficult; see [Williams (1966)].) *Because function and advantage are used synonymously throughout the book, the authors occasionally (and apparently unknowingly) offer interpretations that imply effective group selection*" (my emphasis).

units are taken into account – that the two properties are tightly connected (see Sober and Lewontin, 1982; Wimsatt, 1981).

This brings me to the third major point of this section. It is, indeed, the practice of many biologists to casually assume that any distinctive feature of an organism which is associated with a reproductive advantage (i.e., high expected fitness) for its bearers is an adaptation in the engineering sense. Once one realizes how strong the implied claim is, one recognizes that it deserves serious examination. A development of this theme should shed some light on the recent, many-faceted dispute over the proportion and importance of prominent and biologically interesting features of organisms that are not (in our onerous sense) adaptations. In a controversial paper, Gould and Lewontin (1979) attack the "adaptationist program," arguing that it is overly optimistic to suppose that virtually any prominent or biologically significant feature of an organism is an adaptation, that is, not only that it serves a function, but also that it was brought about by the action of natural selection operating on variants of that feature. Many subsidiary controversies, in fact, turn on whether or not particular biologically significant features are adaptations in this sense. At the genetic level there is the dispute about "selfish" DNA (i.e., DNA that is of no value to any organism, but which hitches a ride on the reproductive machinery of organisms) and the continuing controversy about selectively neutral mutations in genes coding for enzymes. (It may yet turn out that selectively neutral point mutations are required to explain the clockwork regularity of change in DNA sequence versus the stop-and-start irregularity of rates of morphological change, speciation, and so on.) At the level of morphological structure and programmed habit there are innumerable detailed controversies about the correctness, epistemic status of, and support required for selectionist "stories" about the origin of particular features.

There are also general concerns about the paleontological evidence tending to show that whole suites of features change relatively suddenly and relatively rapidly in correlated ways; it is not clear whether such changes occur, in general, via mutual co-adjustment, feature by feature, or via systematic developmental change. Of particular importance is the question whether, on a paleontological time scale, macroevolutionary mechanisms supplement natural selection in a way that must be taken into account in order to give a correct account of the presence of (certain) major traits in surviving populations. (See Vrba, 1980, for a useful review of some of the controversial mechanisms under investigation.) This latter worry is central to the debates about whether one can account for macroevolution by extrapolating what we know about microevolution, a topic which I shall address in a companion paper. Also of concern in the debate over macro-evolution is the effectiveness of group and species selection, that is, selection operating at higher levels than classical natural selection. The resolution of such debates, as should be obvious, requires *far* more than a conceptually adequate account of adaptation, but in all these cases, resolution of the controversy has been impeded by the sort of conceptual confusions about adaptation that I have been illustrating.

My final point deals with problems regarding the units of selection. Classical natural selection operates, however effectively or ineffectively, one generation at a time by means of a systematic statistical bias in the reproductive rate of phenotypically distinct individuals. Yet one finds a great variety of claims in the literature to the effect that units such as genes, gene complexes, cellular organelles, and so on, on the one hand; and groups, populations, and species on the other are units of selection. On the present occasion I will deal briefly with only one of these.

Like many others, Williams (1966, and elsewhere) is inclined to genic reductionism. He holds that virtually everything one wishes to say about natural selection can be said by treating the single gene as the unit of selection. His definition of a gene is a bit unusual. It includes anything in the genome "which segregates and recombines with appreciable frequency" (1966, p. 24), and so includes, for example, whole inversion loops in the chromosomes of certain *Drosophilae* as genes. (But this definitional sleight of hand makes no difference to the fundamental issue.) Specifically, he holds that the central processes of natural selection can all be represented in terms of the mean selective effects of each gene, taken singly. Williams grants that, in different (conspecific) organisms, a given allele has different effects, both because of genetic differences in the organisms and because of the complexities of epigenesis. (These complexities mean that environmental and developmental differences may enable different copies of the same allele, even when it is in the same genetic context, to have different effects.) Nonetheless, Williams argues, each allele has an *average* phenotypic effect that will determine the net strength of the selective processes acting upon it. So far as the genetical theory of natural selection is concerned, the heart of the matter is the *net selective value* of each gene, taken singly.

There are two closely related arguments showing that such extreme genetic reductionism is wrong. Both arguments are based on the claim that in quite common and specifiable circumstances the mean selective value of a gene, were it available, would not work as a construct for making evolutionary predictions and explanations. The more elegant argument, recently articulated in Wimsatt (1980, 1981), traces back to controversies between Fisher and Wright and between Dobzhansky and Muller, though Wimsatt is developing it out of some work by Lewontin. It points out that in very many (arguably quite typical) circumstances, the selective value of an allele per generation is a function of the allelic frequencies at *other* loci. If this is so, and if random or selective pressures are affecting those other loci independently of the allele being considered, the selective value of that allele will change in each generation as the frequency distribution of alleles at the other loci changes. Any system of bookkeeping which assigns fitnesses or selective values to individual alleles will have to take account of allele frequencies *at other loci* and will, thus, only appear to treat each locus independently of the others. In Wimsatt's terminology, even though the mean selective value of the allele may be determinate in every generation, it is a "local" not a "global" value in the genetic state space and so *extrapolations based on that selective value will have no predictive or explanatory value.*[22] In my terminology, the net selective value of a gene in a given generation, like realized fitness, is a *description* of the reproductive outcome *in that generation only*, not an objective propensity or an enduring property with explanatory value.

The second argument is based on a slightly more direct consideration of the way genes affect phenotypes. To the extent that phenotypes are determined by additive interactions among genes, the mean selective value of a gene is determinate. To the extent that genic interactions are nonadditive (and they are not additive in a great many, mostly ill-understood cases, including those involving frequency-dependent interactions, changes in the developmental schedule, and so-called regulatory interactions), the mean phenotypic effect of a single gene is unstable and, hence, ill defined.

As these parallel arguments show, there are deep, substantial, and probably false assumptions that would have to be true if everything we wish to say about natural selection were to be captured, even in principle, in a treatment based on mean phenotypic effects and the associated selective values of genes taken singly. There are thus good

---

[22] Sober and Lewontin (1982) point out that genic fitnesses can be parametrized so as to take account of the genetic background (and other disturbing factors). They argue that although such devices allow one to calculate the actual change in gene frequencies from appropriate information, the relevant causal factors generating the selective forces are *genotypes*, not genes (and so on in other cases), so that the *explanatory* use of fitnesses *in the relevant cases* (one is mentioned in Note 19) occurs at the genotypic (or appropriate other) level, *not* the genic level. Wimsatt (1981) makes a very similar point.

grounds for holding that the gene is almost certainly not *the* ultimate unit of selection. This result (which may be generalized to cover a broad class of purported reductions) undercuts the idea that we can work our way back up from single genes to locate the "correct" phenotypic categories in terms of which to describe those features of organisms that have been directly shaped by natural selection. In other words, the atomistic study of single genes and their phenotype effects will not, in general, reveal which features of organisms are adaptations arrived at in the course of natural selection.

The general moral of this story for the problem of the units of selection is not that the gene is *not* a unit of selection. (It is.) Rather, the moral is that there are units of selection at many hierarchical levels and that we must take into account interactions that cross these levels. The result will be a considerable, but inevitable, increase in the complexity of evolutionary theory, for it means that we will have to examine the strength of the interactions between selective processes at various levels (at least from the level of the single gene to the level of species selection). I, for one, will be very much surprised if it turns out (as, of course, it might) that classical natural selection, operating between phenotypically distinct organisms in virtue of their design, is the only type (or the highest level type) of selection we need to take into account in reconstructing the history of life. But a full examination of the controversies over the units of selection will have to wait another day.

References

Ayala, F., and J. Valentine. 1979. *Evolving*. Menlo Park, Calif.: Benjamin/ Cummings.

Brandon, R. N. 1978. "Adaptation and evolutionary theory." *Studies in the History and Philosophy of Science* 9:181–206.

Burian, R. "On some controversies concerning macroevolution." (Ms.)

Darwin, C. 1859. *On the Origin of Species by Means of Natural Selection*. London: Murray. [Rprt. Cambridge, Mass.: Harvard Univ. Press, 1964.]

Dawson, P. S. 1969. A conflict between Darwinian fitness and population fitness in *Tribolium* "competition" experiments. *Genetics* 62:413–19.

Dobzhansky, Th. 1951. *Genetics and The Origin of Species*, 3rd ed. New York: Columbia Univ. Press.

1970. *Genetics of the Evolutionary Process*. New York: Columbia Univ. Press.

Gould, S. J. 1971. Darwin's retreat. [Review of P. J. Vorzimmer, *Charles Darwin: The Years of Controversy*]. *Science* 172:677–8.

1980. Is a new and general theory of evolution emerging? *Paleobiology* 6:119–30.

1981. But not Wright enough: reply to Orzack. *Paleobiology* 7:131–4.

Gould, S. J., and R. C. Lewontin. 1979. The spandrels of San Marco and the

Panglossian paradigm: a critique of the adaptationist programme. *Proc. Roy. Soc. (Lond. B)* 205:581–98.

Gould, S. J., and E. S. Vrba. 1982. Exaptation – a missing term in the science of form. *Paleobiology* 8:4–15.

Lewontin, R. C. 1974. *The Genetic Basis of Evolutionary Change.* New York: Columbia Univ. Press.

Mayr, E., and W. B. Provine (eds.) 1980. *The Evolutionary Synthesis: Perspectives on the Unification of Biology.* Cambridge, Mass.: Harvard Univ. Press.

Mills, S., and J. Beatty. 1979. The propensity interpretation of fitness. *Philos. Sci.* 46:263–86.

Orzack, S. H. 1981. The modern synthesis is partly Wright. *Paleobiology* 7:128–31.

Ospovat, D. 1981. *The Development of Darwin's Theory.* Cambridge: Cambridge Univ. Press.

Prout, T. 1969. The estimation of fitness from population data. *Genetics* 63:949–67.

  1971. The relation between fitness components and population prediction in *Drosophila.* I and II. *Genetics* 68:127–49; 151–67.

Rosenberg, A. "Fitness." (Ms.)

Rosenberg, A. 1982. On the propensity interpretation of fitness. *Philos. Sci.* 49:268–73.

Ruse, M. 1973. *The Philosophy of Biology.* London: Hutchinson Univ. Library.

Scriven, M. 1959. Explanation and prediction in evolutionary theory. *Science* 130:477–82.

Simpson, G. G. 1949. *The Meaning of Evolution.* New Haven: Yale Univ. Press.

Sober, E., and R. C. Lewontin. 1982. Artifact, cause, and genic selection. *Philo. Sci.* 49:157–80.

Thornhill, R. 1979. [Review]: *Insect Behavior,* by R. W. Matthews and J. R. Matthews. *Q. Rev. Biol.* 54:365–6.

Vrba, E. 1980. Evolution, species and fossils. How does life evolve? *South Afr. J. Sci.* 76:61–84.

Williams, G. C. 1966. *Adaptation and Natural Selection.* Princeton: Princeton Univ. Press.

Williams, M. 1973. The Logical Status of the Theory of Natural Selection and Other Evolutionary Controversies. In *The Methodological Unity of Science,* ed., M. Bunge, pp. 84–102. Dordrecht: Reidel.

Wimsatt, W. 1980. Reductionistic Research Strategies and Their Biases in the Units of Selection Controversy. In *Scientific Discovery: Case Studies,* ed. T. Nickles, pp. 213–59. Dordrecht: Reidel.

  1981. The Units of Selection and the Structure of the Multi-level Genome. In *PSA 1980,* vol. 2, ed. P. Asquith and R. Giere, pp. 122–83. East Lansing, Mich.: Philosophy of Science, Association.

Evolutionary theory and its consequences for the concept of adaptation

D. S. PETERS

Introduction

The concept of adaptation has a long history not only in biology but also in philosophy and other sciences. Since ancient times, adaptations taken for teleological expediences were substantial arguments in philosophical speculations about the arrangement of the universe. More recently the meaning of biological adaptation was differentiated with respect to modern evolutionary theories; on one hand, being considered an important link in evolutionary reasoning, on the other, being used as an argument against evolution (Wilder-Smith, 1978; Scheven, 1979). We shall consider adaptation here only from the evolutionary point of view.

The main innovation of Darwin's and of Wallace's theory was the introduction of selection as a canalizing factor of development. Because of this concept of selection the new theory gained its convincing power in contrast to Lamarckism. Because the canalizing effect of selection as to the positively selected variants could result only in adaptations, the concept of adaptation was connected with the Darwinian theory very strictly from the beginning. Nevertheless, there seems to be no unanimity concerning the concept of adaptation among biologists, even among those taking evolution for granted. A new discussion is arising.

The concept of selection reconsidered

Possibly, the different opinions expressed are not so different as they might seem at a first glance. Very often terms are used equivocally thus obscuring the author's intention. This might be the case also with selection, the most important concept in our context. Do we know to what extent the meanings of the word "selection" are identical in different publications? But it is not my task to analyze this problem. I shall rather express my own view with two assertions adding some explanations thereafter: First, selection is the inevitable consequence of diversification. Second, selection involves the total organism (living system), and not only its relations to the environment (*Umwelt*). For this reason there is no qualitative difference between adaptation to environment and adjustment of internal structures.

With respect to the first point, consider the following four cases:

1. Organisms that are identical in every regard are living in a totally homogeneous world. No mutations occur.

   Unnecessary to explain, there will be no evolution.

2. Identical organisms as in case 1 are living now in a diversified world. No mutations occur.

   Again, no evolution can take place. At the most the distribution of the organisms will achieve a differentiated pattern because they might prefer some places now and avoid others. Even if populations were isolated from each other and reproducing at different rates no evolutionary change can be expected.

3. Let us imagine the contrary situation: A homogeneous world and identical organisms as in case 1, but from a certain moment mutations can occur. There seem to be different opinions as to the possible consequences of that situation.

If we assume that the environment is the decisive factor of selection there should be no further development expected in spite of mutations. Environment would operate stabilizing the original condition of organisms. Nevertheless, this would be a process of selection by means of different reproductive rates, the new variants being repressed in behalf of the old ones.

To be sure, I am convinced that stabilizing selection is not the only possible consequence. To my knowledge there is no sound reason for refusing the assumption that positively selected mutations may occur even under stable environmental conditions. Any change improving external adaptation or internal efficiency has to be an advantage in selection process.

Indeed, "perfect" organisms cannot be improved. But total perfection, which means "omnipotent" organisms, is not realistic. Even if organisms were perfectly adapted or constructed with respect to certain conditions, they would be never adapted to everything. No organism can realize all relations that are theoretically possible. Provided that this is true, then every mutation inaugurating a new relation will change the whole situation, including the significance of already existing adaptations.

I want to stress at this point, that it is not necessary to assume macromutations. Small changes such as slight improvements of sensorial sensitivity or digestive enzymes, may change slightly the selective optimum, thus inducing other innovations. Evolution can start in our fanciful homogeneous world! Of course, in a strict sense this world will not be homogeneous any longer if inhabited by different organisms, because the organisms are mutually parts of their environment.

Arguing on the "metatheoretical level" (Tuomi, 1981) we cannot deduce, of course, the kind of specific development. The specific evolutionary events depend on the specific conditions. Assuming, for instance, that our organisms are uniparental we would have no difficulty imagining that several anagenetically different evolutionary pathways could emerge and proceed simultaneously, because no genetic exchange would happen. Taking into consideration biparental organisms, their cladogenesis, which means the splitting of genealogical branches, depends on whether conditions admit sympatric or only allopatric speciation, whether the initial population was distributed already all over the world or only over a certain area, so that new isolated founder populations can establish themselves, and so on.

But that is beside the point. We have to focus our attention on the fact that the change in the genetical constitution of populations is the inevitable consequence of diversification as soon as diversification influences the rates of reproduction (Mollenhauer, 1976; Van Valen, 1976; Peters, 1978).

The question is, When are such influences realized? As outlined earlier, mutation (diversification) and selection (the nonrandom differential perpetuation of varying genomes) are the two sides of the same coin. In this context the definition of selection seems to be a very tautological statement. But we should take into account that the reproductive success is by no means an independent feature. The capacity to donate genes to future generations is connected rigidly with the other features and capacities of the organism, which, as such, is restricted in a twofold manner. In the first place, its own capacities are specific, that is finite; second, it is living in a finite world with limited resources. Restricted by those two kinds of limitations, survival and reproduction of organisms prove to be a question of competition for gathering and converting matter and energy. Thus an improvement of reproductive success can be gained only by the optimization of the organism's efficiency with regard to given specific frame conditions, and vice versa, each change of the organism's efficiency influences its rate of reproduction. I postpone the further discussion on this point until later.

4. On behalf of completeness only I would like to mention the fourth
   case, which is characterized by a differentiated world and varying
   organisms. The situation is identical with our real world. There is
   no doubt evolution can proceed here even better than in case 3.

With respect to my second point, Gould and Lewontin (1979) attempted recently in a remarkable essay to show that the "adaptationist program" is wrong if not ridiculous. The "spandrels of San
Marco" serve as examples in their argumentation. These spandrels
are "by-products of mounting a dome on rounded arches," and they
contain on their surface designs that are perfectly fitted into their
triangular spaces. In spite of the harmonious impression we should
not be tempted to view the design "as the starting point of any analysis, as the cause in some sense of the surrounding architecture,"
because the spandrels are the necessary effect of architectural constraints. To view the spandrels as "adaptations" for the design, reminds Gould and Lewontin of the ridiculous opinions of Voltaire's
Dr. Pangloss. We can add as Dr. Pangloss' German counterpart H.
Löns' "purposeful Meier" (*Der zweckmässige Meier*). Both gentlemen
– and Meier's opponent in Löns' story, too – are simpleminded indeed, but is this a support for Gould and Lewontin's view? What are
architectural constraints? Are there no such constraints in a piece of
linen or in a wooden table prepared especially for painting?

There is no doubt about it, that St. Mark's Cathedral in Venice was
erected according to a plan based on an architectural idea. The material used for the building, of course, had the usual attributes of
matter, for example, extension and gravity. Yet the cathedral's specific
singularity of shape and configuration is not the effect of those attributes. The specific quality of the cathedral as a building consists
in the realization of the architectural idea, in the way the different
parts were arranged by the bricklayers and carpenters. These people
had to obey natural laws while doing their labor, but there was no
natural law enforcing the characteristic construction of San Marco.
On the other side, if they wanted to realize the ground-plan, the
solution of their task was limited very narrowly. The material, the
parts of the building, could fulfill its function only in case it was
"adapted" to both physical constraints and the intended artistic
expression. The spandrels have to be considered as being involved
in this dialectic relation. Notwithstanding their natural material attributes, the spandrels are in the special shape of their material body
"adapted" to the structure of the whole. In view of the special treatment of their surfaces they are adapted even as the ground for the
design.

Gould and Lewontin are right in stating that the spandrels were not created primarily as painting-grounds, but in my opinion they are wrong in stating that the spandrels are not adapted at all. Transferring this metaphor to biology again, I would like to say that I agree completely with Gould and Lewontin insofar as it is nonsense to interpret organisms merely as accumulations of environmental adaptations. But I cannot agree with the conclusion that the internal constraints of the organismic construction have nothing to do with selection.

The authors would be right, of course, if only the mere material attributes were interpreted as "constraints," but this would be a trivial statement. To be sure, the hydrochloric acid in our stomach has the same chemical properties as the synthetic one we buy in a drugstore. From the viewpoint of evolutionary biology, however, not the chemical properties as such but the integration of an acid with these properties in a living system should be the interesting problem. It is obviously not enough to know only the properties of collagen-fibers or bones, we want to understand rather how these elements are fitting in the construction of organisms.[1] There is no doubt that this "manner of fitting in," this mutual limiting of the involved parts is heavily influenced by selection. The organismic structure is exposed to selection first with respect to the minimum of integration, beyond which the construction does not work; and second, with respect to a continuous improvement toward the optimal state, the paradigm (Rudwick, 1964). We should keep in mind that the construction as well as the activity of organismic structures require matter and energy that are only available in limited amounts. As we mentioned before, this is the reason for competition in evolutionary process.

We found that selection can be defined as the effect of diversification of organisms on the differentiation of their reproductive success. If that meets reality, selection can operate only through an eval-

[1] During discussion, Gould mentioned gravity acting on flying fishes as another example of a constraint not accessible for the adaptationist program. This illustrates very well the misunderstandings in the antiadaptationist approach. Of course gravity as such is not exposed to selection because it is a general attribute of matter. But the consequences of gravity for the organism, the systemic interrelations of weight, shape, and size of body and fins, power of muscles, and so forth, are certainly exposed to selection and undergo a process of adaptation with respect to the actually given frame conditions. The subject of biology is organisms as living systems. Organismic structures and processes have to meet the limitations given by the general natural laws; and biological research has for its goal the understanding of these organismic responses, not the explanation of natural laws. It is not interesting at all to know that a flying fish is not weightless, but it may be interesting to know that it is not too heavy to take off and not too light to be blown away by the wind.

uation of the different inclusive efficiencies of the varying organisms. Both internal and external relations of organisms are involved in this process. In many cases it is impossible to make a distinction between internal and external conditions. Imagine a doe with a cardiac defect. It becomes an easy prey of the wolf. Did it perish because of environmental influence or owing to internal deficiency? And what do we answer, if the doe perishes without being chased by the wolf?

The strong selective pressure acting on the internal construction is demonstrated for instance by the fate of teratological cases. Thus the distinction between "environmental adaptation" and "architectural constraint" is fading away. It vanishes even more if we take into account that both "adaptation" and *Bautechnik* are, at least partially, based on genetic information. Analogically to the spandrel example we surely cannot maintain that the chemical character of hydrochloric acid, the toughness of collagen-fibers, or the properties of calcite are inherited features, but the fact that these substances are part of (some) organisms, as well as the pattern of their structural and functional integration are clearly influenced genetically, whatsoever the course of epigenetical process might be.

It is important to understand that the specific architectural constraints are dependent on the systemic character of the whole construction. General natural laws are necessary conditions for these constraints but they do not yield the sufficient explanation. Neither the spandrels of San Marco nor any organismic constructions can be derived or deduced from natural laws. There is no law for a knee joint or the mimetic coloration of a nightjar. Also in this respect no differences between environmental adaptation and internal constraints (no matter whether in a static or developmental, ontogenetical, sense) can be found. Consequently, all features, the construction and the developmental mechanisms, become selective factors for every structural or functional change. The construction of the organism itself participates in the canalizing limitation of further evolution.

Our considerations also indicate that there is no adaptation and no optimization in the abstract (Bonik, Gutmann, and Peters, 1977). It depends on the circumstances whether a feature is interpreted as an adaptation. The hare and the partridge are hunted by animals that rely upon their visual orientation. That is the only reason why the coloration of the hare and the partridge is thought to be an adaptation enabling them to escape – at least sometimes. Were the legs of tetrapods not part of a very specialized locomotory apparatus, nobody would consider their joints to be consequences of architectural constraints.

Before recognizing adaptations or architectural constraints in a concrete specific case, it is necessary to analyze very carefully this case, its conditions, and relations. One of those conditions, often neglected, is the history of the structures in question. Every construction had its forerunners that, as mentioned earlier, were part of the selective conditions leading to the present state. The quantity of four legs in tetrapods, for instance, has its reason in conditions swimming fishes are confronted with, and not in conditions connected with terrestrial life, although many other features in the tetrapod limbs are connected with the technical constraints of walking (Peters and Gutmann, 1976, 1978).

Another condition is probably even more important and indispensable. It is the dependence of all organismic structures and processes on energetic costs. Regarding energetic costs there are no neutral solutions possible, or else we believe in the perpetuum mobile. Keeping this in mind, we are enabled, or even forced, to explain reductions of organismic structures also as changes canalized by selection. There has been much confusion about this point in the past. Discussing the problem of the diminutive front legs of Tyrannosaurus, Gould and Lewontin (1979) maintain a "nonadaptive" explanation by means of allometric correlations. I have nothing against allometric methods, but they yield, at most, descriptive results requiring additional explanations. Gould and Lewontin are arguing against "untestable speculations based on secondary utility." They are quite right as to that, but why should we assume that "the primary evolutionary reason" is nonadaptive? I think that the energetic base of life is the primary reason for every evolutionary change. Reduction of structures means inevitably reduction of expenses, other functions notwithstanding. According to the specific frame conditions such reductions may be canalized very narrowly as in the case of limb reduction in some reptiles (Lande, 1978); or if strong constraints are lacking, a broad variety of regressive phenomena result as in the case of the eyes of some cave-dwelling fishes (Wilkens, 1972; Peters, Scholl, and Wilkens, 1975). In the first case we may find allometric correlations, whereas in the second, we will not. In any case, every reduction fitted in the frame construction may be interpreted as being an adaptation with respect to the energetic budget of the organism.

As demonstrated in a previous paper (Peters and Gutmann, 1971), the energetic principle of evolution has been understood, though to a different degree, by many authors in the past. Recently, programs assembled under the heading of "sociobiology" seem to be based on the energetic aspect, although this is rarely expressed as explicitly as in the following: "Presumably, natural selection will usually act to maximize the amounts of matter and energy gathered per unit time; the problem is to understand how this matter and energy are partitioned among somatic and reproductive tissues and activities" (Pianka, 1970). The numerous models elaborated by sociobiologists would be deprived of stringency at once if organisms, the "survival machines" (Dawkins, 1976), could spend matter and energy without selective relevancy.

Dullemeijer (1980) states correctly that the energetic principle is a very theoretical claim in the context of evolutionary arguments because of the difficulty of obtaining the complete energetic balance of any organism. Indeed the energetic principle was not formulated on an inductive base, it was derived from the assumption that the universal natural laws, among them the rule of energy conservation, are valid also in organisms without exception. Hence this principle is testable.

Before closing, another misunderstanding remains to be settled. It arises apparently from an overinterpretation of the parsimony aspect of the concept of optimization. Probably some formulations in previous publications nourished the misunderstanding. This could be the case for instance with our (Peters and Gutmann, 1971) reference to the definition of adaptation given by Bock and v. Wahlert (1965). It reads: "Evolutionary adaptation, the process, is defined as any evolutionary change which reduces the amount of energy required to maintain successfully a synerg, or the niche as the case may be, toward the minimum possible amount." A "synerg" is defined as ". . . a link between the organism and its umwelt." Indeed, one is tempted to take evolution for a kind of strategy for saving energy. However, it is difficult from such a point of view to explain how organisms can evolve that are bigger, heavier, more complicated, in a word, more expensive than their predecessors. One may approach the solution of the problem by combining the contents of both, this definition and Pianka's sentence cited previously.

The starting point of all considerations has to be the fact that reproduction, that is, the number of offspring per unit time, is the measure of evolutionary success. As Fisher (1930) pointed out it is more adequate in this case to evaluate the grandchildren because the children are still part of the expenditures. But for the sake of comprehension let us assume that all children of an individual have equal chances; and then we may neglect the grandchildren.

All capacities of an organism can be summarized in two faculties. First, an organism must be capable of *gathering* matter and energy; second, it must be capable of *converting* matter and energy. Hence, there are two main pathways to gain improvements of reproduction.

The first one consists in reducing the expenditures of the "converting machinery" in such a way that a surplus of matter and energy is left to produce more offspring. This advantage can be achieved with different strategies, for example, reduction of useless parts, improvement of mechanical or chemical structures to make them work at lower costs, and so forth. The second pathway consists of the improvement of the "gathering machinery." In this case, too, there are many possible approaches; for example, improvement of catching mechanisms, greater efficiency of digestion, improvement of the means of defense and resistance, and so forth. The results of the second pathway, of course, may be more expensive, when evaluated in "joules," than the competitive alternatives. Nevertheless, they will enable organisms affected by them to gain more reproductive success even if their children are produced at greater cost, provided that these children are more numerous and/or in other respects "better" than other children. In reality both pathways may occur together, as in business investment and economy do not exclude each other. The general outcome of this process will be an optimization of the whole system called "organism." The specific character of an optimization, however, has to be studied anew in every case.

Consequences

Terminology

The preceding arguments suggest a reconsideration of certain terms frequently used in discussion. Hitherto, the meaning of "adaptation" was almost exclusively the meaning of environmental "adaptation." Even the quoted definition (Bock and von Wahlert, 1965) is no exception in spite of making allowance for the energetic aspect. Such being the case we could let things remain as they were, and Gould and Lewontin's distinction between "adaptive" (related to environment) and "nonadaptive" (intraorganismic) would be clear. However, the concept of adaptation was related as well to the concept of selection, thus evoking a curious circularity of argument. Because, in most cases, the connection of adaptation and environment was understood very exclusively, adaptations to environment were thought, consequently, to be the only field of evolutionary events influenced by selection. Hence, all processes or states that according to the mentioned distinction appeared to be nonadaptive were regarded as neutral with respect to selection. The distinction between Darwinian and non-Darwinian evolution is one of the consequences. It seems to be at least problematic; in my opinion it is unnecessary and even wrong.

We cannot escape from the simple fact that every feature, be it internal or external, that improves reproductive success will be positively selected by means of this success. If we maintain that any feature that had an advantage in selection is an adaptation we have to accept not only external but internal adaptations as well. By these criteria the concept of adaptation becomes very broad and unwieldy.

It is not my intention to unravel the problem by proposing a new terminology. Probably we shall continue to use the term "adaptation" for such features as can be immediately connected with some properties of the environment. I do not raise any objection to that as long as it is accepted that not only these features are selected. From the viewpoint of evolution it is indifferent too whether mechanical, chemical, and other internal adjustments are termed "adaptations" or not, as long as one keeps in mind that these features are subjected to selection. The usage of the terms "internal" and "external" adaptation may be practicable, though it is arbitrary too. But we should not exaggerate this terminological problem. The comprehension of evolutionary theory and its implications for the analysis of organismic structures is much more important than a neatly defined vocabulary, which may be used equivocally anyhow. But this is not an argument against any attempt to establish a new adequate terminology.

Phylogenetic reconstructions

Much more radical are the consequences of a strict concept of selection for every serious attempt to reconstruct phylogeny. Some basic considerations concerning this aim were expounded by members of "Senckenbergische Arbeitsgruppe Phylogenetik" (SAP) in previous papers (e.g., Peters and Gutmann, 1971; Peters, 1972; Gutmann and Peters, 1973; Peters et al., 1975; Gutmann and Bonik, 1981). The postulates published there were supported by Bock (1981), Boy (1981), Dullemeijer (1980), Leisler (1975, 1981), Mayr (1974), Starck (1978), Vogel (1979), and others. Recently also some studies within the Special Research Division "Paleoecology" (Reif, 1982) are in agreement with these postulates, thus contradicting their own initial methodological concept (Seilacher, 1970), which with regard to "constructional aspects" (Bautechnik) anticipated rather Gould and Lewontin's point of view.

The keynote of SAP's considerations consists in the methodological claim to analyze organisms from a biotechnical standpoint. Organisms have to be studied as machine-like systems, which, of course, have a genealogical history. In this context the comparison of form and shape as such gains a preliminary character. Although environmental relations must not be neglected, the technical, functional properties of organismic structures have to be understood in the first place.[2] As explained previously, they are restricted by both natural laws and specific frame conditions which are to be analyzed very carefully. If these limitations are examined properly we may be able to outline in a model the evolutionary pathways that presumably led to the organisms in question. This approach deviates in many aspects from other current methods; it utilizes the essence of the evolutionary concept of selection and the systemic concept of organisms.

The differences are by no means purely academic as can be demonstrated by the results obtained. The biotechnical approach revealed so many features, limitations, relations, internal and external constraints that were overlooked or misunderstood. The insistence on the logical consequences of a strict concept of selection was sometimes mistaken for dogmatism (Reif, 1982). In my opinion, a neglect of these consequences abandons phylogeny to ambiguity.

Outlook

I hope you will not conclude from my statements that I am a reductionist. My insistence on natural laws, on the machine-like constitution of organisms had for its goal to demonstrate that biologists are research workers acting on the Cartesian field of natural sciences. I believe that the subject of biology is organisms, not life (Peters, 1979). This is an important distinction. To my mind, life seems to be rather a philosophical category. We meet with life in organisms of course, but while investigating them we investigate by reductionistic means. To be sure, this is the only way to obtain results that can be formulated in the language of natural science. Nevertheless, the organism is not reducible to such results. Each particular investigation deals with special aspects of things. The "whole" is beyond the scope of such research.

If I am right, one of the reasons for arranging this conference was a kind of uneasiness evoked by the multitude of new questions and by the ambiguity of many pretended answers in evolutionary biology. The feeling that the current theory of evolution should be revised or defined anew is symptomatic. I share this feeling, and that is why I argued so emphatically in favor of a logical connection between the theoretical reasoning and the practical approach. This would facilitate

---

[2] We learned from G. von Wahlert's contribution to our discussion at the Bad Homburg conference that ecological considerations are very important. As far as I understand, however, we have, indeed, to start from studying the construction of organisms. If we do not know anything about the structures of the organisms involved we are unable to understand their synecological interrelations discussed by G. von Wahlert.

the detection of inconsistent concepts and methods, thus promoting the revision. At the same time the limits of biological science would be ascertained more precisely, thus clarifying the field for philosophy, which certainly will have to play the leading part in this enterprise.

In his retrospect to one of the most extraordinary books of the past fifteen years, W. H. Thorpe (1969) stated with regard to the organismic concept: ". . .we are required to believe in a source of value added to, or injected into, the natural process as complexity develops which we are totally unable to understand." For the present I still feel in agreement with this remark.

References

Bock, W. J. 1981. Functional–adaptive analysis in evolutionary classification. *Amer. Zool.* 21:5–20.

Bock, W. J., and G. von Wahlert 1965. Adaptation and the form–function complex. *Evolution* 19:269–99.

Bonik, K., W. F. Gutmann, and D. S. Peters. 1977. Optimierung und Ökonomisierung im Kontext von Evolutionstheorie und phylogenetischer Rekonstruktion. *Acta Biotheor.* 26:75–119.

Boy, J. A. 1981. Zur Anwendung der Hennigschen Methode in der Wirbeltierpaläontologie. *Paläont. Z.* 55 1:87–107.

Dawkins, R. 1976. *The Selfish Gene.* New York: Oxford Univ. Press.

Dullemeijer, P. 1980. Functional morphology and evolutionary biology. *Acta Biotheor.* 29:151–250.

Fisher, R. A. 1930. *The Genetical Theory of Natural Selection.* Oxford: Clarendon Press.

Gould, S. J., and R. C. Lewontin. 1979. The spandrels of San Marco and the Panglossian paradigm: a critique of the adaptationist programme. *Proc. Roy. Soc. Lond. B* 205:581–98.

Gutmann, W. F., and K. Bonik. 1981. *Kritische Evolutionstheorie.* Hildesheim: Gerstenberg.

Gutmann, W. F., and D. S. Peters. 1973. Das Grundprinzip des wissenschaftlichen Procedere und die Widerlegung der phylogenetisch verbrämten Morphologie. *Aufs. Reden SNG* 24:7–25.

Lande, R. 1978. Evolutionary mechanisms of limb loss in tetrapods. *Evolution* 32:73–92.

Leisler, B. 1975. Die Bedeutung der Fussmorphologie für die ökologische Sonderung mitteleuropäischer Rohrsänger (Acrocephalus) und Schwirle (Locustella). *J. Ornithol.* 116(2):117–53.

1980. Okomorphologische Freiland- und Laboratoriumsuntersuchungen. *Acta XVII Congr. Intern. Ornithol.* 1:202–8.

Mayr, E. 1974. Cladistic analysis or cladistic classification? *Z. Zool. Syst. Evolutionsforsch.* 12:94–128.

Mollenhauer, D. 1976. Systemtheorie und botanische Systematik. Drei Betrachtungen. *Aufs. Reden SNG* 28:32–68.

Peters, D. S. 1972. Das Problem konvergent entstandener Strukturen in der anagenetischen und genealogischen Systematik. *Z. Zool. Syst. Evolutionsforsch. 10*(3):161–73.

1978. Voraussetzungen einer selektionistischen Evolution. *Aufs. Reden SNG* 29:159–62.

1979. Biologie – Wissenschaft vom Leben? *Nat. Museum 109*(2):385–93.

Peters, D. S., J. L. Franzen, W. F. Gutmann, and D. Mollenhauer. 1974. Evolutionstheorie und Rekonstruktion des stammesgeschichtlichen Ablaufs. *Umschau 74*(16):501–6.

Peters, D. S., and W. F. Gutmann. 1971. Über die Lesrichtung von Merkmals- und Konstruktionsreihen. *Z. Zool. Syst. Evolutionsforsch 9*(4):237–63.

1976. Die Stellung des "Urvogels" Archaeopteryx im Abstammungsmodell der Vögel. *Nat. Museum 106*(9):265–75.

1978. Ausgangsform und Entwicklungszwange der Gliedmassen landlebender Wirbeltiere. *Nat. Museum 108*(1):16–21.

Peters, N., A. Scholl, and H. Wilkens. 1975. Der Micos-Fisch, Hohlenfisch in statu nascendi oder Bastard. *Z. Zool. Syst. Evolutionsforsch. 13*:110–24.

Pianka, E. R. 1970. On r- and K-selection. *Am. Natural. 104*:592–7.

Reif, E.-W., 1982. Functional morphology on the procrustean bed of the neutralism–selectionism debate. Notes on the constructional morphology approach. *N. Jahrb. Geol. Paläont. Abhandl. 164*:46–59.

Rudwick, M. J. S. 1964. The inference of function from structure in fossils. *Brit. J. Phil. Sci. 15*:27–40.

Scheven, J. 1979. Daten zur Evolutionslehre im Biologieunterricht. Neuhausen, Stuttgart: Hanssler-Verlag.

Seilacher, A. 1970. Arbeitskonzept zur Konstruktions-Morphologie. *Lethaia* 3:393–6.

Starck, D. 1978. *Vergleichende Anatomie der Wirbeltiere.* Berlin, Heidelberg, New York: Springer-Verlag.

Thorpe, W. H. 1969. Retrospect. In *Beyond Reductionism,* ed. A. Koestler and J. R. Smythies, pp. 428–34.

Tuomi, J. 1981. Structure and dynamics of Darwinian evolutionary theory. *Syst. Zool. 30*(1):22–31.

Van Valen, L. 1976. Domains, deduction, the predictive method, and Darwin. *Ecol. Theory 1*:231–45.

Vogel, K. 1979. Efficiency of biological constructions and its relation to selection and rate of evolution (General remarks). *Paläogeogr. Paläoclim. Paläoecol. 28*:315–19.

Wilder, Smith, A. E. 1978. *Die Naturwissenschaften kennen keine Evolution.* Basel, Stuttgart: Schwabe.

Wilkens, H. 1972. Zur phylogenetischen Rückbildung des Auges Cavernicoler: Untersuchungen an *Anoptichthys jordani* (= *Astyanax mexicanus*), Characidae, Pisces. *Am. Speleol. 27*:411–32.

INDEX

adaptation, 2–3, 6–8, 11–12, 44–6, 52–3, 56, 63–7, 99–100, 106, 115, 119–20, 123–4, 129–30, 133–4, 144–5, 148, 163, 165–6, 233, 315–24
   controversies, 276, 280–1, 283–6, 306–13
   Darwin's theory, 47–9, 287–98
   ecological perspective, 99
   English evolutionists, 86
   mimicry, 106–8, 110, 114, 116–18, 122, 130–1, 158
   neo-Darwinism, 298–306
   patterns of thought, 236
   revolt against adaptationism, 9–10
   shifting balance theory, 45–6, 61
   speciation, 102
   synthetic theory, 74–91, 153
   typostrophism theory, 180–1, 187–8
adaptedness, 287–99
adaptogenesis, 190
Alberch, P., 257, 259, 264
Albert, H., 215
Allee, W. C., 113, 115
Allen, E., 160
allometry, 58–60, 86–7, 321
Alverdes, P., 38
American evolutionists, 86
anagenesis, 78–9
analogy, 176, 206, 224
aristogenesis, 35
Aristotle, 11, 218–19
asexual species, 154–5, 279
atomism, 10–11
Ayala, F. L., 9, 299, 303

Baer, K. E. von, 235
Bailey, V. A., 121
Baldwin, J. M., 39
Balme, D., 11
Bambach, R. K., 251, 254–5, 257

Barker, M., 161
Bates, H. W., 97, 102–5, 109, 112, 133
Batesian mimicry, 107–8, 133, 136, 147, 150, 165
Bateson, W., 19–20, 23, 27, 44, 49, 74, 98, 100–2, 105, 108, 132–3, 135, 153, 155, 160
Beatty, J., 8, 300–1
Berg, L. S., 34
Berry, R. J., 162
Berry, W. B. N., 246
Bertalanffy, L. von, 235, 246
Beurlen, K., 35, 174, 180–2, 184–5, 188–9, 197
biometric–Mendelian debate, 19, 23, 132, 136–7
biometry, 23–5
Bock, W. J., 262, 322–4
body size, changes in, 58
Böker, H., 34–5
Bonik, K., 320, 324
Boucot, A. J., 244, 246
Box, J. F., 20–2, 26
Boy, J. A., 324
Brandon, R. N., 10–11, 288, 300
Braun, A., 176
Bresch, C., 195
Bretsky, P. W., 244, 256
Bretsky, S. S., 256
Brooks, D. R., 8
Brown, H., 2
Brown, K. S., Jr., 156
Brunswik, E., 207, 215, 219
Buchner, P., 34
Bumpus, H. C., 52
Burian, R. M., 289

Cain, A. J., 86, 89, 276
Carpenter, G. D. H., 113
Carr, T. R., 256

329